T4-AJF-623

Molecular Biology of the Skin

Molecular Biology of the Skin

The Keratinocyte

Edited by

Michel Darmon
Institut National de la Sante et de la Recherche Medicale
Université de Nice
Parc Valrose
France

Miroslav Blumenberg
Department of Dermatology
New York University
New York, New York

Academic Press, Inc.
A Division of Harcourt Brace & Company

San Diego New York Boston London Sydney Tokyo Toronto

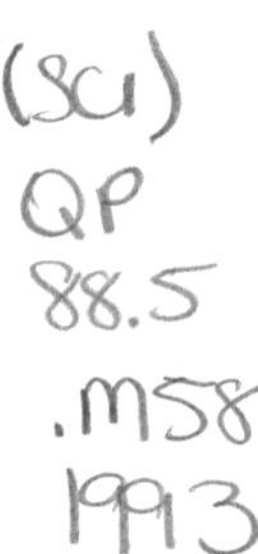

Front cover photograph: Morphological appearance of a DMBA/TPA-induced mouse skin squamous cell carcinoma. See Chapter 2, Figure 16.

This book is printed on acid-free paper. ∞

Academic Press, Inc.
1250 Sixth Avenue, San Diego, California 92101-4311

United Kingdom Edition published by
Academic Press Limited
24–28 Oval Road, London NW1 7DX

Library of Congress Cataloging-in-Publication Data

Molecular biology of the skin : the keratinocyte / edited by Michel Darmon, M. Blumenberg.
p. cm.
Includes bibliographical references and index.
ISBN 0-12-203455-4
1. Skin--Molecular aspects. 2, Skin--Differentiation. 3. Keratinocytes. I. Darmon, Michel. II. Blumenberg, Miroslav Lj. [DNLM: 1. Epidermis--cytology. 2. Keratinocytes--physiology. 3. Molecular Biology. WR 101 M718 1993]
QP88. 5.M58 1993
612.7'9--dc20
DNLM/DLC
for Library of Congress

92-48406
CIP

PRINTED IN THE UNITED STATES OF AMERICA
93 94 95 96 97 98 BC 9 8 7 6 5 4 3 2 1

Contents

3 Phenotypic Expression and Processing of Filaggrin in Epidermal Differentiation

Beverly A. Dale, Richard B. Presland, Philip Fleckman, Ephraim Kam, and Katheryn A. Resing

4 The Cornified Envelope: A Key Structure of Terminally Differentiating Keratinocytes

Uwe Reichert, Serge Michel, and Rainer Schmidt

5 Loricrin

Daniel Hohl and Dennis Roop

6 Retinoic Acid in Epithelial and Epidermal Differentiation

Michel Darmon and Miroslav Blumenberg

Contributors

Numbers in parentheses indicate the pages on which the authors' contributions begin.

Bruno A. Bernard (207), L'OREAL, Centre de Recherche C. Zviak, 92583, Clichy Cedex, France

Miroslav Blumenberg (1, 181), Ronald O. Perelman Department of Dermatology and Department of Biochemistry, New York University Medical Center, New York, New York 10016

Pascale Briand (245), Génétique et Pathologie Expérimentale, Institut Cochin de Génétique Moléculaire, CJF INSERM 90-03, F-75014, Paris, France

Joseph M. Carroll (269), Department of Oral Biology and Pathology, School of Dental Medicine, State University of New York at Stony Brook, Stony Brook, New York 11794

Catherine Cavard (245), Génétique et Pathologie Expérimentale, Institut Cochin de Génétique Moléculaire, CJF INSERM 90-03, F-75014, Paris, France

Beverly A. Dale (79), Departments of Oral Biology, Periodontics, Biochemistry, and Medicine/Dermatology, University of Washington, Seattle, Washington 98195

Michel Darmon (181), Institut National de la Sante et de la Recherche Medicale, Université de Nice, INSERM U 273, Parc Valrose, B.P. 71, 06108, Nice Cedex 2, France

Elizabeth S. Fenjves (269), Department of Oral Biology and Pathology, School of Dental Medicine, State University of New York York at Stony Brook, Stony Brook, New York 11794

Philip Fleckman (79), Department of Medicine/Dermatology, University of Washington, Seattle, Washington 98195

Jonathan A. Garlick (269), Department of Oral Biology and Pathology, School of Dental Medicine, State University of New York York at Stony Brook, Stony Brook, New York 11794

Daniel Hohl (151), Service de Dermatologie, Centre Hospitalier Universitaire Vaudois–Hôpital Beaumont, Lausanne, 1011 Switzerland

Ephraim Kam (79), Department of Oral Biology, University of Washington, Seattle, Washington 98195

Serge Michel (107), Centre International de Recherches Dermatologiques Galderma, (CIRD Galderma), F-06902 Sophia Antipolis, France

Richard B. Presland (79), Department of Oral Biology, University of Washington, Seattle, Washington 98195

Uwe Reichert (107), Centre International de Recherches Dermatologiques Galderma, (CIRD Galderma), F-06902 Sophia Antipolis, France

Katheryn A. Resing (79), Department of Chemistry and Biochemistry, University of Colorado, Boulder, Colorado 80309

Dennis Roop (151), Departments of Dermatology and Cell Biology, Baylor College of Medicine, Houston, Texas 77030

Rainer Schmidt (107), Centre International de Recherches Dermatologiques Galderma, (CIRD Galderma), F-06902 Sophia Antipolis, France

Jürgen Schweizer (33), Research Program II, German Cancer Research Center, D-6900 Heidelberg, Germany

Lorne B. Taichman (269), Department of Oral Biology and Pathology, School of Dental Medicine, State University of New York York at Stony Brook, Stony Brook, New York 11794

Alain Zider (245), Génétique et Pathologie Expérimentale, Institut Cochin de Génétique Moléculaire, CJF INSERM 90-03, F-75014, Paris, France

Foreword

By anatomical definition, the skin is a peripheral organ, and for generations of students and researchers it has also been a bit in the periphery of general interest—both in basic research and in medicine. For the ambitious biochemist and molecular biologist, the liver, for example, and the exocrine pancreas were in the limelight of global attention. Research on the skin was left to a small sect of specialists, dedicated afficionados, and dermatology professors.

This view of the skin—and of the epidermis in particular—has changed increasingly over the past two decades, and today the different kinds of cells and the mechanisms that govern their differentiation and functions are of central interest in general cell and molecular biology. Researchers studying the molecular and genetic aspects of the skin and its diseases have been recognized as pioneers of medical research. In recent years, results of skin research have been discussed repeatedly in editorials of major journals in science and medicine and have made it to the front pages and cover pictures.

Many individual advancements in methodology and molecular characterization have contributed importantly to the growth and development of skin research. For example, methods for culturing almost all cell types of normal and diseased skin have developed, and even complex processes of cell differentiation can now be studied in well-defined systems *in vitro*. Moreover, the keratinocytes of the epidermis and related epithelia have been both a source and a research object in studies of diverse groups of cytoskeletal molecules, from keratins to desmosomal proteins, not to mention the major proteins associated with the terminal differentiation of epidermal cells and the special molecules contributing to hair and nail formation. Research on skin cells has also made significant contributions to the recognition of the roles of growth factors, of oncogenes and tumor promoters, and of diverse molecular defense mechanisms—from short protective peptides to true immune responses.

Recently, for example, major autoantigenic molecules of certain blistering autoimmune diseases of the skin have been identified and mutational molecular defects have been associated with two forms of hereditary human epidermolytic diseases.

In such exciting times of an exponential increase of molecular information and insight into mechanisms, *Molecular Biology of the Skin*, which brings together experts in different domains of basic research, is most timely and topical. Consequently, the wide scope of chapters in this book also reflects the development and differentiation of the field of skin research, ranging from valuable reviews of the ever growing family of keratins and of the genomic arrangements of their genes and pseudogenes (Chapters 1 and 2, by Blumenberg and Schweizer, respectively) to a most comprehensive and detailed treatment of the major molecular components characteristic of terminal differentiation of epidermal cells (Chapters 3, 4, and 5, by Dale *et al.*, Hohl and Roop, and Reichert *et al.*, respectively). An important principle in the regulation of epithelial differentiation, that is, the control of the activities of certain genes by retinoic acid, is presented in Chapter 6 by Darmon and Blumenberg. The importance of studies using transgenic animals to understand molecular functions and the development of skin diseases is summarized in Chapter 8, by Cavard *et al.* A logical consequence of applicability, the advantage of the skin and of the keratinocyte as its predominant cell type as an obvious route for somatic gene therapy is outlined in an exciting review from Lorne Taichman's laboratory (Chapter 9, Carroll *et al.*). Finally, as the development of molecular biology of the skin has been paralleled by the research revealing the role of human papillomaviruses in the formation of epithelial tumors, malignant carcinomas included, this group of viruses is appropriately discussed in Chapter 7 by Bernard.

With rapid development of contemporary research, books summarizing the evidence and knowledge accumulated are needed at certain well-chosen intervals, for both the researchers and the broader community of scientists and medical researchers. Clearly, *Molecular Biology of the Skin* appears at the right time. And because of the breath-taking growth rate of scientific information, authors of such reviews on "hot" research subjects must be not only competent but also courageous. In this spirit, the present book is doubtless a great service to the field of molecular research on the skin, and authors as well as editors are to be congratulated for this accomplishment.

W. W. Franke

Preface

Skin plays protective (and esthetic) roles. In contrast to other organs, it cannot be compared to a filter, a pump, a factory, or a computer. For this reason the physiology and pathology of skin do not depend upon anatomical organization, but concern mainly basic cellular processes. For example, the protective role of the epidermis stems directly from the terminal differentiation of keratinocytes into corneocytes. As a result, a biologist willing to study skin is obliged to address his or her questions directly in terms of cellular and molecular biology. Despite the "simplicity" of skin as an organ, dermatological diseases are paradoxically very complex: They affect cellular pathways that are poorly understood and many of these diseases alter epidermal differentiation itself. Thus, researchers in dermatology also address their questions in the scope of molecular and cellular research. For example, the extensive study of the biology of keratins in cultured cells, and in transgenic animals has led to the information necessary to unravel the molecular basis of several human genetic skin disorders.

The use of a common language by dermatologists and molecular biologists is obvious when one attends scientific meetings or reads journals, whether specialized in dermatology or of wider scope. But a book emphasizing this harmony of ideas and reviewing essential topics about the molecular biology of the keratinocyte for both the medical and the scientific communities was still lacking. We hope that this book will play this role. The time is particularly appropriate for this volume because the studies from several divergent areas, as described here, converge to create a critical mass that will, we are certain, attract a large number of dermatologists to basic and clinical research.

We hope that this book will be useful not only for people interested in dermatology but also for people with wider interests, such as developmental biologists, because keratinization represents a prototype of terminal differentiation and also offers a very attractive model for the study of retinoids; surgeons and gene therapists, because human keratinocytes, unlike most cell types, can be grown *in vitro* for many generations without losing their normal phenotype and thus can reform *in vitro* a functional epidermis, which can be grafted back to patients; virologists, because human papillomaviruses and other viruses whose target is exclusively the keratinocyte still represent a major medical challenge; and, finally, pharmacologists, because effective drugs without dangerous side effects are still expected in dermatology.

Michel Darmon and Miroslav Blumenberg

1

Molecular Biology of Human Keratin Genes

Miroslav Blumenberg

Introduction

Ten years have passed since the pioneering reports of cloning and sequencing of human keratin genes (Hanukoglu and Fuchs, 1982). Murine, bovine, and ovine genes soon followed (Steinert *et al.*, 1983; Jorcano *et al.*, 1984; Powell *et al.*, 1986), heralding an unabating stream of keratin gene sequences. Of the 19 keratins described in the Moll catalog (Moll *et al.*, 1982), all but one

TABLE 1 **Keratin Sequence Determinations**

Keratin	Reference source
K 1	Johnson *et al.*, 1985
	Steinert *et al.*, 1985 (Mouse)
K 2	Not available
K 3	Klinge *et al.*, 1987
K 4	Hanukoglu and Fuchs, 1983
	Leube *et al.*, 1988
	Knapp *et al.*, 1986 (Mouse)
K 5	Lersch and Fuchs, 1988
K 6	Tyner *et al.*, 1985
K 7	Glass *et al.*, 1985
K 8	Glass and Fuchs, 1988
	Leube *et al.*, 1988
	Yamamoto *et al.*, 1990
	Krauss and Franke, 1990
	Oullet *et al.*, 1988 (Mouse)
	Tamai *et al.*, 1991
	Hsieh *et al.*, 1992 (Rat)
	Franz and Franke, 1986 (Toad)
	Giordano *et al.*, 1990 (Goldfish)
K 9	N.A.
K 10	Darmon *et al.*, 1987
	Zhou *et al.*, 1988
	Steinert *et al.*, 1983 (Mouse)
	Rieger *et al.*, 1985 (Bovine)
K 11	Mischke and Wild, 1987 Allelic variant of K 10
K 12	N.A.
K 13	Kuruc *et al.*, 1989
	Mischke *et al.*, 1989
	Winter *et al.*, 1990 (Mouse)
K 14	Hanukoglu and Fuchs, 1982
	Marchuk *et al.*, 1985
	Knapp *et al.*, 1987 (Mouse)
K 15	Bader *et al.*, 1988
	Leube *et al.*, 1988
K 16	RayChaudhury *et al.*, 1986[a]
K 17	Savtchenko *et al.*, 1990[b]
	Knapp *et al.*, 1987 (Mouse)
K 18	Kulesh and Oshima, 1989
	Oshima *et al.*, 1986
	Romano *et al.*, 1986
	Ichinose *et al.*, 1988 (Mouse)
K 19	Stasiak and Lane, 1987
	Bader *et al.*, 1988
	Bader *et al.*, 1986 (Bovine)
K 20	W. W. Franke, 1991 pers. com.
K 21	Chandler *et al.*, 1991 (Rat)
KHa1	Bertolino *et al.*, 1988 (Murine)
KHa2	Bertolino *et al.*, 1990 (Murine)
	Wilson *et al.*, 1988 (Ovine)

(*continued*)

TABLE 1 (*Continued*)

Keratin	Reference source
KHa3	
KHa4	
KHb1	
KHb2	Yu *et al.*, 1991
KHb3	
KHb4	Tobiasch *et al.*, 1992 (pers. com.)
70 KDa	Rentrop *et al.*, 1987 (Mouse)

[a]Misidentified as K17 at the time of publication.
[b]A human K17 pseudogene locus, not the functional gene.

(K12) have been cloned and sequenced from various organisms, and several previously unidentified keratins have subsequently been cloned and at least partially sequenced. A tangle of hair keratins has apparently been unsnarled, yielding a neat group of four keratins in each "family" (Heid *et al.*, 1988). Many of these have subsequently been cloned as well (Powell *et al.*, 1986; Bertolino *et al.*, 1988; Wilson *et al.*, 1988; Yu *et al.*, 1991). A compilation of sequences made available in one decade is given in Table 1.

This wealth of sequence information has been used to analyze the molecular structure of keratins, their evolution, chromosomal mapping, genetics, relationship with other intermediate filament (IF) proteins, filament assembly, etc. The availability of genomic clones has led to major efforts to elucidate the molecular mechanisms that regulate the expression of keratin genes, but although neither the "trees" nor the "forests" of regulation can yet be seen, the keratin genes are already being used for very sophisticated molecular biological studies. Examples include efforts to deliver marker genes to specific tissues in transgenic and somatic gene therapy model experiments.

Gene Structure

Protein Sequences

The sequences of keratin proteins conform to the "common blueprint" for all IF proteins—an approximately 310 amino acid central alpha-helical rod domain, bracketed by end domains highly variable in sequence and structure (reviewed by Steinert and Roop, 1988). The alpha-helical conformation of the central rod domain is disrupted at conserved sites by nonhelical linkers. The alpha-helical subdomains are highly conserved in sequences and in lengths, while the linkers vary considerably among different keratins both in sequence and in length (Fig. 1).

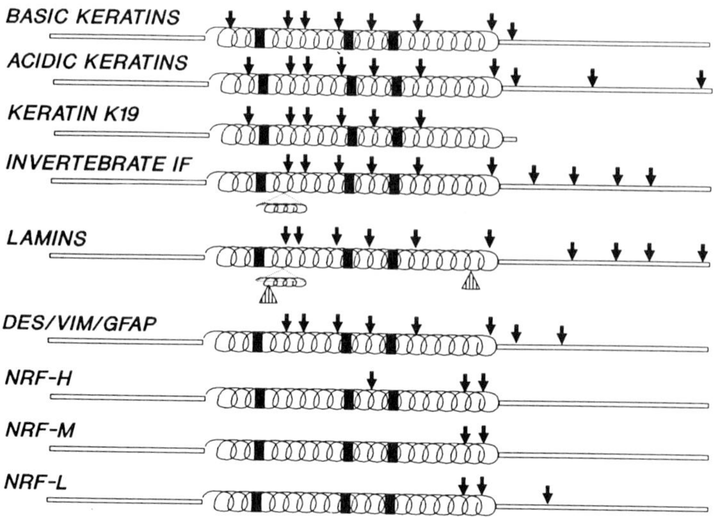

FIGURE 1 Structure of Keratin Proteins and Positions of Introns. The central alpha-helical domains are represented with spirals, the nonhelical linkers with black boxes. Open bars represent the terminal domains, but note that while the lengths of the central domains are conserved, those of the terminal domains can vary greatly, as shown for keratin K19. Arrows mark the positions of introns. Lamins and invertebrate IF proteins have alpha-helical inserts in the central domains, represented with smaller spirals. The *Drosophyla* lamin gene has only two introns; these are marked with triangles. The three neurofilament genes, NRF-H, -M and -L have fewer introns, whereas the type III genes, desmin, vimentin, and GFAP have introns in the positions equivalent to those in keratin genes.

The strict conservation of the arrangement of alpha-helical and linker subdomains was initially thought to be essential for the assembly, structure, and function of keratin filaments. However, recent mutagenesis experiments indicate that this assumption may not be fully warranted (Letai *et al.*, 1992). When the central domain of the K14 keratin is mutated to include proline (an amino acid incompatible with alpha-helical conformation), the resulting proteins form normal-looking filaments, provided the prolines are placed deep within the central domain. Conversely, removal of helix-breaking amino acids from the nonhelical linkers (thus potentially allowing linkers to assume alpha-helical conformation), also leads to normal-looking filaments. The structure of the central domain appears to be tolerant to significant alterations: both disrupting the alpha helices and making linkers alpha-helical is compatible with filament formation and structure.

However, disruption of the ends of the central domain destroys the filaments. The ends are the most highly conserved sequences and are preserved in all intermediate filament proteins. The amino-proximal end of IF central domains is invariably LNDR $\substack{\mathrm{L}\\\mathrm{F}}$ AX $\substack{\mathrm{YIE}\\\mathrm{FLD}}$ K whereas the carboxy-proximal end is TYRXLLEGEE, except for acidic type keratins, which have

TYRXLLEGED. So highly are these sequences conserved that a modestly trained eye can identify a protein sequence as belonging to the IF family by recognizing one of the two end sequences.

The importance of the end sequences of the central domain was clarified by deletion analysis (Albers and Fuchs, 1989). Proteins with deleted end sequences completely fail to assemble into filaments. However, such deletions also remove terminal domains, suggesting that deletion of terminal domains could be responsible for filament disruption. This was later, however, disproven: point mutations that substitute proline for individual amino acids in end sequences of central domains (preserving terminal domains) also completely disrupt filaments (Letai *et al.*, 1992).

The ends of the central domain, however, are more important than the middle portions either (1) because their sequences provide an essential function or (2) because mutations that disrupt the alpha-helical conformation at the ends would result in a shortening of the alpha-helical rod, thus preventing its proper alignment with its neighbors. The second hypothesis has been supported by unpublished data from D. R. Roop whose laboratory engineered a deletion within the sequence coding for an internal alpha-helical segment (thus shortening the rod domain) that did not alter the end sequences. The resulting shorter protein resulted in a dominant negative phenotype and failed to properly assemble into filaments.

Interestingly, proteins with mutations that shorten the rod domain not only themselves fail to assemble into filaments, but destroy even the preexisting filament network (Albers and Fuchs, 1989). This means that shortened central domains retain some assembly function. In the simplest model, the truncated rod remains could form an abortive association with the filaments, causing misalignment of proteins within the filaments and ultimately the collapse of the network. Alternatively, the truncated rod domains could associate with, and thus sequester, an accessory protein required for filament assembly or maintenance and in doing so indirectly cause filament disassembly. The first alternative seems, from current perspective, more likely, because intermediate filaments can assemble *in vitro* from purified components not requiring any accessory factors.

The carboxy-proximal end of the alpha-helical domain is encoded by a single exon which, curiously, is sometimes found orphaned, (i.e., without the remaining gene sequence). Such orphan exons have been found in both ovine and human genomes (Powell *et al.*, 1986; M. Blumenberg, unpublished results). Their role is unclear and it should be pointed out that while the amino acid sequences have been strictly conserved, the corresponding DNA sequences diverged as far as allowed by code degeneracy. This makes it unlikely that the sequences encoding the carboxy-proximal ends have a meta-function (e.g., regulatory) at the DNA level.

Other members of the IF protein family share with keratins not only the similarities in the ends of the central domain, but also the architecture of the alpha-helical and linker subdomains (see Fig. 1). Lamins are the exception, because they have a 42 amino-acid alpha-helical insert following the first linker segment (McKeon *et al.*, 1986). Unlike keratins, most other IF proteins can form homopolymeric filaments; the reason for this difference is not yet known.

The conserved central domains are responsible for the characteristics common to all keratins (such as assembly, interactions with associated proteins, etc.). The terminal domains, however, are not conserved and may vary greatly thus, presumably, giving each keratin its individual "character." Epidermal keratins have glycine-rich termini, hair keratins cysteine-rich, etc. One common characteristic of the carboxy-terminal sequence is a potential Ca^{2+} binding site (Franke, 1987). This site has been identified by similarity with other Ca^{2+}-binding sequences. The binding of $Ca^{2+}$to keratins at this or any other site, however, has not yet been demonstrated.

Mapping of Keratin Genes

Soon after several keratin genes were cloned, it was apparent that they were not dispersed throughout the genome but clustered in linkage groups. This was apparent for human (Blumenberg and Savtchenko, 1986; RayChaudhury *et al.*, 1986), ovine (Powell *et al.*, 1986), and bovine genes (Blessing *et al.*, 1987). Ovine basic-type keratin genes seem arranged in head-to-head orientation (Powell *et al.*, 1986); other linkage groups contain head-to-tail arrangements. The patterns of expression do not correlate with the arrangement of keratin genes on the chromosome (Savtchenko *et al.*, 1988; Rosenberg *et al.*, 1991).

The basic- and acidic-type keratin genes are not linked to each other as was originally suggested (Blumenberg and Savtchenko, 1986). Basic-type human keratin genes are found on chromosome 12q11–q13; acidic-type genes are found on chromosome 17q12–q21 (Lessin *et al.*, 1988; Romano *et al.*, 1988; Rosenberg *et al.*, 1988; Popesku *et al.*, 1989). Interestingly, syntenic regions in the murine genome contain several additional loci that affect skin and hair: *Re* (kinky hair), *Den* (denuded), and *Bsk* (bareskin) are linked to the acidic-type keratin genes on murine chromosome 11, while basic-type keratin genes are linked on chromosome 15 to *Ca* (caracul, curled hair), *Sha* (shaven), and *Ve* (velvet). This potentially includes the keratin genes into "epidermis-specific" chromosomal loci (Compton *et al.*, 1991).

Keratins K8 (basic) and K18 (acidic) are found in early embryos and simple epithelia. Judging from their sequence, they are the most distantly related to keratins of the same type and may thus be the most similar to respective ancestral keratins. Their gene location is somewhat controversial. Heath *et al.* (1990) localized a gene for K18 to chromosome 17p11–p12 by *in situ* hybridization; Waseem *et al.* (1990a,b) mapped both K8 and K18 genes to chromosome 12 using polymerase chain reaction. Because Heath *et al.* (1990) used a cDNA probe that can crossreact with numerous processed pseudogenes of K18 (Kulesh and Oshima, 1988), whereas Waseem *et al.* (1990a,b) used intron-specific sequences that are specific for the functional gene, the localization of the K18 gene to chromosome 12 is more likely. However, this implies that K18, unlike all other known acidic-type keratin genes, is in linkage with a basic-type keratin gene locus. Indeed, this emphasizes the unique character of the K8 and K18 genes. The exact location of the K8 and K18 genes on chromosome 12 is not yet known: they may be closely linked to the basic-type linkage group, distant from the linkage group but close to each other, distant from the linkage group and from each other, or the K8 gene may be with its "family" while K18 is located at a distance.

Similar linkage groups of keratin genes have been found in both murine and bovine genomes. As described earlier, murine acidic-type keratin genes are found on chromosome 11 (Nadeau *et al.*, 1989); basic-type genes are found on chromosome-15 (Compton *et al.*, 1991). The bovine equivalents are located on chromosome 19q16q–*ter* and 5q14–q23, respectively. In addition, several pseudogenes are found in bovine (Fries *et al.*, 1991), murine, and human genomes (Vasseur *et al.*, 1985; Oshima *et al.*, 1988). The myriad of K8- and K18-processed pseudogenes reflect their expression in early embryonic development (Oshima *et al.*, 1983; Oshima *et al.*, 1986). Genes expressed only in somatic tissue would not be expected to give rise to processed pseudogenes present in the germline. This makes the finding of a K19-processed pseudogene particularly interesting because it indicates that K19 could also be expressed in early embryonic or in germline cells (Savtchenko *et al.*, 1988).

Evolution

The human keratin gene-linkage group of chromosome 17 evidently underwent a duplication involving many, if not all, of the acidic-type genes of the original locus. The genes of the duplicated locus, which map at 17p11–p12, were then inactivated by mutation soon after the duplication (Savtchenko *et al.*, 1988; Rosenberg *et al.*, 1988). Subsequently, the pseudogene locus was reduplicated *in situ*, so that today there are three parallel linkage groups of acidic-type keratin genes in the human genome. This reduplication is specific for human lineage because it occurred approximately 3.8 million years ago, after the separation of human and chimpanzee ancestors. This represents the most recent known duplication of a human gene (Savtchenko *et al.*, 1990).

Intron–Exon Boundaries

While in many other proteins intron–exon boundaries demarcate the boundaries of functional domains, the domains in keratins are not encoded by separate exons (see Fig. 1). The positions of introns in the sequences encoding the terminal domains vary, reflecting independent evolutionary histories of keratin terminal domains (Klinge *et all.*, 1987; Steinert *et al.*, 1985).

The positions of introns in the central sequences are largely conserved both within the two families and between them. The first intron is the exception; it is posed somewhat more upstream in the basic-type keratin genes than in the acidic ones. Occasionally an intron will "creep"; in some keratin genes it is found a codon or two away from its exact consensus site (Rieger *et al.*, 1985). The last intron of the central domain-coding sequence is missing in the K19 gene (Bader *et al.*, 1988). This gene does not have a carboxy-terminal domain and ribosomes simply read through to the first termination codon, thus extending the alpha-helical domain by a few amino acids. With few exceptions, the position of introns corresponds to that of type III IF genes and lamin genes. This presumably means that the keratin, lamin, and type III IF genes had a common ancestor with most introns already in place (see Fig. 1).

Extraordinarily, the neurofilament genes have a completely different intron–exon arrangement: they have a pair of introns near the end of the central domain coding region but are missing all the introns the other IF genes have in common (Lewis and Cowan, 1986; Julien *et al.*, 1987; Levy *et al.*, 1987; Myers *et al.*, 1987). It has been suggested that the neurofilament genes arose from an mRNA transposition intermediate that has lost all of its introns (Lewis and Cowan, 1986). Subsequently, but before the duplications that created the three neurofilament proteins, the two introns were inserted. Later, the NF–H subunit gene received a third intron that was inserted in the vicinity of one of the consensus intron sites in all IF genes (Julien *et al.*, 1988).

The processed mRNA intermediate may have played a part in the evolution of the *Drosophyla* lamin gene as well, independently of neurofilament gene evolution, because (as indicated in Fig. 1) this gene also has only two introns, in distinct positions, but none of the consensus introns found in other IF genes (Osman *et al.*, 1990). Thus, two independent mRNA retrotranspositional events in the evolutionary history of the IF gene family must have occurred to account for the current position of introns (Fig. 2).

The relative rarity of retrotranspositional events has prompted speculation about a different evolutionary history (Julien *et al.*, 1987). It has been suggested that the common ancestor to all IF genes had a large set of introns, corresponding to all those present today in various genes. During evolution the introns were lost, most of them in the neurofilament and *Drosophyla* lamin genes; fewer in vertebrate lamin genes and nonneurofilament ancestor genes. According to this suggestion, the common ancestor of neurofilament genes was the first to separate from the common ancestor of the other IF genes (i.e., the evolutionary tree of IF genes is rooted between neurofilament genes and the remainder; Fig. 2).

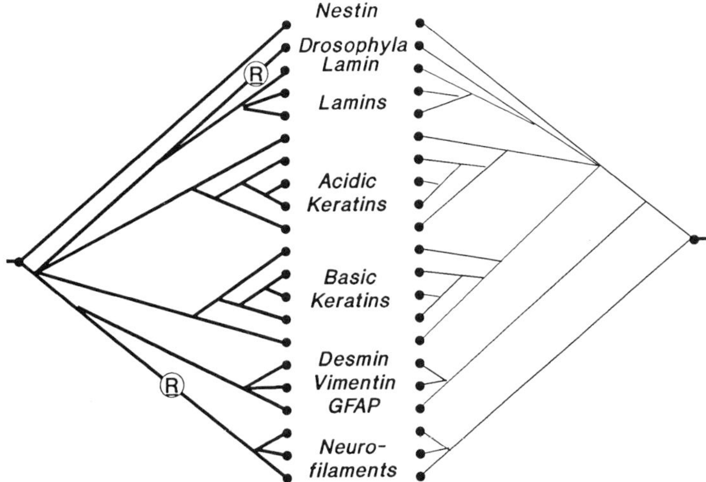

FIGURE 2 Evolutionary Trees of IF Protein Sequences. The tree on the left, with thicker branches, is based on sequence comparison and the assumption of a steady molecular clock. The two retrotranspositional events required to construct this tree are marked with a circled *R*. The tree on the right is based on the assumption that retrotranspositions did not occur and that the introns were lost gradually.

Two lines of evidence argue against such a scenario. First, the primary sequences of neurofilaments are most similar to type III IF proteins, desmin, vimentin, and GFAP. For this to occur, the mutation rate of neurofilaments and type III IF proteins must be significantly slower than of the other IF proteins, yet, in comparing the orthologous sequences from mammalian species, no significant difference in mutation rate among any of the IF proteins is evident (Blumenberg, 1989). Second, the invertebrate IF gene has the introns in the consensual positions (Dodemont *et al.,* 1990). Therefore, inasmuch as the invertebrate IF gene relates to the ancestral gene, it appears that the ancestor had the introns in the consensus positions as well.

Retrotranspositional events are so rare that we cannot at present assess their evolutionary contribution. The gradual accumulation of point mutations, on the other hand, creates a relatively steady molecular clock, and using this clock to reconstruct events from the past, it appears that the common ancestor of all IF proteins gave rise within a relatively short time period to four families, one of which later divided into the neurofilament and the type III IF families (Blumenberg, 1989).

Central Domains

Whereas the intron–exon boundaries do not provide a reliable picture of IF evolution, the comparison of primary sequences among IF proteins does. However, only the central alpha-helical domains derive from a common ancestor and so only these can be analyzed using available methodology. The ends of the central alpha-helical domains, as mentioned already, are the most highly conserved sites. The nonhelical linkers are the least conserved, both within and among IF families. The entire lengths of the alpha-helical domains are highly conserved within each family, but, except for the ends, they are much less conserved between families (Blumenberg, 1989; Levin, M., in preparation). No site can be particularly designated as characteristic of acidic or basic keratin families.

Acidic-type and basic-type keratin families are not more closely related to each other than to other IFs. Within each keratin family, the first gene duplications are likely to have separated the embryonic keratin genes from the two ancestors of remaining keratins. In this respect, as in their chromosomal location, keratin K8 and K18 genes are unique within their respective families. Soon afterward, bursts of gene duplications created the large number of keratin genes present in our genome today. The most recent duplications created the two hair keratin subfamilies and the subfamilies of basal-cell–specific and hyperproliferation-associated keratins. Remarkably, the bursts of gene duplications occurred in parallel in the acidic- and the basic-type families indicating that there exists a mechanism for sensing or coordinating gene duplication activities in two loci, unlinked and on two differ- chromosomes (Blumenberg, 1988).

Terminal Domains

The terminal domains do not derive from common ancestors. Their sizes and sequences are generally dissimilar and intron positions vary from gene to gene. Even those keratins that are extremely glycine rich—K1, K3, K9, and K10—have different, individual evolutionary histories. The glycine-rich end domains evolved by repeated, accordion-like expansions of short glycine-containing peptides (Steinert *et al.*, 1985). The initial peptide sequence varies from keratin to keratin and even within the same keratin; the amino-terminal and carboxy-terminal domains likely arose by amplification of a different initial peptide sequence (Klinge *et al.*, 1987). The amplification process most likely involved unequal crossovers between alleles, followed by selection of the longer and longer variants, until the optimal length had been achieved. The length and glycine content seem to have been more important than the actual primary sequence, judging from the comparison of human and murine orthologous sequences (Rieger and Franke, 1988).

Why do epidermal and corneal keratins have long glycine-rich terminal domains? Amino acids are usually divided, according to the affinity of their side chains for water, into hydrophilic and hydrophobic. This is a conceptual misnomer: "hydrophobic" side chains have a relatively low affinity for water, but glycine does not even have a side chain, has no affinity for water at all, and is thus the ultimate in hydrophobicity. Furthermore, polyglycine, because it lacks side chains, can pack extremely tightly into paracrystaline, hydrophobic aggregates, which are resistant to proteolysis (Crick and Rich, 1985). These are the characteristics of the materials one would use to make skin.

Hair keratins, on the other hand, have very high cysteine content, which facilitates crosslinking to the sulfur-rich components of the hair matrix. Although the structure of the hair keratin terminal domain is replete with cysteine and proline amino acids, the domains do not appear to have arisen by an accordion-like mechanism. It is more likely that the process occurred by a gradual accumulation of those point mutations that progressively increased the number of cysteine codons (Wilson *et al.*, 1988; Bertolino *et al.*, 1988).

As previously mentioned, the K19 keratin does not have a carboxy-terminal domain. This domain may have been lost by a disruption of a 5′ splice site, because the K19 gene is the only one that lacks an intron near the central domain/carboxy-terminal domain boundary-coding region (Bader *et al.*, 1988).

Mutations

Polymorphism

The heterogeneity of epidermal keratins, their modification by phosphorylation and proteolytic degradation, and their relative insolubility have made very difficult the full characterization of each member of the set of epidermal keratins. It is still controversial whether K2 is a separate keratin or a degradation product of K1. However, meticulous analysis of expressed keratins in a large population and in families has indicated that keratins K1, K4, K5, and K10 are polymorphic (Mischke and Wild, 1987; Mischke *et al.*, 1990). The polymorphism is attributed to size differences of the various proteins, which normally contain two alleles that segregate in a Hardy–Weinberg equilibrium. The lack of linkage disequilibrium between K4 and K5 loci indicates that the polymorphism is ancient and, while both genes are linked on chromosome 12, they may be quite distant from each other.

The accordion-like terminal domains would be expected to be polymorphic and indeed, using PCR, polymorphism was detected in the K10 gene (Korge *et al.*, 1992). Interestingly, only the carboxy-terminal domain was found to be polymorphic, although both terminal domains are glycine-rich and would be expected to undergo accordion-like expansions and contractions due to unequal crossovers. Similarly, the carboxy-terminal glycine-rich region of the K1 gene is polymorphic. The two alleles differ in the presence of a seven-amino acid segment, whereas in the amino-terminal glycine-rich domain a polymorphism involving a single glycine codon has been characterized (P. Steinert, personal communication).

Deleterious Mutations

Whereas polymorphism in the terminal domains need not necessarily lead to disease, mutations in the central alpha-helical domains of keratins would be expected to be more severe. Furthermore, because keratins are structural proteins, deleterious mutations would be expected to have a dominant phenotype. With this in mind, Bonifas *et al.* (1990) set out to investigate linkage between Epidermolysis Bulosa Simplex (EBS) and the keratin loci in several affected families.

EBS is a severe blistering disease. Several subtypes with dominant inheritance have been described; the most common are Weber-Cockayne (EGS-WC) and Koebner (EBS-K), which are associated with acral and generalized distribution of blisters, respectively. The blisters appear early after birth, heal without scarring, and can be induced by physical causes, such as heat or mild trauma. The blisters form in the basal layer of the epidermis and other stratified epithelia, (i.e., in cells that express K5 and K14 keratins). In one family with nine EBS-K patients, a close linkage between the disease phenotype and a marker on chromosome 17 was established (Bonifas *et al.*, 1991). Furthermore, a restriction site polymorphism within the K14 gene itself was associated with the disease. This mutation induces a Leu⟶Pro change in the helical domain near the carboxy-proximal end; it could be traced to the founder of the pedigree and has a 100% penetrance. Thus, a point mutation that disrupts the alpha-helical conformation of the K14 keratin rod domain results in the EBS-K phenotype.

In another family, with 26 EBS-WC patients, linkage to chromosome 17 was excluded, but the disease locus is clearly associated with the basic-type keratin gene family on chromosome 12. This EBS-WC phenotype was linked to, and is probably caused by, a mutation in the K5 keratin gene (Bonifas *et al.*, 1991).

Similarly, a mutation in the K5 keratin gene (specifically in the carboxy-proximal end of the alpha-helical domain) caused EBS in another large family (Lane *et al.*, 1992). A severe, Dowling-Meara subtype (EBS-DM) was linked to a Glu⟶Gly mutation in the TYRXLLEGE sequence that was conserved in all intermediate filament proteins, as described earlier. This mutation resulted in a blistering phenotype confirming that deleterious mutations either in K5 or in K14 can lead to EBS.

Another study, using a different approach, came to similar conclusions. First, the *in vitro* mutagenized keratin gene that results in disrupted filaments was inserted into the genome of transgenic mice. The resulting animals exhibited a dominant blistering phenotype with a fragile basal layer, reminiscent of EBS (Vassar *et al.*, 1991). Prompted by these studies and the results of Bonifas *et al.* (1991), Coulombe *et al.* (1991a) examined the sequences of K14 genes of two patients with EBS-DM and found two different mutations in the very same codon for a highly conserved arginine in the vicinity of the amino-proximal end. Further studies showed that the genes carrying a mutation in the conserved arginine encode proteins that are unable to form filaments *in vitro* and disrupt the endogenous keratin network when transfected into human keratinocytes *in vivo* (Coulombe *et al.*, 1991b). Therefore, it is apparent that the DM subtype of EBS is a clinical category related to the severity of the disease (i.e., a phenotype rather than a genotype designation) because mutations either in the K5 or in the K14 genes can cause EBS-DM.

A similar approach was taken by the research group of D. Roop (personal communication), who created transgenic mice carrying the internal deletion in the human K1 gene. This dominant mutation created a severe (occasionally fatal) phenotype with suprabasal blistering and scarring in the epidermal layers that produce the K1 keratin. The phenotype of affected animals closely resembled the symptoms of Epidermolytic Hyperkeratosis (EHK) thus confirming the reports of abnormal keratin proteins in this disease (Ogawa *et al.*, 1979; Yoshike *et al.*, 1983). Keratin K1 is coexpressed with K10 in the suprabasal layers of the epidermis. As reported by E. Fuchs (at the Keystone symposium, April 1992), a mutation in the K10 gene also leads to EHK-like symptoms in transgenic animals. Thus, whereas mutations in either of the basal-cell–specific keratins, K5 and K14, lead to EBS, mutations in either of the suprabasal keratins, K1 and K10, lead to EHK.

Both EBS and EHK thus seem to be caused by dominant mutations in keratin genes. The primary effect in both diseases is the breakdown of three-dimensional tissue architecture due to the collapse of the intracellular skeleton. This points to the important role of keratins in the epidermis, which is to create and maintain the proper structure and organization of the three-dimensional multicellular sheet.

This notion, that keratins are necessary for establishment and maintenance of three-dimensional epithelial structures, is confirmed by a combination of seemingly contrasting findings. These involved the mutations engineered into the K8 and K18 keratin genes, which are expressed in murine embryonic and simple epithelial cells. Targeted inactivation of both alleles of K8 (the basic-type keratin gene) in murine embryonic stem cells did not affect formation of embryoid bodies after differentiation of these cells. Even without any keratin IFs, functional polarized epithelia formed with normal microvilli and desmosomes (Abe and Oshima, 1990; Baribault and Oshima, 1991). The lack of keratin is thus compatible with normal epithelial differentiation, structure, and function. In contrast, dominant negative mutation in the K18 gene collapsed the endogenous keratin cytoskeleton in differentiating F9 cells, causing a disruption of the desmosomes and preventing the formation of a functional epithelium (Trevor, 1990). The simplest way to reconcile these two findings is to suggest that, while total lack of keratins in the cell is permissive for epithelium formation, a collapsed keratin network is highly detrimental.

Regulation of Keratin Expression

Mutual Costabilization

One of the more interesting questions in the regulation of keratin synthesis is the achievement of equimolarity between the acidic- and basic-type keratins within a cell. Because 1 : 1 interactions are involved in the early steps of the filament assembly *in vitro,* and in most cases equimolar amounts are found *in vivo,* the cell must somehow equilibrate the levels of acidic- and basic-type keratins. As it turns out, the mechanism that ensures equimolarity is refreshingly simple: the two keratin types stabilize each other; any excess of either type is rapidly degraded (Domenjoud *et al.,* 1988; Kulesh and Oshima, 1988; Kulesh *et al.,* 1989).

The presence of only one type of keratin does not induce the expression of the complementary type (Trevor *et al.,* 1987; Kulesh and Oshima, 1988; Blessing *et al.,* 1989). The early report suggesting such an induction (Giudice and Fuchs, 1987) probably misinterpreted the data due to the exquisitely powerful costabilization of keratin proteins: transfection of a basic-type gene on an expression vector results in production of a basic-type keratin, which allows pairing with (and thus preservation of) the endogenous acidic-type protein. The recipient cells express low levels of an endogenous acidic-type keratin mRNA; its translation product is undetectable in untransfected cells because of the efficient proteolysis. Although no evidence of induction of transcription of the acidic-type gene has been found, the surprising appearance of an acidic-type keratin protein epitope was assumed to be a result of induction of the acidic-type gene by a basic-type one (Giudice and Fuchs, 1987).

Regulation of Transcription

The regulation of expression of keratin genes, except for the costabilization, appears to be entirely at the level of transcription initiation (Roop *et al.,* 1988). An early report of posttranscriptional regulation was not confirmed by other researchers and may be peculiar to the K6 (hyperproliferation-associated) keratin, the transcription of which may be induced by preoperative handling of patients' skin (Tyner and Fuchs, 1986). To date, no alternative splicing of keratin mRNAs has been reported, and each protein is a product of its own gene. In one report, two transcription starts have been described for the K8 gene in PCC4 embryonal carcinoma cells, with the alternative promoter at a significant distance upstream from the coding region (Sémat *et al.,* 1988). There has been no evidence that the alternative mRNA is translated into protein.

The Role of DNA Methylation

A frequent mechanism for extinguishing transcription (or keeping extinguished) is by DNA methylation. Although significant counterexamples exist, the methylated genes are silenced and demethylation can elicit their transcription. Similar situations exists in the murine and human K8 and K18 genes. Apparently, during early embryogenesis DNA methylation is involved in silencing the K8 and K18 genes, which are already expressed at the two-cell stage and need to be silenced in the mesenchymal cells. Demethylation using 5-azacytidine induces either K8 or K18 expression at a relatively high frequency. However, in view of the instability of individual keratin proteins (*vide supra*), it takes a much rarer event (which occurs at the square of the frequency for a single activation) to demethylate both K8 and K18 genes, resulting in costabilization and filament formation (Darmon, 1985; Darmon *et al.*, 1984; Sémat *et al.*, 1986).

A different type of DNA methylation governs the expression of the murine K13 gene. Unlike K8 and K18, K13 is a differentiation marker specific for nonkeratinizing stratified squamous epithelia (e.g., esophagus). While methylation of K8 and K18 DNA appears to be general, methylation of the K13 gene is much more specific and involves the C–G dinucleotide located 2.3 kb upstream from the transcription initiation site (Winter *et al.*, 1990). The methylation of this dinucleotide strictly and inversely correlates with the transcriptional availability of the K13 gene in various epithelia and tumors. The unmethylated K13 gene is expressed in all cells, with the exception of those basal cells that are destined to express it once they leave the basal compartment. The methylatable dinucleotide is found within a DNA-se hypersensitive region that, similar to "lack of methylation," strictly correlates with the expression of the K13 gene.

Nuclear Transcription Factors

Study of regulation of transcription of keratin genes is in its infancy and several laboratories are actively pursuing the elucidation of molecular mechanisms of the various keratin genes. It currently appears that the factors involved in keratin gene transcription may be divided into three broad categories: (1) general transcription factors, components of the transcriptional machinery; (2) cell-type–specific and keratin-gene–specific factors; and (3) modulators of the level of transcription. A discussion of this categorization follows.

General transcriptional factors are considered to be those sequences (and the proteins that recognize them) that are present in many different genes and function in many different cell types. A compilation and analysis of those can be found in Wingender (1990).

The cell-type– and keratin-gene–specific factors are those that are functional only in cell types in which a given keratin is normally expressed. Such protein transcription factors have not as yet been purified and molecularly characterized.

The modulators are the transcription factors and recognition element sequences that are affected by various environmental signals, such as hormones, vitamins, growth factors, and possibly extracellular matrix components and cell–cell adhesion molecules. The modulators in turn affect the levels of transcription of various genes, including the keratin genes. A paradigm of such modulators is the retinoid receptor described in detail in Chapter 6.

The two embryonic keratins, K8 and K18, are in many ways quite different from the other keratins and some of their peculiarities will be described separately, as will the *Xenopus* keratins. As we learn more about the details of keratin gene transcription and its regulation, the boundaries between the three categories may blur, and new boundaries may appear.

General Transcription Factors

The upstream sequences for the following human keratin genes are known: K1, K3, K5, K6, K7, K8, K10, K14, K16, K17, K18, K19, and a hair keratin (Fig. 3; see Table 1). All promoters have a canonical TATA box, or a variant of it. Thus, the transcription factor TFIID, which binds the TATA box, is essential for transcription initiation. This is apparently the only transcription factor common to all keratin genes. In addition to the TATA box,

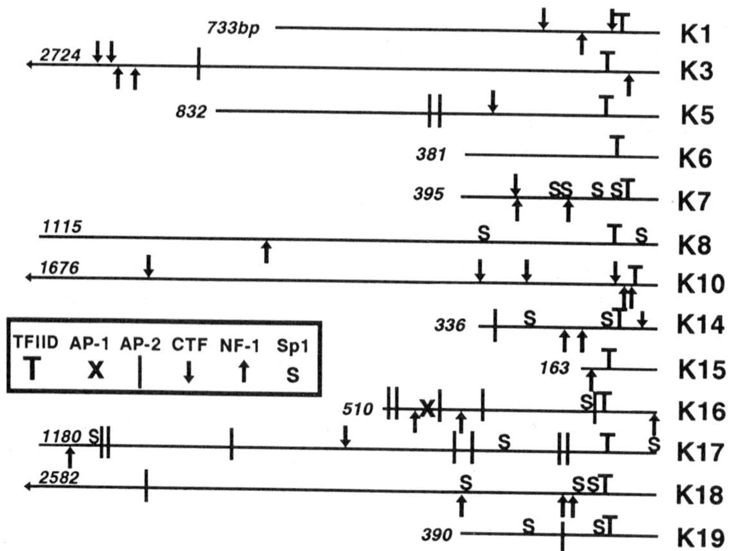

FIGURE 3 Promoters of Human Keratin Genes. The numbers represent the lengths of the sequences known; the arrowheads in the K3, K10, and K18 promoters indicate that additional sequences (*not represented here*) are available. The consensus sequences recognized by transcriptional factors are marked on each promoter DNA (*see box*). In all cases, transcription proceeds to the right.

keratin promoters contain several consensus sequences recognized by other transcriptional factors. For example, K6, K7, and K8 as well as K14, K16, K17, K18, and K19 contain one or more Sp1 sites. Interestingly, the differentiation specific murine keratin genes, K1, K3, K10, and K13, do not. The CAAT box which binds CTF protein is present (although not in the vicinity of the TATA box) in K1, K6, K7, K10, and K19 genes in the upstream region. NF-1 half-sites are present in K1, K3, K5, K8, K10, K14, K16, and K18 genes.

It is generally accepted that TATA boxes and TFIID factors play a role in transcription of keratin genes, but it is important to note that the interactions with Sp1, CTF, and NF-1 proteins are implied from the DNA sequences alone, and that functional interactions have not yet been established.

In contrast, functional interactions have been demonstrated for two general transcription factors: AP-2 with epidermal keratin genes and AP-1 with K18. The consensus sequence for AP-2 is GCCNNNGCC and similar sequences have been found in K1, K5, K6b, K10, K14, and K16 genes (Leask *et al.*, 1991). AP-2 is expressed in epithelial cells including those in which the above keratin genes are not expressed. It is also expressed in a variety of nonepithelial cells. Purified AP-2 proteins can bind both K5 and K14 DNA, and binding of an AP-2–like protein to the K14 site in crude nuclear extract has been demonstrated. The importance of the AP-2 site in regulating keratin gene transcription is not fully understood, because disruption of the site in the K14 gene promoters results in only a twofold reduction of transcriptional activity (Leask *et al.*, 1990).

On the other hand, the AP-1 site in the first intron of human and murine K18 genes is very important for expression, because disruption of the AP-1 site results in a significant, fivefold reduction in transcription (Oshima *et al.*, 1990). The role of AP-1 is particularly interesting in view of the fact that K8 and K18 genes are induced in F9 cells during differentiation. The FOS and JUN proteins, which constitute the AP-1 factor, are also induced during F9 cell differentiation. This may provide one of the causal molecular links between differentiation and induction of K8 and K18 genes. Note that a functional AP-1 site has also been described in the K8 gene (Tamai *et al.*, 1991), and that in the first intron of the K18 gene, the AP-1 site is a component of an enhancer (a necessary but not sufficient element) for increasing the level of transcription (Oshima *et al.*, 1990).

Specific Elements

The most salient characteristic of keratin gene expression is its epithelial specificity; the molecular mechanisms that confer it are, however, unknown. Keratin synthesis is downregulated in keratinocyte X fibroblast cell hybrids, indicating that it is under negative control in nonepithelial cells (Peehl and

Stanbridge, 1981). In contrast, in hybrids between different epithelial cell types, the high level of keratin expression is inherited dominantly from the more differentiated parent. This indicates that within epithelial cell types, keratin expression may be positively regulated (Berdichevsky and Taylor-Papadimitriou, 1991).

It is still a mystery why and how different types of epithelia express different keratins. For instance, why are hair keratins expressed in certain regions of the tongue (Dhouailly *et al.*, 1989)? What determines the abrupt change in the expression of the corneal keratins at the limbus–conjunctiva border (Kolega *et al.*, 1989)? What determines the varieties of keratins expressed in different epithelia of the oral cavity (Ouhayoun *et al.*, 1985)? The simplest hypothesis (the basis for much research effort) states that various cell types have correspondingly unique transcription factors or a combination of factors that specifically regulate different keratin genes.

The first cell-type–specific element in keratin genes was described by Blessing *et al.* (1989) for the bovine K6 gene. An enhancer exists within a 650-bp sequence upstream from the gene and functions in proliferating cells of stratified epithelial origin. It does not function in simple epithelial cells. This enhancer may be specific for either stratified or hyperproliferating cells, or both. It can confer enhancement of transcription to other promoters including those for K1 and K10 keratins, which are expressed only in differentiating cells (and thus in conjunction with the K6 gene enhancer) in cultured cells.

The promoter region of the human keratin K14 gene, originally thought to be promiscuously expressed in nonepithelial cells (Vassar *et al.*, 1989), is actually specific for epithelial cells (Jiang *et al.*, 1990). In transgenic mice, it faithfully mimics the expression of the endogenous, murine K14 gene in the basal layers of stratified epithelia (Vassar *et al.*, 1989). In transfection experiments, a relatively short, 300-bp, promoter region, was not expressed in fibroblasts and melanocytes, but was expressed in cells derived both from simple and stratified epithelia. It thus contains the epithelial-specific, but not the stratified epithelial-specific determinants. This construct does not have the AP-2 site described earlier. Expression in stratified epithelial cells is extremely high, and, in these cell types, surpasses even the strong viral promoters of SV40 and RSV (Jiang *et al.*, 1990).

When we performed functional comparison analyses with the promoter regions of the human K5, K6, and K10 genes, we found that, while none of them was expressed in nonepithelial fibroblast cells, only K14 was expressed at high levels in transformed epithelial (HeLa) or simple epithelial cells [i.e., those derived from rabbit or mouse mesothelium (Jiang *et al.*, 1991)]. The K5 promoter was functional, albeit at relatively low levels, in HeLa cells. Thus, while all epidermal keratin genes appear epithelial-specific, K6 and K10 gene promoters also contain the stratified epithelial-specific determinants.

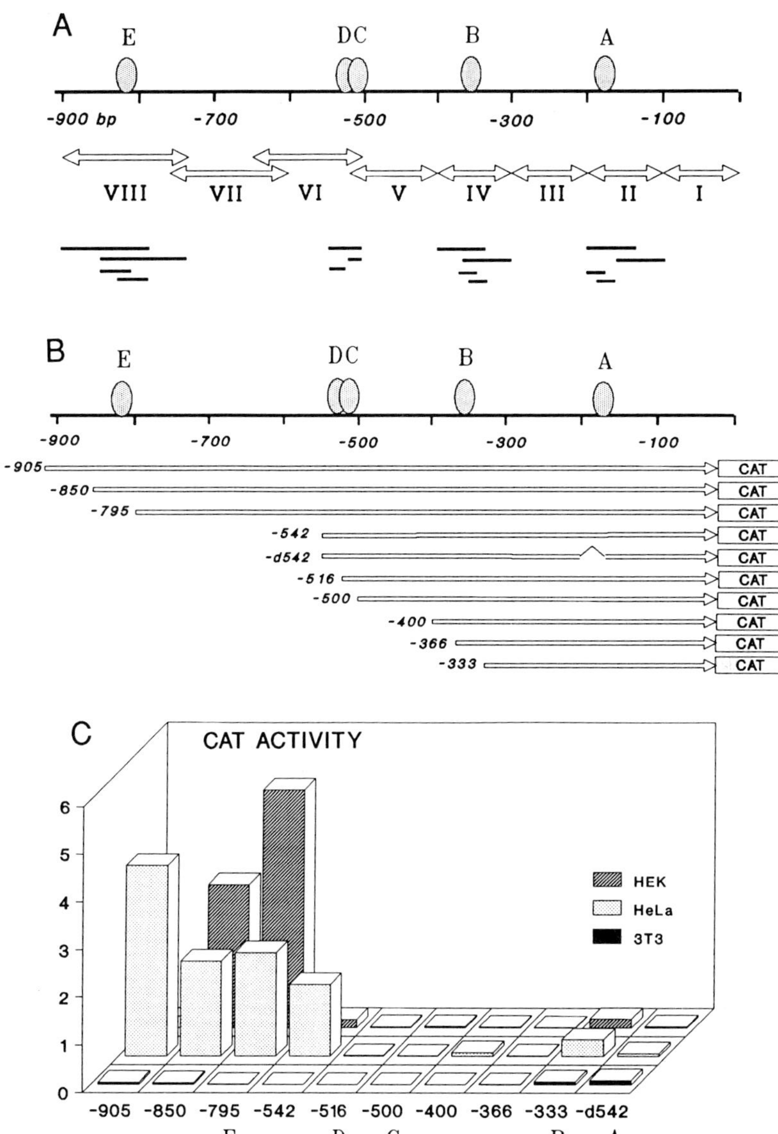

FIGURE 4 Upstream Region of the K5 Keratin Gene. (*A*) The top solid black line represents the K5 gene DNA with the ovals representing the five proteins, *A–E*, bound to their corresponding sites. The double-headed arrows represent segments used in gel shift experiments and the filled-in bars below them DNAs used as competitors. (*B*) Arrows represent deletion constructs linked to the reporter gene CAT and functionally tested by transfection into 3T3 fibroblasts, HeLa cells, and human epidermal keratinocytes (HEK). (*C*) The panel shows the results of the transfection experiments in those three cell types. Note that the *d*542 construct has an internal deletion of site *A*. Reprinted with permission from Blumenberg *et al.*, (1992) Elsevier, *J. Invest. Dermatol.* **99**, 206–215.

The complex regulation of expression of the K5 keratin prompted us to investigate the molecular interactions between nuclear proteins and DNA upstream from the human K5 gene. Both positive and negative factors play a role in regulation of K5 expression. This is as expected from the induction of K5 keratin in the basal layer of stratified epithelia and its subsequent supression in the suprabasal layers. Similarly, the "turn-on" in urothelium in culture, and the "shut-off" *in vivo* (Surya *et al.*, 1990) as well as other similar regulations is multifaceted. The upstream regulatory region of the K5 keratin gene binds at least five different nuclear proteins from HeLa cells. The five binding sites are marked *A–E* in Figure 4. At least two sites bind positive regulators (*A* and *D*); one appears to bind a negative regulator (site *B*); for one the role is not determined (site *C*); and the most distal (site *E*) does not seem to have a function under the conditions tested (Ohtsuki *et al.*, 1992). We have chosen HeLa cells because they are a prototype of epithelial cell lines and our previous results indicate that the cloned K5 promoter is active in HeLa cells (Jiang *et al.*, 1991). Of course, other cell types (e.g., epidermal keratinocytes) may use a different, possibly overlapping set of nuclear proteins to regulate K5 expression.

Gel retardation assays preferentially detect abundant nuclear binding proteins. Such proteins are most commonly, but not always, those affecting transcriptional regulation. The nuclear proteins from HeLa cells do not bind the AP-2 site of the K5 promoter, although the purified AP-2 protein (as described earlier) does. We have not detected significant competition among the five binding sites, which implies that each protein binds separately to its own site. The proteins can be separated physically. Even the proteins that bind adjacently to each other (at sites *C* and *D*) do not seem to cooperate in binding. The simplest model then has five proteins binding at five sites to perform five tasks.

The Modulators

Because of their biologic, pharmacologic, and cosmetic importance, the effects of retinoids are described in a separate chapter (Chapter 6). Briefly, epidermal keratin genes contain clusters of binding sites for the retinoic acid receptor monomer. Several receptors can simultaneously bind to these clusters to suppress transcription (Tomic-Canic *et al.*, 1992).

The same clusters of monomer binding sites are also recognized by the thyroid hormone receptor, which can also, in a ligand-dependent way, suppress the expression of keratin genes (Tomic *et al.*, 1990). In contrast, the vitamin D_3 receptor has no direct effect on epidermal keratin gene transcription (Tomic *et al.*, 1991a). This is unexpected because vitamin D_3 is the "skin vitamin," and is produced in the epidermis. Keratinocytes have receptors for vitamin D_3. Keratinocytes respond dramatically to vitamin D_3 by turning into cornified envelopes by the first suprabasal layer and thus virtually completing their differentiation (Regnier and Darmon, 1991). Therefore, the effect of vitamin D_3 is diametrically opposed to the RA effect. In addition, the vitamin D_3 receptor can recognize and bind to the same consensus monomer binding site to which RA and T3 receptors bind. Yet, vitamin D_3 and its receptor do not affect keratin gene expression directly, but by indirect modulation of the differentiation state of the keratinocyte. The analysis of the effects of additional nuclear receptors (i.e., estrogen, progesterone, androgens, and glycocorticoids) is also being investigated.

An interesting detail in the modulation of keratin gene expression is the induction of K17 by UV light in human keratinocytes (Kartasova *et al.*, 1987). The induction occurs at the mRNA level, but the mechanism has not yet been explored in detail.

A very important (but as yet unexplored at the molecular level) modulation effect is provided by the cytokines and growth factors. These molecules are both produced by keratinocytes and affect their physiology, including modulation of keratin expression. Their interactions with ECM, as well as their intercellular interactions, may also prove to be modulators of keratin gene expression. Pathologic states are also known to affect expression of keratin genes, but the corresponding molecular mechanisms are as yet unknown.

The Specific Case of K8 and K18 Genes

As previously mentioned, the K8 and K18 keratins are in many ways exceptional when compared to other keratins: they are the most distantly related members of their respective families, and their genes are linked together. Also, K8 and K18 are the first keratins to be expressed during embryonic development and are detectable after the first cell division of the zygote (Oshima *et al.*, 1983, Lehtonen *et al.*, 1983). K8 and K18 are the only keratins induced, instead of suppressed, by retinoic acid (Tomic-Canic *et al.*, 1992, unpublished). In adult organisms, K8 and K18 are only found in simple epithelia (Moll *et al.*, 1982), but stratified epithelial cells, when oncogenically transformed, can re-express K8 and K18, thus replacing the stratification-specific keratins (Markey *et al.*, 1991). Because of these unique characteristics, the regulation of expression of K8 and K18 keratins has received special attention.

In undifferentiated embryonal carcinoma cells (PCC4), K8 keratin expression is negatively regulated (i.e., inhibited by a protein with a short half-life). Blocking of total protein synthesis depletes cells of this inhibitor protein and induces K8 gene transcription (Crèmisi and Duprey, 1987). The site of action of the inhibitor protein is not known: it could be in the promoter region, but it could also act on the enhancer that has been found three kilobases downstream from the gene (Takemoto *et al.*, 1991). The enhancer consists of six tandemly repeated elements unrelated to AP-1 sites. An AP-1 site is found in the first intron of the murine gene, but its function is thus far unknown (Tamai *et al.*, 1991). On the other hand, K18 requires positive regulatory factors in different embryonal carcinoma cells (F9), including at least the retinoic acid receptor RARβ and the FOS and JUN proteins of the AP-1 factor (Oshima *et al.*, 1990). The receptor acts in the 5′-upstream region (M. Tomic-Canic, personal communication), whereas the AP-1 site is a part of an enhancer located in the first intron. Thus, both K8 and K18 genes require sequences downstream from the transcription initiation site for their function; they have enhancers in the 3′ region, and in the first intron, respectively.

The unique expression of keratins K8 and K18 is perhaps most striking in fishes: even nonepithelial tissues, such as endothelia, in smooth muscle, as well as fibroblasts, chondrocytes, and glia express K8 and K18 as a predominant or sole IF protein (Markl and Franke, 1988; Giordano *et al.*, 1990). In mammals, these cell types express vimentin IFs; yet, even in mammals, K8 and K18 are the only keratins that can be found in nonepithelial cells such as the cells of vascular smooth muscle (Bader *et al.*, 1988).

Keratin Expression in Xenopus laevis

In amphibian *X. laevis,* oocytes contain the K8 keratin transcript (Franz and Franke, 1986). K18 keratin is found expressed in the notochord of the developing embryo (LaFlamme *et al.*, 1988). Staged succession of expression of some other keratins can be precisely correlated to specific developmental stages in the tadpole (Winkles *et al.*, 1985; Miyatani *et al.*, 1986). These keratins are not present in the adult animal. The expression of keratins found in the epidermis is transcriptionally regulated (Jonas *et al.*, 1989). A specific protein that binds to a site in the promoter region has been characterized (Snape *et al.*, 1990).

The changes in keratin gene expression that accompany transformation are under thyroid hormonal regulation (Mathisen and Miller, 1989). The anuran keratins thus resemble mammalian keratins, although it has not yet been demonstrated in *Xenopus* that keratins are directly regulated by thyroid hormones, i.e., by direct interaction of the receptor with keratin genes.

Perspectives

Since the cloning of keratin genes, a tremendous wealth of knowledge has been derived about the genes themselves and as paradigms of differentially regulated large gene families. Still, many unanswered questions remain that will be the source of extensive study in the future.

Regarding keratin protein structure, it is still unclear why and how they are obligate heteropolymers. Perhaps even more intriguing are the questions of structural interaction with associated proteins, desmosomal and hemidesmosomal components, filaggrin, cytoskeletal proteins, matrix proteins in hair, etc. These questions are especially important because one demonstrated function of keratins is to maintain the three-dimensional tissue architecture. Keratins can accomplish this only by interacting with additional structural proteins. Keratin proteins are extensively modified both by proteolysis and by phosphorylation (and possibly by glycosylation as well), but the mechanism, structure, and function of the modifications is at present unknown. Predictably, these modifications will assume higher importance as they become better understood.

At the gene level, as the complete maps of the keratin loci are revealed, they will further illuminate evolutionary history. Maps will also help analyze the regulation of transcription of keratin genes because, as pointed out by E. Epstein at the S.I.D. meeting held in Seattle in 1991, "The 5′ of one keratin gene is the 3′ of another." Thus keratin genes may be coordinately regulated, as well as compete for transcription factors. The sites described in Figures 3 and 4 are merely a small fraction of the regulatory elements of keratin genes. It is doubtful that the complete regulatory circuities will be known for quite some time.

Among the unanswered questions of transcriptional regulation are the mechanisms that ensure cell type specificity and the mechanisms of pair-wise expression. Why is K1 coexpressed with K10; K5 with K14; and K6 with K16? Whereas K17 and K19 do not have basic-type partners, why can both K7 and K8 be coexpressed with K18, and why is K8 often coexpressed with K18 and K19? An understanding of the regulation of keratin gene expression will lead to a precise targeting of genes into epidermis. This promises enormous benefits in future somatic gene therapy, because the accessible epidermal keratinocytes are a very convenient target vector.

Finally, exciting discoveries of the role keratins play in severe epidermal diseases indicates that they may perhaps play a similar role (albeit in more subtle ways) in other cutaneous disorders. Here milder symptoms may be caused by inappropriate interactions among keratins or between keratins and associated proteins. Such diseases may be amenable to pharmacological intervention using drugs that modulate keratin gene expression or post-transcriptional modification. Therefore, we can expect that upcoming research on keratin genes will continue to grow and attract the interest and talents of many scientists and clinicians.

Dedicated to Charley Yanofsky

Acknowledgments

I would like to thank all those who worked with me in this field, particularly Irwin Freedberg, as well as Ekaterina Savtchenko, Chuan-Kui Jiang, and Marjana Tomic-Canic. Our research has been supported by National Institutes of Health grants AR30682 and AR39176; the NYU Skin Disease Research Center Grant AR39749; and the R. L. Baer Foundation. M. B. is a recipient of the Irma T. Hirschl Career Scientist Award.

References

Albers, K., and Fuchs, E. (1989). Expression of mutant keratin cDNAs in epithelial cells reveals possible mechanisms for initiation and assembly of intermediate filaments. *J. Cell Biol.* **108**, 1477–1493.

Bader, B. L., Jahn, L., and Franke, W. W. (1988). Low level expression of cytokeratins 8, 18 and 19 in vascular smooth muscle cells of human umbilical cord and in cultured cells derived therefrom, with an analysis of the chromosomal locus containing the cytokeratin 19 gene. *Euro. J. Cell. Biol.* **47**, 300–319.

Baribault, H. and Oshima, R. G. (1991). Polarized and functional epithelia can form after the targeted inactivation of both mouse keratin 8 alleles. *J. Cell Biol.* **115**, 1675–1684.

Berdichevsky, F. and Taylor-Papadimitriou J. (1991). Morphological differentiation of hybrids of human mammary epithelial cell lines is dominant and correlates with the pattern of expression of intermediate filaments. *Exp. Cell Res.* **194**, 267–274.

Bertolino, A. P., Checkla, D. M., Notterman, R., Sklaver, I., Schiff, T. A., Freedberg, I. M., and DiDona, G. J. (1988). Cloning and characterization of a mouse type I hair keratin cDNA. *J. Invest. Dermatol.* **91**, 541–546.

Bertolino, A. P., Checkla, D. M., Heitner, S., Freedberg, I. M. and Yu, D. -W. (1990). Differential expression of type I hair keratins. *J. Invest. Dermatol.* **94**, 297–303.

Blessing, M., Zentgraf, H., and Jorcano, J. L. (1987). Differentially expressed bovine cytokeratin genes. Analysis of gene linkage and evolutionary conservation of 5′-upstream sequences. *EMBO J.* **6**, 567–575.

Blessing, M., Jorcano, J. L., and Franke, W. W. (1989). Enhancer elements directing cell-type–specific expression of cytokeratin genes and changes of the epithelial cytoskeleton by transfections of hybrid cytokeratin genes. *EMBO J* **8**, 117–126.

Blumenberg, M. (1988). Concerted gene duplications in the two keratin gene families. *J. Mol. Evol.* **27**, 203–211.

Blumenberg, M. (1989). Evolution of homologous domains of cytoplasmic intermediate filament proteins and lamins. *Mol. Biol. Evol.* **6**, 53–65.

Blumenberg, M., and Savtchenko, E. S. (1986). Linkage of human keratin genes. *Cytogen. Cell Genet.* **42**, 65–71.

Bonifas, J. M., Rothman, A. L., and Epstein, E. (1990). Linkage of epidermolysis bullosa simplex to probes in the region of keratin gene clusters on chromosomes 12q and 17q. *J. Invest. Dermatol.* **96**, 550.

Bonifas, J. M., Rothman, A. L., and E. P. Epstein, Jr. (1991). Epidermolysis bullosa simplex: evidence in two families for keratin gene abnormalities. *Science* **254**, 1202–1205.

Chandler, J. S., Calnek, D. and Quaroni, A. (1991). Identification and characterization of rat intestinal keratins. *J. Biol. Chem.* **266**, 11932–11938.

Compton, J. G., Ferrara, D. M., Yu, D., Recca, V., Freedberg, I. M., and Bertolino, A. P. (1991). Chromosomal localization of mouse hair keratin genes. In *The Molecular and Structural Biology of Hair* (K. S. Stenn, A. G. Messenger, and H. P. Baden, eds.). pp. 32–43. Ann. N.Y. Acad. of Sci.: New York.

Coulombe, P. A., Hutton, M. E., Letai, A., Hebert, A., Paller, A. S., and Fuchs, E. (1991a). Point mutations in human keratin 14 genes of epidermolysis bullosa simplex patients: genetic and functional analyses. *Cell* **66**, 1301–1311.

Coulombe, P. A., Hutton, M. E., Vassar, R., and Fuchs, E. (1991b). A function for keratins and a common thread among different types of epidermolysis bullosa simplex diseases. *J. Cell Biol.* **115**, 1661–1674.

Crèmisi, C. and Duprey, P. (1987). A labile inhibitor blocks *endo* A gene transcription in murine undifferentiated embryonal carcinoma cells. *Nucleic Acids Res.* **15**, 6105–6116.

Crick, F. H. C. and Rich, A. (1985). Structure of polyglycine II. *Nature* **176**, 780–782.

Darmon, M. (1985). Coexpression of specific acid and basic cytokeratins in teratocarcinoma-derived fibroblasts treated with 5-azacytidine. *Dev. Biol.* **110**, 47–52.

Darmon, M., Nicolas, J.-F., and Lamblin, D. (1984). 5-Azacytidine is able to induce the conversion of teratocarcinoa-derived mesenchymal cells into epithelial cells. *EMBO J.* **3**, 961–967.

Darmon, M. Y., Sémat, A., Darmon C. D., and Vasseur, M. (1987). Sequence of a cDNA encoding human keratin NO. 10 selected according to structural homologies of keratins and their tissue-specific expression. *Mol. Biol. Rep.* **12**, 277–283.

Dhouailly, D., Xu, C., Manabe, M., Schermer, A., and Sun, T.-T. (1989). Expression of hair-related keratins in a soft epithelium: subpopulations of human and mouse dorsal tongue keratinocytes express keratin markers for hair-, skin- and esophageal-types of differentiation. *Exp. Cell Res.* **181**, 141–158.

Dodemont, H., Riemer, D., and Weber, K. (1990). Structure of an invertebrate gene encoding cytoplasmic intermediate filament (IF) proteins: implications for the origin and the diversification of IF proteins. *EMBO J.* **9**, 4083–4094.

Domenjoud, L., Jorcano, J. L., Breuer, B., and Alonso, A. (1988). Synthesis and fate of keratins 8 and 18 in nonepithelial cells transfected with cDNA. *Exp. Cell Biol.* **179**, 352–361.

Ehlen, T. and Dubeau, L. (1989). Detection of *ras* point mutations by polymerase chain reaction using mutation-specific inosine-containing oligonucleotide primers. *Biochem. Biophys. Res. Commun.* **160**, 441–447.

Franke, W. W. (1987). Homology of a conserved sequence in the tail domain of intermediate filament proteins with the loop region of calcium binding proteins. *Cell Biol. Int. Rep.* **11**, 831.

Franz, J. and Franke, W. W. (1986). Cloning of cDNA and amino acid sequence of a cytokeratin expressed in oocytes of *Xenopus laevis*. *Proc. Nat. Acad. Sci. U.S.A.* **83**, 6475–6479.

Fries, R., Threadgill, D. W., Hediger, R., Gunawardana, A., Blessing, M., Jorcano, J. L., Stranzinger, G., and Womack, J. E. (1991). Mapping of bovine cytokeratin sequences to four different sites on three chromosomes. *Cytogenet. Cell Genet.* **57**, 135–141.

Giordano, S., Hall, C., Quitschke, W., Glasgow, E., and Schecter, N. (1990). Keratin 8 of simple epithelia is expressed in glia of the goldfish nervous system. *Differentiation* **44**, 163–172.

Giudice, G. J. and Fuchs, E. (1987). The transfection of epidermal keratin genes into fibroblasts and simple epithelial cells: evidence for inducing a type I gene by a type II gene. *Cell* **48**, 453–463.

Glass, C. and Fuchs, E. (1988). Isolation, sequence, and differential expression of a human K7 gene in simple epithelial cells. *J. Cell Biol.* **107**, 1337–1350.

Glass, C., Kim, K. H., and Fuchs, E. (1985). Sequence and expression of a human type II mesothelial keratin. *J. Cell Biol.* **101**, 2366–2373.

Hanukoglu, I. and Fuchs, E. (1982). The cDNA sequence of a human epidermal keratin: divergence of sequence but conservation of structure among intermediate filament proteins. *Cell* **31**, 243–252.

Hanukoglu, I. and Fuchs, E. (1983). The cDNA sequence of a type II cytoskeletal keratin reveals constant and variable structural domains among keratins. *Cell* **33**, 915–924.

Heath, P., Elvin, P., Jenner, D., Gammack, A., Morten, J. and Markham, A. (1990). Localisation of a cDNA clone for human cytokeratin 18 to chromosome 17p11–p12 by *in situ* hybridisation. *Hum. Genet.* **85**, 669–670.

Heid, H. W., Moll, I., and Franke, W. W. (1988). Patterns of expression of trichocytic and epithelial cytokeratins in mammalian tissues. I. Human and bovine hair follicles. *Differentiation* **37**, 137–157.

Hsieh, J.-T., Zhau, H. E., Wang, X.-H., Liew, C.-C., and Chung, L. W. K. (1992). Regulation of basal and luminal cell-specific cytokeratin expression in rat accessory sex organs. *J. Biol. Chem.* **267**, 2303–2310.

Ichinose, Y., Morita, T., Zhang, F., Srimahasongcram, S., Tondella, M. L. C., Matsumoto, M.,

Nozaki, M., and Matsushiro, A. (1988). Nucletide sequence and structure of the mouse cytokeratin *endo B* gene. *Gene* **70**, 85–95.

Jiang, C.-K., Epstein, H. S., Tomic, M., Freedberg, I. M., and Blumenberg, M. (1991). Functional comparison of the upstream regulatory DNA sequences of four human epidermal keratin genes. *J. Invest. Dermatol.* **96**, 162–167.

Jiang, C.-K., Epstein, H. S., Tomic, M., Freedberg, I. M., and Blumenberg, M. (1990). Epithelial-specific keratin gene expression: identification of a 300 base-pair controlling segment. *Nucleic Acids Res.* **18**, 247–253.

Johnson, L. D., Idler, W. W., Zhou, X.-M., Roop, D. R., and Steinert, P. M. (1985). Structure of a gene for the human epidermal 67-kDa keratin. *Proc. Nat. Acad. Sci. U.S.A.* **82**, 1896–1900.

Jonas, E. A., Snape, A. M., and Sargent T. D. (1989). Transcriptional regulation of a *xenopus* embryonic epidermal keratin gene. *Development* **106**, 399–405.

Jorcano, J. L., Franz, J. K., and Franke, W. W. (1984). Amino acid sequence diversity between bovine epidermal cytokeratin polypeptides of the basic (type II) subfamily as determined from cDNA clones. *Differentiation* **28**, 155–163.

Julien, J.-P., Grosveld, F., Yazdanbaksh, K., Flavell, D., Meijer, D., and Mushynski, W. (1987). The structure of a human neurofilament gene (NF-L): a unique exon–intron organization in the intermediate filament gene family. *Biochem. Biophys. Acta* **909**, 10–20.

Julien, J.-P., Coté, F., Beaudet, L., Sidky, M., Flavell, D., Grosveld, F., and Mushynsky, W. (1988). Sequence and structure of the mouse gene coding for the largest neurofilament subunit. *Gene* **68**, 314–322.

Kartasova, T., Cornelissen, B. J., Belt, P., and van de Putte, P. (1987). Effects of UV, 4-NQO and TPA on gene expression in cultured human epidermal keratinocytes. *Nucleic Acids Res.* **15**, 5945–5962.

Klinge, E. M., Sylvestre, Y. R., Freedberg, I. M., and Blumenberg, M. (1987). Evolution of keratin genes: different protein domains evolve by different mechanisms *J. Mol. Evol.* **24**, 319–229.

Knapp, B., Rentrop, M., Schweizer, J., and Winter, H. (1986). Nonepidermal members of the keratin multigene family, cDNA sequences and *in situ* localization of the mRNAs. *Nucleic Acids Res.* **14**, 751–763.

Knapp, B., Rentrop, M., Schweizer, J., and Winter, H. (1987). Three cDNA sequences of mouse type I keratins. *J. Biol. Sci.* **262**, 938–945.

Kolega, J., Manabe, M., and Sun, T.-T. (1989). Basement membrane heterogeneity and variation in corneal epithelial differentiation. *Differentiation* **42**, 54–63.

Korge, B. P., Gan, S.-Q., McBride, O. W., Mischke, D., and Steinert, P. M. (1992). Extensive size polymorphism of the human keratin 10 chain resides in the C-terminal *V2* subdomain due to variable numbers and sizes of glycine loops. *Proc. Natl. Acad. Sci. U.S.A.* **89**, 910–914.

Krauss, S. and Franke, W. W. (1990). Organization and sequence of the human gene encoding cytokeratin 8. *Gene* **86**, 241–249.

Kulesh, D. A. and Oshima, R. G. (1988). Cloning of the human keratin 18 gene and its expression in nonepithelial mouse cells. *Mol. Cell. Biol.* **8**, 1540–1550.

Kulesh, D. A. and Oshima, R. G. (1989). Complete structure of the gene for human keratin 18. *Genomics* **4**, 339–347.

Kulesh, D. A., Cecena, G., Darmon, Y. M., Vasseur, M., and Oshima, R. G. (1989). Posttranslational regulation of keratins: degradation of mouse and human keratins 18 and 8. *Mol. Cell. Biol.* **9**, 1553–1565.

Kuruc, N., Leube, R. E., Moll, I., Bader, B. L., and Franke, W. W. (1989). Synthesis of cytokeratin 13, a component characteristic of internal stratified epithelia, is not induced in human epidermal tumors. *Differentiation* **42**, 111–123.

LaFlamme, S. E., Jamrich, M., Richter, K., Sargent, T. D., and Dawid, I. B. (1988). *Xenopus endo B* is a keratin preferentially expressed in the embryonic notochord. *Genes & Devel.* **2**, 853–862.

Lane, E. B., Rugg, E. L., Navsaria, H., Leigh, I. M., Heagery, A. H. M., Ishida-Yamamoto, A. I., and Eady, R. A. J. (1992). A mutation in the conserved helix termination peptide of keratin 5 in hereditary skin blistering. *Nature* **356,** 244–246.

Leask, A., Rosenberg, M., Vassar, R., and Fuchs, E. (1990). Regulation of a human epidermal keratin gene: sequences and nuclear factors involved in keratinocyte-specific transcription. *Genes & Devel.* **4,** 1985–1998.

Leask, A., Byrne, C., and Fuchs, E. (1991). Transcription factor AP2 and its role in epidermal-specific gene expression. *Proc. Natl. Acad. Sci. U.S.A.* **88,** 7948–7952.

Lehtonen, E., Lehto, V.-P., Vartio, T., Badley, R. A., and Virtanen, I. (1983). Expression of cytokeratin polypeptides in mouse oocytes and preimplantation embryos. *Devel. Biol.* **100,** 158–165.

Lersch, R. and Fuchs, E. (1988). Sequence and expression of a type II keratin, K5, in human epidermal cells. *Mol. Cell. Biol.* **8,** 486–493.

Lessin, S. R., Huebner, K., Isobe, M., Croce, C. M., and Steinert, P. M. (1988). Chromosomal mapping of human keratin genes: evidence of nonlinkage. *J. Invest. Dermatol.* **91,** 572–578.

Letai, A., Coulombe, P. A., and Fuchs, E. (1992). Do the ends justify the mean? Proline mutations at the ends of the keratin coiled–coil rod segment are more disruptive than internal mutations. *J. Cell Biol.* **116,** 1181–1195.

Leube, R. E., Bader, B. L., Bosch, F. X., Zimbelmann, R., Achtstaetter, T., and Franke, W. W. (1988). Molecular characterization and expression of the stratification-related cytokeratins 4 and 15. *J. Cell Biol.* **106,** 1249–1261.

Levin, M., and Blumenberg, B. In preparation.

Levy, E., Liem, R. K. H., D'Eustachio, P., and Cowan, N. J. (1987). Structure and evolutionary origin of the gene encoding mouse NF-M, the middle-molecular-mass neurofilament protein. *Eur. J. Biochem.* **166,** 71–77.

Lewis, S. A. and Cowan, N. J. (1986). Anomalous placement of introns in a member of the intermediate filament multigene family: an evolutionary conundrum. *Mol. Cell. Biol.* **6,** 1529–1534.

Marchuk, D., McCrohon, S., and Fuchs, E. (1984). Remarkable conservation of structure among intermediate filament genes. *Cell* **39,** 491–498.

Marchuk, D., McCrohon, S., and Fuchs, E. (1985). Complete sequence of a gene encoding a human type I keratin: sequences homologous to enhancer elements in the regulatory region of the gene. *Proc. Nat. Acad. Sci. U.S.A.* **82,** 1609–1613.

Markey, A. C., Lane, B. E., Churchill, L. J., MacDonald, D. M., and Leigh, I. M. (1991). Expression of simple epithelial keratins 8 and 18 in epidermal neoplasia. *J. Invest. Dermatol.* **97,** 763–770.

Markl, J. and Franke, W. W. (1988). Localization of cytokeratins in tissues of the rainbow trout: fundamental differences in expression patterns between fish and higher vertebrates. *Differentiation* **39,** 97–122.

Mathisen, P. M. and Miller, L. (1989). Thyroid hormone induces constitutive keratin gene expression during *Xenopus laevis* development. *Mol. Cell. Biol.* **9,** 1823–1831.

McKeon, F. D., Kirschner, M. W., and Caput, D. (1986). Homologies in both primary and secondary structure between nuclear envelope and intermediate filament proteins. *Nature* **319,** 463–468.

Mischke, D. and Wild, G. (1987). Polymorphic keratins in human epidermis. *J. Invest. Dermatol.* **88,** 191–197.

Mischke, D., Wachter, E., Hochstrasser, K., Wild, A. G. and Schulz, P. (1989). The N-, but not the C-terminal domains of human keratins 13 and 15 are closely related. *Nucleic Acids Res.* **17,** 7984.

Mischke, D., Wille, G., and Wild, A. G. (1990). Allele frequencies and segregation of human polymorphic keratins K4 and K5. *Am. J. Hum. Genet.* **46,** 548–552.

Miyatani, S., Winkles, J. A., Sargent, T. D., and Dawid, I. B. (1986). Stage-specific keratins in *Xenopus laevis* embryos and tadpoles: the XK81 gene family. *J. Cell. Biol.* **103,** 1957–1965.

Moll, R., Franke, W. W., Schiller, D. L., Geiger, B., and Krepler, R. (1982). The catalog of human cytokeratins: patterns of expression in normal epithelia, tumors, and cultured cells. *Cell* **31**, 11–24.

Myers, M. W., Lazzarini, R. A., Lee, V. M.-Y., Schlaepfer, W. W., and Nelson, D. L. (1987). The human mid-size neurofilament subunit; a repeated protein sequence and the relationship of its gene to the intermediate filament gene family. *EMBO J.* **6**, 1617–1626.

Nadeau, J. H., Berger, F. G., Cox, D. R., Crosby, J. L., Davisson, M. T., Ferrara, D., Fuchs, E., Hart, C., Hunihan, L., Lalley, P. A., Langley, S. H., Martin, G. R., Nichols, L., Phillips, S. J., Roderick, T. H., Roop, D. R., Ruddle, F. H., Skow, L. C., and Compton, J. G. (1989). A family of type I keratin genes and the Homeobox-2 gene complex are closely linked to the *rex* locus on mouse chromosome 11. *Genomics* **5**, 454–462.

Ogawa, H., Hattori, M., and Ishibashi, T. (1979). Abnormal fibrous protein isolated from the stratum corneum of a patient with bullous congenital ichthyosiform erythroderma (BCIE). *Arch. Dermatol. Res.* **266**, 109–116.

Ohtsuki, M., Tomic-Canic, M., Freedberg, I. M. and Blumenberg, M. (1992). Nuclear proteins involved in transcription of the human K5 keratin gene. *J. Invest. Dermatol.* **99**, 206–215.

Oshima, R. G., Howe, W. E., Klier, F. G., Adamson, E. D., and Shevinsky, L. H. (1983). Intermediate filament protein synthesis in preimplantation murine embryos. *Dev. Biol.* **99**, 447–455.

Oshima, R. G., Millan, J. L., and Cecena, G. (1986). Comparison of mouse and human keratin 18: A component of intermediate filaments expressed prior to implantation. *Differentiation* **33**, 61–68.

Oshima, R. G., Trevor, K., Shevinsky, L. H., Ryder, O. A., and Cecena G. (1988). Identification of the gene coding for the *endo B* murine cytokeratin and its methylated, stable inactive state in mouse nonepithelial cells. (1988). *Genes & Devel.* **2**, 505–516.

Oshima, R., Abrams, L., and Kulesh, D. (1990). Activation of an intron enhancer within the keratin 18 gene by expression of *c-fos* and *c-jun* in undifferentiated F9 embryonal carcinoma cells. *Genes & Devel.* **4**, 835–848.

Osman, M., Paz, M., Landesman, Y., Fainsod, A., and Gruenbaum, Y. (1990). Molecular analysis of the *Drosophila* nuclear lamin gene. *Genomics* **8**, 217–224.

Ouhayoun, J. -P., Gosselin, F., Forest, N., Winter, S. and Franke W. W. (1985). Cytokeratin patterns of human oral epithelia: differences in cytokeratin synthesis in gingival epithelium and the adjacent alveolar mucosa. *Differentiation* **30**, 123–129.

Ouellet, T., Levac, P., and Royal A. (1988). Complete sequence of the mouse type-II keratin *Endo A:* its amino-terminal region resembles mitochondrial signal peptides. *Gene* **70**, 75–84.

Peehl, D. M. and Stanbridge, E. J. (1981). Characterization of human keratinocyte X HeLa somatic cell hybrids. *International Journal of Cancer* **27**, 625–635.

Popescu, N. C., Bowden, P. E., and DiPaolo, J. A. (1989). Two type II keratin genes are localized on human chromosome 12. *Hum. Genet.* **82**, 109–112.

Powell, B. C., Cam, G. R., Fietz, M. J., and Rogers, G. E. (1986). Clustered arrangement of keratin intermediate filament genes. *Proc. Nat. Acad. Sci. U.S.A.* **83**, 5048–5052.

RayChaudhury, A., Marchuk, D., Lindhurst, M., and Fuchs, E. V. (1986). Three tightly linked genes encoding human type I keratins: conservation of sequence in the 5′-untranslated leader and 5′-upstream regions of coexpressed keratin gene. *Mol. Cell Biol.* **6**, 539–548.

Regnier, M. and Darmon, M. (1991). 1,25-Dihydroxyvitamin D_3 stimulates specifically the last steps of epidermal differentiation of cultured human keratinocytes. *Differentiation* **47**, 173–188.

Rentrop, M., Nischt, R., Knapp, B., Schweizer, J., and Winter, H. (1987). An unusual type-II 70-kDa keratin protein of mouse epidermis exhibiting postnatal body-site specificity and sensitivity to hyperproliferation. *Differentiation* **34**, 189–200.

Rieger, M. and Franke W. W. (1988). Identification of an orthologous mammalian cytokeratin gene. *J. Mol. Biol.* **204**, 841–856.

Rieger, M., Jorcano, J. L., and Franke, W. W. (1985). Complete sequence of a bovine type I cytokeratin gene: conserved and variable intron positions in genes of polypeptides of the same cytokeratin subfamily. *EMBO J.* **4,** 2261–2267.

Romano, V., Hatzfeld, M., Magin, T. M., Zimbelmann, R., Franke, W. W., Maier, G., and Ponstingl, R. (1986). Cytokeratin expression in simple epithelia. I. Identification of mRNA coding for human cytokeratin no. 18 by a cDNA clone. *Differentiation* **30,** 244–253.

Romano, R., Bosco, P., Rocchi, M., Costa, G., Leube, R. E., Franke, W. W., and Romeo, G. (1988). Chromosomal assignments of human type I and type II cytokeratin genes to different chromosomes. *Cytogenet. Cell Genet.* **48,** 148–151.

Roop, D. R., Krieg, T. M., Mehrel, T., Cheng, C. K., and Yuspa, S. H. (1988). Transcriptional control of high molecular weight keratin gene expression in multistage mouse skin carcinogenesis. *Cancer Res.* **48,** 3245–3252.

Rosenberg, M., RayChaudhury, A., Shows, T. B., Le Beau, M. M., and Fuchs, E. (1988). A group of type I keratin genes on human chromosome 17: characterization and expression. *Mol. Cell. Biol.* **8,** 722–736.

Rosenberg, M., Fuchs, E., Le Beau, M. M., Eddy, R. L., and Shows, T. B. (1991). Three epidermal and one simple epithelial type II keratin genes map to human chromosome 12. *Cytogenet. Cell Genet.* **57,** 33–38.

Rothnagel, J. A., Dominey, A. M., Dempsey, L. D., Longley, M. A., Greenhalgh, D. A., Gagne, T. A., Huber, M., Frenk, E., Hohl, D. and Roop D. R. (1992). Mutations in the rod domains of keratins 1 and 10 in epidermolytic hyperkeratosis. *Science* **257,** 1128–1130.

Savtchenko, E. S., Schiff, T. A., Jiang, C.-K., Freedberg, I. M., and Blumenberg, M. (1988). Embryonic expression of the human 40-kD keratin: evidence from a processed pseudogene sequence. *Am. J. Hum. Genet.* **43,** 630–637.

Savtchenko, E. S., Tomic, M., Ivker, R., and Blumenberg, M. (1990). Three parallel linkage groups of human acidic keratin genes. *Genomics* **7,** 394–407.

Sémat, A., Duprey, P., Vasseur, M., and Darmon, M. (1986). Mesenchymal-epithelial conversions induced by 5-azacytidine: appearance of cytokeratin *Endo A* messenger RNA. *Differentiation* **31,** 61–66.

Sémat, A., Vasseur, M., Maillet, L., Brulet, P., and Darmon, Y. M. (1988). Sequence analysis of murine cytokeratin *endo A* (n°8) cDNA. Evidence for mRNA species initiated upstream of the normal 5′ end in PCC4 cells. *Differentiation* **37,** 40–46.

Snape, A. M., Jonas, E. A., and Sargent, T. D. (1990). KTF-1, a transcriptional activator of *Xenopus* embryonic keratin expression. *Development* **109,** 157–165.

Stasiak, P. C. and Lane, E. B. (1987). Sequence of cDNA coding for human keratin 19. *Nucleic Acids Res.* **15,** 10058.

Steinert, P. M. and Roop, D. R. (1988). Molecular and cellular biology of intermediate filaments. *Ann. Rev. Biochem.* **57,** 593–625.

Steinert, P. M., Rice, R. H., Roop, D. R., Trus, B. L., and Steven, A. C. (1983). Complete amino acid sequence of a mouse epidermal keratin subunit and implications for the structure of intermediate filaments. *Nature* **302,** 794–799.

Steinert, P. M., Parry, D. A. D., Idler, W. W., Johnson, L. D., Steven, A. C., and Roop, D. R. (1985). Amino acid sequences of mouse and human epidermal type II keratins of M_r 67,000 provide a systematic basis for the structural and functional diversity of the end domains of keratin intermediate filament subunits. *J. Biol. Chem.* **260,** 7142–7149.

Stick, R. (1988). cDNA cloning of the developmentally regulated lamin L_{III} of *Xenopus laevis*. *EMBO J.* **7,** 3189–3197.

Surya, B., Yu, J., Manabe, M., and Sun, T.-T. (1990). Assessing the differentiation state of cultured bovine urothelial cells: elevated synthesis of stratification-related K5 and K6 keratins and persistent expression of uroplakin I. *J. Cell Sci.* **97,** 419–432.

Takemoto, Y., Fujimura, Y., Matsumoto, M., Tamai, Y., Morita, T., Matsushiro, A., and Nozaki, M. (1991). The promoter of the *endo A* cytokeratin gene is activated by a 3′ downstream enhancer. *Nucleic Acids Res.* **19,** 2761–2765.

Tamai, Y., Takemoto, Y., Matsumoto, M., Morita, T., Matsushiro, A., and Nozaki, M. (1991). Sequence of the *Endo A* gene encoding mouse cytokeratin and its methylation state in the CpG-rich region. *Gene* **104**, 169–176.

Tomic, M., Sunjevaric, I., Savtchenko, E. S., and Blumenberg, M. (1990). A rapid and simple method for introducing specific mutations into any position of DNA leaving all other positions unaltered. *Nucleic Acids Res.* **18**, 1656.

Tomic-Canic, M., Jiang, C.-K., Connolly, D., Freedberg, I. M., and Blumenberg, M. (1991). Vitamin D3, its receptor, and regulation of epidermal keratin gene expression. **1**, 70–75.

Tomic-Canic, M., Sunjevaric, I., Freedberg, I. M., and Blumenberg, M. (1992). Identification of the retinoic acid and thyroid hormone receptor responsive element in the human K#14 keratin gene. *J. Invest Dermatol.* **99**, 842–847.

Trevor, K. T. (1990). Disruption of keratin filaments in embryonic epithelial cell types. *New Biologist* **2**, 1004–1014.

Trevor, K., Linney, E., and Oshima, R. G. (1987). Suppression of *endo B* cytokeratin by its antisense RNA inhibits the normal coexpression of *endo A* cytokeratin. *Dev. Biol.* **84**, 1040–1044.

Tyner, A. L. and Fuchs, E. (1986). Evidence for posttranscriptional regulation of the keratins expressed during hyperproliferation and malignant transformation in human epidermis. *J. Cell Biol.* **103**, 1945–1955.

Tyner, A. L., Eichman, M. J., and Fuchs, E. (1985). The sequence of a type II keratin gene expressed in human skin: Conservation of structure among all intermediate filament genes. *Proc. Natl. Acad. Sci. USA* **82**, 4683–4687.

Vassar, R., Rosenberg, M., Ross, S., Tyner, A., and Fuchs, E. (1989). Tissue-specific and differentiation-specific expression of a human K14 keratin gene in transgenic mice. *Proc. Natl. Acad. Sci. U.S.A.* **86**, 1563–1567.

Vassar, R., Coulombe, P. A., Degenstein, L., Albers, K., and Fuchs, E. (1991). Mutant keratin expression in transgenic mice causes marked abnormalities resembling a human genetic skin disease. *Cell* **64**, 365–380.

Vasseur, M., Duprey, P., Brulet, P., and Jacob, F. (1985). One gene and one pseudogene for the cytokeratin *endo A. Proc. Natl. Acad. Sci. U.S.A.* **82**, 1155–1159.

Waseem, A., Alexander, C. M., Steel, J. B., and Lane, E. B. (1990a). Embryonic simple epithelial keratins 8 and 18: chromosomal location emphasizes difference from other keratin pairs. *New Biologist* **2**, 464–478.

Waseem, A., Gough, A. C., Spurr, N. K., and Lane, E. B. (1990b). Localization of the gene for human simple epithelial keratin 18 to chromosome 12 using polymerase chain reaction. *Genomics* **7**, 188–194.

Wilson, B. W., Edwards, K. J., Sleigh, M. J., Byrne, C. R., and Ward, K. A. (1988). Complete sequence of a type-I microfibrillar wool keratin gene. *Gene* **73**, 21–31.

Wingender, E. (1990). Transcription-regulating proteins and their recognition sequences. *Eukaryotic Gene Expression* **1**, 11–48.

Winkles, J. A., Sargent, T. D., Parry, D. A. D., Jonas, E., and Dawid, I. B. (1985). Developmentally regulated cytokeratin gene in *Xenopus laevis. Mol. Cell Biol.* **5**, 2575–2581.

Winter, H., Rentrop, M., Nischt, R., and Schweizer, J. (1990). Tissue-specific expression of murine keratin K13 in internal statified squamous epithelia and its aberrant expression during two-stage mouse skin carcinogenesis is associated with the methylation state of a distinct CpG site in the remote 5′-flanking region of the gene. *Differentiation* **43**, 105–114.

Yamamoto, R., Kao, L.-C., McKnight, C. E., and Strauss III, J. F. (1990). Cloning and sequence of cDNA for human placental cytokeratin 8. Regulation of the mRNA in trophoblastic cells by cAMP. *Mol. Endocrinology* **4**, 370–374.

Yoshike, T., Negi, M., Hattori, M., and Ogawa, H. (1983). Fractionation and characterization of the epidermal stratum corneum in bullous congenital ichthyrosiform erythroderma (BCIE). *J. Dermatol.* **10**, 427–431.

Yu, D., Pang, S. Y.-Y., Checkla, D. M., Freedberg, I. M., Sun, T.-T., and Bertolino, A. P. (1991). Transient expression of mouse hair keratins in transfected HeLa cells: interactions between "hard" and "soft" keratins. *J. Invest. Dermatol.* 97, 354–363.

Zhou, X.-M., Idler, W. W., Steven, A. C., Roop, D. R., and Steinert, P. M. (1988). The complete sequence of the human intermediate filament chain keratin 10. *J. Biol. Chem.* 263, 15584–15589.

Addendum

Since this chapter was written, several important and interesting developments have occurred, fulfilling the expectations in the last sentence.

Several additional laboratories have confirmed simultaneously that the dominant mutations in the K1 or K10 genes cause Epidermolytic Hyperkeratosis (Chipev *et al.*, 1992; Cheng *et al.*, 1992). It is now expected that, knowing what to look for, additional inherited diseases will be associated with mutations in other keratin genes, such as K9, the palmoplantar keratin gene (Reis *et al.*, 1992; E. Epstein, personal communication), or hair keratin genes.

Keratin K2 takes its rightful place in the family of basic-type keratins. It has been cloned and, furthermore, it appears that several different genes, one expressed in the oral cavity, another in the epidermis, constitute a K2 keratin subfamily (Collin *et al.*, 1992a,b).

Details of transcription regulation have also received considerable attention. Basonuclin, a putative transcription factor with multiple zinc fingers, has been cloned and characterized (Tseng and Green, 1992), but its direct involvement in expression of keratins has not been demonstrated yet. Our own experiments characterized binding sites for Sp1, Ap1, and AP2 as well as additional transcription factors in the promoters of K6 and K16 genes (Bernerd *et al.*, 1993, *Gene Expression,* in press; Magnaldo *et al.*, 1993, *DNA and Cell Biology*, in press). More importantly, we have characterized the EGF-responsive elements in the K6 and K16 promoters and discovered a nuclear protein that binds this factor (Jiang *et al.*, 1993, *Proc. Natl. Acad. Sci. USA*, in press).

Addendum References

Chipev, C. C., Korge, B. P., Markova, N., Bale, S. J., DiGiovanna, J. J., Compton, J. G., and Steinert, P. M. (1992). A leucine–proline mutation in the H1 Subdomain of keratin 1 causes epidermolytic hyperkeratosis. *Cell* 70, 821–828.

Cheng, J., Snyder, A. L., Yu, Q. -C., Letai, A., Paller, A. S., and Fuchs, E. (1992). The genetic basis of epidermolytic hyperkeratosis: a disorder of differentiation-specific epidermal keratin genes. *Cell*, 70, 881–891.

Collin, C., Moll, R., Kubicka, S., Ouhayoun, J. -P., and Franke, W. W. (1992). Characterization of human cytokeratin 2, an epidermal cytoskeletal protein synthesized late during differentiation. *Exp. Cell. Res.*, 202, 132–141.

Collin, C., Ouhayoun, J. -P., Grund, C., and Franke, W. W. (1992). Suprabasal marker proteins distinguishing keratinizing squamous epithelia: Cytokeratin 2 polypeptides of oral masticatory epithelium and epidermis are different. *Differentiation* **51**, 137–148.
Reis, A., Küster, W., Eckardt, R., and Sperling, K. (1992). Mapping of a gene for epidermolytic palmoplantar keratoderma to the region of the acidic keratin gene cluster at 17q12-q21. *Hum. Genet.*, **90**, 113–116.
Tseng, H. and Green, H. (1992). Basonuclin: a keratinocyte protein with multiple paired zinc fingers. *Proc. Natl. Acad. Sci USA* **89**, 10311–10315.

2

Murine Epidermal Keratins

Jürgen Schweizer

Introduction

With reference to epithelial keratin expression, the mouse is, next to humans, the most frequently studied species. In both species, the epidermis is the most intensively and studied epithelium. This review compiles the current knowledge on differential keratin expression in normal mouse epidermis; the changes in hyperplastic or transformed mouse epidermis; and expression in cultured mouse keratinocytes. Comparisons are made with the respective situation in human epidermis. Because of the comparative character of this article, and following an increasing tendency in keratin research, we will use the human keratin numerical nomenclature according to the catalog of Moll *et al.* (1982) when an orthologous relationship between human and murine keratins is secure. This review does not contain a specific section concerning the structural features and characteristics of keratin pro-

Molecular Biology of the Skin: The Keratinocyte

teins and genes. Some of these features, however, will be discussed in reference to specific practical research questions that arise. The reader who wishes a more detailed description can refer to the appropriate reviews and articles cited. In the same sense, theoretical speculations about the functional significance of keratins and in particular aspects of the regulation of expression of specific keratin genes will be limited as such studies, which have not proceeded as far in the mouse as in man, are similarly limited. Nevertheless, the review is more than a compendium of practical data concerning keratin expression in normal and pathologically changed murine (and human) epidermis. It will also serve as a tool both for researchers who work specifically on keratins and scientists who study keratins as one parameter of epidermal differentiation.

The Basic Pattern of Mouse Epidermal Keratins

The entire integumental epidermis of the adult mouse is a highly inappropriate source for the investigation and elucidation of the basic pattern of epidermal keratins for several reasons. First, both dorsal and ventral epidermis are extremely thin epithelia consisting essentially of a monolayered basal compartment with at most two or three layers of living suprabasal cells (Hanson, 1947; Weiss and Zelikson, 1975). Second, these skin sites are covered by an extremely dense hair coat. Both properties make it difficult to obtain pure epidermis without follicular remnants that express keratin members normally absent from interfollicular epidermis. On the other hand, anatomical regions with a lower frequency of hairs (i.e., ear and tail) or glabrous skin sites (i.e., sole of the foot) from which the epidermis can be properly and more easily removed from the underlying dermis, express body site specific keratins in addition to the basic keratin pattern. Fortunately, newborn mouse epidermis constitutes an ideal source for study of the basic epidermal keratin pattern. At birth, the entire mouse body epidermis is completely mature in terms of morphology (Fig. 1). It no longer contains peridermal residues and, except for the footpads of the sole of the foot, is essentially uniform in thickness and exhibits a flat dermal epidermal junction (Weiss and Zelikson, 1975). In addition (most important when comparing mouse and human epidermal keratins), the proportions of the different cell layers (basal, spinous, granular, and cornified) of newborn mouse epidermis largely correspond to those of human body epidermis. Newborn mouse epidermis can easily be separated from the dermis, e.g., by a short (20 sec) incubation in 56–60°C hot water (Baumberger, 1942; Marrs and Vorhees, 1971). This method is superior to the frequently used separation of the tissues by maceration in cold 1% acetic acid (Cowdry, 1952), as it leaves the still immature hair follicles (Fig. 1) in the dermis so that almost pure interfollicular epidermis is obtained. Moreover, the rapid separation of epidermis from

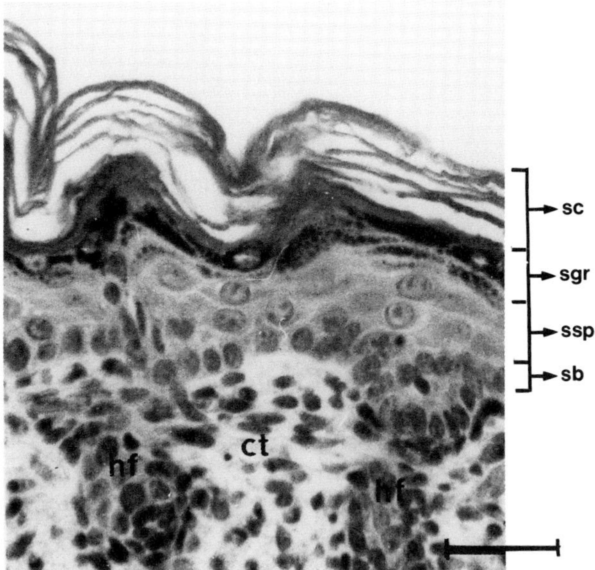

FIGURE 1 Morphological appearance of newborn mouse epidermis (back skin). sb, stratum basale; ssp, stratum spinosum; sgr, stratum granulosum; sc, stratum corneum; ct, connective tissue; hf, hair follicles. Bar = 100 μm.

dermis by heat is a suitable method for the isolation of undegraded epidermal RNA (Schweizer and Goerttler, 1980).

The solubilization of keratins from newborn mouse epidermis by standard protocols and their two-dimensional resolution by nonequilibrium pH gradient gel electrophoresis (NEPHGE) (O'Farrell, 1975; Garrels and Gibson, 1976) leads to the schematic shown in Figure 2a. Basically, eight individual keratin proteins can be detected. They are indicated in Figure 2a by their molecular weights in kilodaltons (kDa) which rely on size determinations performed in different laboratories (Schweizer and Winter, 1982b; Franke *et al.*, 1982; Roop *et al.*, 1983; Bowden *et al.*, 1984; Molloy and Laskin, 1988). Whereas the reported SDS–PAGE size estimates of the three largest keratins (67, 64, and 62kDa) are identical in various laboratories, considerable discrepancies exist with regard to the remaining smaller keratin species (Fig. 2a). As early keratin research used only one-dimensional protein electrophoretic separation methods, it became evident that the precise identification of a distinct murine keratin in the 60- to 47-kDa region as well as a comparison of the mouse keratin pattern with keratin patterns of other species required lengthy descriptions and resulted in confusion. These communication problems were largely overcome as keratin research progressed leading to the general recognition of compartmentalized keratin synthesis in

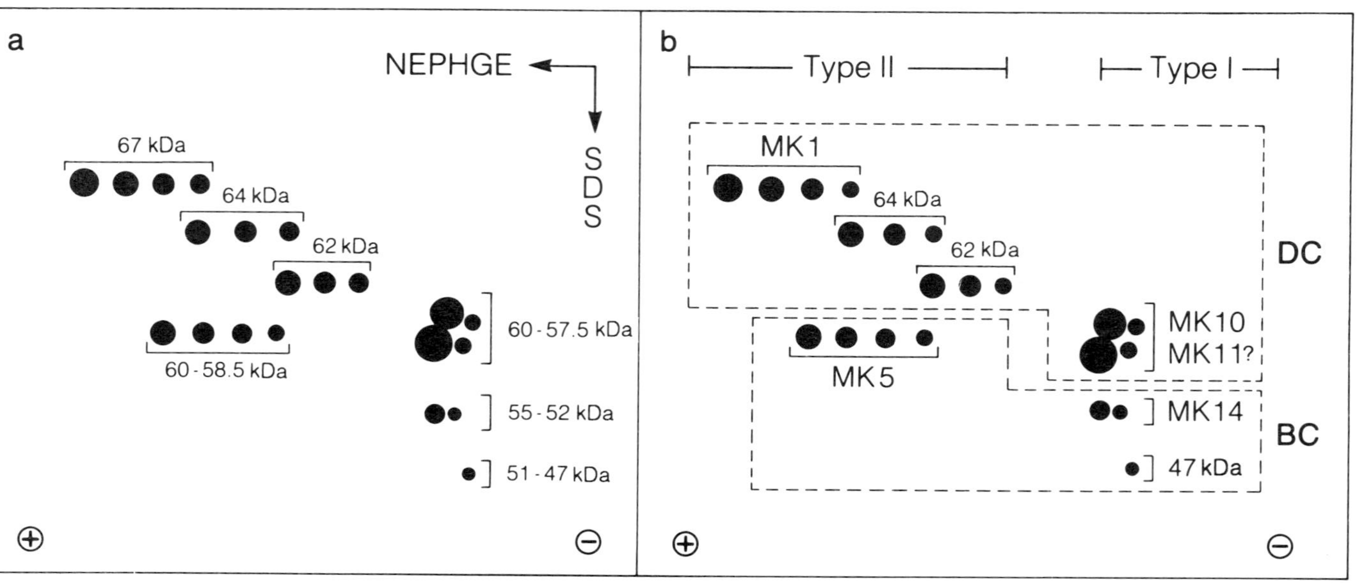

FIGURE 2 Schematic representation of the basic keratin pattern of mouse epidermis as revealed by nonequilibrium pH gradient gel electrophoresis (NEPHGE) in the first dimension and SDS–PAGE in the second dimension. (a) Designation of the keratins according to SDS–PAGE size estimates. (b) Designation of the keratins according to the numerical nomenclature of human keratins (Moll *et al.*, 1982). BC, basal cells; DC, differentiated cells.

mammalian epidermis proceeding via the sequential expression of basal and suprabasal pairs of basic to neutral (type II) and acidic (type I) keratins as well as the generation of posttranslationally derived keratins in the stratum corneum. (For review see Moll *et al.*, 1982; Sun *et al.*, 1983; Sun *et al.*, 1984; Schweizer *et al.*, 1984; Fuchs, 1988; Steinert and Roop, 1988.) However, the most important step toward unequivocal communication between scientists working on keratins of different species, has been the growing tendency to adopt a species-independent numerical keratin nomenclature as originally developed by Moll *et al.* (1982) for human keratins. On the basis of this catalog, the basic keratin pattern of mouse epidermis would read as indicated in Figure 2b. As it stands, keratins designated by an arabic number (i.e., MK1) represent mRNA-encoded proteins, whereas keratins still indicated by the molecular weight values may represent either posttranslationally derived keratin variants or gene-encoded mouse-specific keratins having no homolog in human epidermis.

Keratins Expressed in Basal Epidermal Cells

The main keratin pair expressed in basal cells of mouse epidermis consists of MK5 and MK14 (Fig. 2b). The type II MK5 represents one of the rare epidermal keratins for which no sequence data from cloned epidermal mRNA have been published. [With regard to a postulated MK5 sequence (Steinert *et al.*, 1984), see section, "Murine Hyperproliferation-Associated Keratins MK6 and MK17."] Thus, the actual molecular weight of this keratin cannot be predicted and compared with the varying estimates from SDS–PAGE (Fig. 2a). In contrast, both partial and full length sequence information exists for HK5 (Galup and Darmon, 1988; Lersch and Fuchs, 1988; Eckert and Rorke, 1988; Lersch *et al.*, 1989) whose molecular weight according to SDS–PAGE estimates is 58 kDa (Moll *et al.*, 1982). Interestingly the actual size predicted for HK5 from the two full length clones differs slightly [62.5kDa (Eckert and Rorke, 1988) versus 61.5kDa (Lersch *et al.*, 1989)]. However, both values confirm the observation that molecular weights calculated from sequenced keratins are, with apparantly one exception (Johnson *et al.*, 1985), generally larger than those estimated from SDS–PAGE (for a summary, see Eckert and Rorke, 1988). The apparent occurrence of two related, but distinct, mRNAs for HK5 appears to be in line with the recent demonstration of inter-individual variants of the HK5 protein that differ electrophoretically by about 1kDa (Wild and Mischke, 1986). This phenomenon has been explained by a polymorphism of the HK5 gene, which is supposed to consist of two diverged but codominant alleles (Wild and Mischke, 1986).

In contrast to the situation with K5, sequence data exists for both the basal type I MK14 (Knapp *et al.*, 1987) and HK14 (Marchuk *et al.*, 1984; Marchuk *et al.*, 1985). This laboratory has isolated a cDNA clone, pkSCC52, that codes for MK14 (Knapp *et al.*, 1987). [cDNA clones encoding this keratin have also been reported from other laboratories; however, sequence data were not published (Roop *et al.*, 1984).] Unfortunately, clone pkSCC52 lacks the sequence encoding the amino-terminal region of MK14 so that its actual molecular weight cannot be calculated. The molecular weight predicted for HK14 (51.6kDa; Marchuk *et al.*, 1984; Marchuk *et al.*, 1985), is again larger than the SDS–PAGE estimates of 50kDa (Moll *et al.*, 1982). Sequence comparisons of the known portions of MK14 and HK14 reveal an extremely high homology (> 95%), which at the nucleotide level extends to the 3′ noncoding region (Fig. 3) Remarkably, the nucleotide sequence of MK14 contains a tandem duplicate of 18 bases immediately adjacent to the region encoding the central α-helical domain, a phenomenon occurring only once in HK14 (Fig. 3). It is evident that the resulting additional six amino acid residues in mouse keratin contribute to the observed size difference between MK14 and HK14 on polyacrylamide gels [55–52kDa (Fig. 2a) versus 50 kDa]. Unlike human basal epidermal cells, which express only HK5 and HK14 (Moll *et al.*, 1982; Sun *et al.*, 1984), mouse basal epidermal cells contain an additional small type I keratin of 47kDa (Fig. 2a,b). The molecular weight of this keratin is given according to SDS–PAGE size evaluations in our laboratory. It should be emphasized that the amount of this keratin seen on polyacrylamide gels varies considerably relative to the body site from which the keratins are isolated. Depending on the mass ratio of basal and suprabasal cells, the 47-kDa keratin that appears relatively prominent in extracts of the thin dorsal epidermis is, however, almost undetectable in the thick epidermis covering the sole of the foot. Accordingly, the largest amounts of the 47-kDa keratin are observed in conventionally cultured mouse epidermal keratinocytes in which stratification and differentiation is only rudimentary.

There is strong evidence that the basal 47-kDa keratin represents a genuine keratin encoded by an mRNA of its own. Incubation of neonatal or adult mouse skin in a medium containing ^{35}S-methionine leads to a rapid labeling of the 47-kDa keratin, which appears at a time identical with that of other mRNA-encoded keratins (Schweizer *et al.*, 1987, and unpublished results). Moreover, *in vitro* translation of polyA^{+} RNA isolated from total epidermis or basal epidermal cells clearly reveals the 47-kDa keratin among the translation products (Schweizer *et al.*, 1984; see also Fig. 8a,b). It has previously been suggested that, with the exception of a shorter amino terminus, the 47-kDa keratin is structurally identical to MK14 and might represent the hyperproliferation-associated HK17 (Roop *et al.*, 1985). Both assumptions must, however, be discounted on the grounds that a monospecific antibody recognizing an epitope on the carboxy terminus of MK14 does not react with the 47-kDa keratin in Western blots (Fisher *et al.*, 1987) and that MK17 is distinct from the 47-kDa keratin. Our laboratory has isolated a new murine Type I keratin cDNA, which according to differential dot blot

```
                          1                    2
MK14 GCTCCGCTGCGAGATGGAGCAGCAGAACCAGGAGTACAAAATCCTGCTGGATGTGAAGACAAGGCTGGAGCAGGAGATCGCCACCTACCGCCGTCTGCTG
     ***************************************R***********Y********R *******************************Y******
HK14 GCTCCGCTGCGAGATGGAGCAGCAGAACCAGGAGTACAAGATCCTGCTGGACGTGAAGACGCGGCTGGAGCAGGAGATCGCCACCTACCGCCGCCTGCTG

                         ↓┌────────1────────┐┌────────2────────┐
     GAGGGAGAGGACGCCCACCTTTCATCTTCCCAATTCTCCTCATCCTCTCAATTCTCCTCTGGCTCTCAGTCATCCAGAGATGTGACCTCCACCAACCGCC
     ********************Y                  ** *****Y**R********** ** *************************  ***R*****
     GAGGGCGAGGACGCCCACCTC------------------TCCTCCTCCCAGTTCTCCTCTGGATCGCAGTCATCCAGAGATGTGACCTCCTCCAGCCGCC
                         ↑

     AGATCCGCACCAAGGTCATGGATGTGCACGATGGCAAGGTGGTCTCCACCCACGAGCAGGTCCTGCGCACCAAGAACTAAAGCTGCTACATGCTGCTCAG
     *R**************************************  *********************  *************R*R*****Y  RY *Y*****
     AAATCCGCACCAAGGTCATGGATGTGCACGATGGCAAGGTGGTGTCCACCCACGAGCAGGTCCTTCGCACCAAGAACTGAGGCTGCCCAGCCCCGCTCAG

     GCTTAGGAGGCTCC----GTGGAC-GAGATACTGCTGGAAGAGCCCCTGTATTGTCCCTATAGGCTTCAC------TCTTTACTTGACCCT-TGTCCTTC
     **Y*******Y**      ******  ****  *YR********  *****  *  Y**   **Y*  R  R******       *YY*R***  *****   Y**Y**
     GCCTAGGAGGCCCCCCGTGTGGACACAGATCCCACTGGAAGATCCCCTCTCCTG--CCCAAGCACTTCACAGCTGGACCCTGCTTCACCCTCACCCCCTC

     CTGGCAAGCAATAAAGCTTCTTTTTCTGAGTTGCAC
     ******  *****  ******  **  **********Y
     CTGGCAATCAATACAGCTTCATTATCTGAGTTGCAT
```

FIGURE 3 Nucleotide sequence comparison of the 3′ coding and noncoding regions of MK14 (Knapp *et al.*, 1987) and HK14 (Marchuk *et al.*, 1985). The arrowheads indicate the end of the region coding for the α-helical domains. The respective stop codons and polyadenylation signals are underlined. R indicates A–G changes, Y indicates C–T changes. "1" and "2" denote a tandem duplicated sequence in MK14 that occurs only once in HK14. (With permission from Knapp *et al.*, 1987.)

analysis with mRNAs from various skin sites and from cultured mouse keratinocytes, may encode the 47-kDa keratin. Moreover, preliminary data indicate that the sequence encoded by this clone is unrelated to any of the known type I mouse keratins. The data also disclose a possible orthologous relationship with HK15 [i.e., a type I keratin which is expressed in basal and also in suprabasal cells of human epidermis (Leube *et al.*, 1988)]. Thus it appears that the 47-kDa keratin has no counterpart in human epidermis. Further characterization of this new cDNA clone will hopefully lead to the elucidation of the true nature of the 47-kDa keratin.

Keratins Expressed in Suprabasal Epidermal Cells

The oppositely charged keratin pair expressed in suprabasal, living cells of mouse epidermis consists of MK1 and MK10 (Fig. 2b). Steinert *et al.* (1985) have published a cDNA clone encoding the type II 67-kDa MK1 which, however, lacks the penultimate part of the amino-terminus. Therefore, the actual molecular weight for this murine keratin cannot be predicted. In contrast, both partial and full length sequence data are available for HK1 (Steinert *et al.*, 1985; Johnson *et al.*, 1985). Inter-individual variants have also been demonstrated for HK1 (Mischke and Wild, 1987) in a fashion similar to basal HK5. However, there is only evidence for one mRNA species encoding this keratin. According to these data, HK1 has a calculated molecular weight of 66.4 kDa (Johnson *et al.*, 1985), which in contrast to other epidermal keratins is slightly smaller than its SDS–PAGE size estimates. MK1 seems to be somewhat larger than HK1, since its carboxy-terminal region exceeds that of the human analog by 11 amino acid residues (Steinert *et al.*, 1985).

Partial and full length sequence data exist for both MK10 (Steinert *et al.*, 1983; Krieg *et al.*, 1985; Quellet *et al.*, 1986; Knapp *et al.*, 1987) and HK10 (Darmon *et al.*, 1987; Zhou *et al.*, 1988; Rieger and Franke, 1988). Similar to the basal HK5, the two full length clones of HK10 code for two highly related but distinct proteins that differ in their calculated molecular weights [59.5 kDa (Rieger and Franke, 1988) versus 57.1 kDa (Zhou *et al.*, 1988)]. This again is in agreement with a possible allelic polymorphism of the HK10 gene (Mischke and Wild, 1987) that might ultimately be responsible for the synthesis of HK10 and HK11. There is evidence that the HK10 locus is even more complex since a partial clone published by Darmon *et al.* (1987) appears to code for a MK10 species different from those encoded by the two full length clones.

In contrast to the complex situation in humans, there is no evidence for structural divergence among the partial and full length clones of MK10 isolated in different laboratories (Steinert *et al.*, 1983; Krieg *et al.*, 1985; Quellet *et al.*, 1986; Knapp *et al.*, 1987). The molecular weight predicted for MK10 on the basis of these structurally uniform clones would then be

57.8 kDa (i.e., 58.4 kDa if the known content of phosphorylated serine residues of this keratin is considered; Krieg *et al.*, 1985). No clear statement can be made with regard to the existence in mouse epidermis of a genuine analog of HK11. Both incubation studies of mouse skin in a medium containing ^{35}S-methionine and *in vitro* translations of mouse epidermal polyA^{+} RNA are controversial with regard to the generation of only one or two protein entities in the acidic 60–57.5-kDa region (unpublished results). In view of these discrepancies and considering further complications due to differentiation-specific keratin posttranslational modifications it might be appropriate to question the validity of MK11 in Figure 2b as a specific gene product at this stage of research.

Posttranslationally Generated Epidermal Keratins

Evidence for the existence of secondarily derived keratins within the complex pattern of epidermal keratins was first obtained from investigations in fractioned epidermis. Studies in different species revealed the specific occurrence of distinct keratin subunits in the stratum corneum concomitant with the loss or strong reduction of keratins expressed in the subjacent strata, thus indicating a precursor–product relationship at the transition from living to dead cell layers (Fuchs and Green, 1980; Banks-Schlegel, 1982; Schweizer and Winter, 1982b; Skerrow and Skerrow, 1983; Breitkreutz *et al.*, 1983; Bowden *et al.*, 1984). The identification of these secondarily derived keratins in mouse epidermis was initially hampered by an early report from our laboratory that suggested that virtually all keratins extractable from epidermis could be translated *in vitro* (Schweizer and Goerttler, 1980). However, subsequently, this observation could not be substantiated (Roop *et al.*, 1983; Schweizer *et al.*, 1984) and additional studies clearly showed that the abundant 64-kDa and 62-kDa proteins (Fig. 2a,b) were indeed part of those keratins that were specifically associated with the cornified cell layer and therefore not labeled during short term *in vitro* incubation of skin or by *in vitro* translation of epidermal polyA^{+} RNA. Comparative peptide mapping experiments revealed that both proteins were stratum corneum equivalents of MK1 (Bowden *et al.*, 1984). Interestingly, HK1 is apparently processed into only one stratum corneum variant in human epidermis (Bowden *et al.*, 1984). The type I MK10 is also subject to posttranslational modifications, however, the fate of this keratin at the transition from living to dead cell layers is less precisely known than that of its type II partner. Whereas the latter is completely degraded in the stratum corneum into the 64-kDa and 62-kDa keratins, conflicting results exist with regard to the disappearance of MK10 from cornified cells (Schweizer and Winter, 1982b; Bowden *et al.*, 1984). Moreover, the understanding of secondary MK10 modifications is obscured by the uncertainty about the existence of a genuine MK11 species. What appears to be clear, however, from two independent studies is that, unlike MK1, MK10 is processed into stratum corneum equivalents that are both smaller and larger than their mRNA-encoded precursor (Schweizer and Winter, 1982b; Bowden *et al.,* 1984). Similar results were obtained for HK10 (Bowden *et al.,* 1984). Considering the secondary modification of MK10 as a unique case, it may otherwise be generally stated that the main modifications to which MK1 and MK10 are subject during the transition from living to dead cell layers consist in highly ordered degradation processes. Only recently has more insight been gained into mechanistic aspects of this complex scenario.

General Remarks

In the preceeding sections, the individual members of the mouse epidermis basic keratin pattern were presented according to their sequential generation in the three main layers of this terminally keratinizing epithelium. As already mentioned, this assignment was originally based either on investigations in fractioned epidermis or on studies in developing embryonic epidermis in which the different strata are built up consecutively according to a rigid timescale (Schweizer and Winter, 1982b; Banks-Schlegel, 1982; Quellet *et al.,* 1990). In contrast to human epidermis in which differential keratin expression was also extensively studied by means of monoclonal keratin antibodies (Eichner *et al,* 1984; Sun *et al.,* 1984), immunological investigations with conventional polyclonal keratin antibodies were performed to a much lesser extent in mouse epidermis (Schweizer *et al.,* 1984). However, the elucidation of the structural features of keratin genes and proteins beginning in the early eighties by means of recombinant DNA techniques opened new perspectives for the generation of both highly specific DNA probes and antibodies that confirmed and refined previous observations on differential epidermal keratin expression. Sequence comparisons of keratin cDNAs of one species reveal that the nucleotide sequences that are virtually unique for an individual keratin encompass those coding for the penultimate carboxy-terminal region as well as the adjacent 3′-noncoding region. Therefore, DNA probes containing only these sequence areas proved to be highly efficient and reliable tools for investigation of chromosomal localization, gene copy number, and site and extent of expression of a distinct keratin member.

The use of specific 3′ fragments of keratin cDNAs in mouse epidermis has confirmed that the synthesis of the MK14 mRNA is indeed confined to a monolayered basal cell compartment, but occurs in a two-layered compartment in some internal stratified epithelia (Knapp *et al.,* 1987; Fig. 4a,b). It was also determined that the expression of suprabasal keratin mRNAs occurs initially in some basal cells obviously in transit to the parabasal cell row (Fig. 4c), and generally preceeds the expression of other mouse epidermal differentiation markers (i.e., loricrin and filaggrin; Nischt *et al.,* 1987). DNA probes for MK1 and MK10 are also helpful in unequivocally confirm-

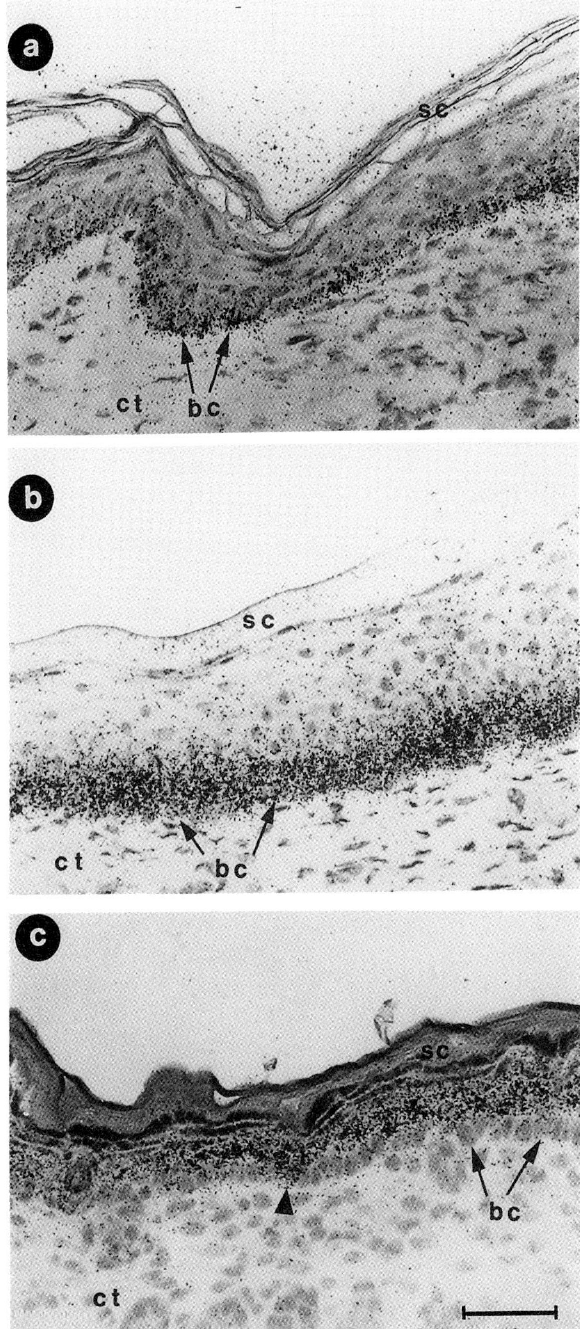

FIGURE 4 *In situ* hybridization with ^{35}S-labeled specific 3′ probes derived from cDNA clones coding for (a) MK14 (adult tail skin); (b) MK14 (adult palate); (c) MK10 (neonatal back skin; the arrowhead indicates a labeled basal cell). bc, basal cells; sc, stratum corneum; ct, connective tissue. Bar = 100 μm.

Cell type	Type	Keratin	Sequence
Basal epidermal cells	Type II	MK5	
		HK5	- S S S S V K F V S T T S S S R K S F K S
	Type I	MK14	- M D V H D G K V V S T H E Q V L R T K N
		HK14	- M D V H D G K V V S T H E Q V L R T K N
Hyperproliferative epidermal cells	Type II	Murine 60kd keratin(MK6)	- G G S S T I K Y T T T S S S - K K S Y R Q
		HK6	- G G S S T I K Y T T T S S S S R K S Y K H
	Type I	MK16	
		HK16	- Q T W P I L K E Q S S S S F S Q G Q C S
		MK17	- E E V Q D G K V I S S R E Q V H Q T T R
		HK17	- E E V Q D G K V I S S R E Q V H Q T T R
Differentiated epidermal cells	Type II	MK1	- K S S G S S T V K F V S T S Y S R Q T K
		HK1	- S S G G S S S V R F V S T S Y S G V T R
	Type I	MK10	- S G G F S G T S G G G D Q S S K G P R Y
		HK10	- G H K S S S S G S V G E S S S K G P R Y
Differentiated cells of internal stratified epithelia	Type II	MK4	- Y G G S G S K I T S S A T I T K R S P R
		HK4	- S V S G S S S S K I I S T T T L N K R R
	Type I	MK13	- T T T S N G G S P S N S G R P D F R K Y
		HK13	- A S V T T T S N A S G R R T S D V R R P

ing the synthesis of these keratins in some internal stratified epithelia (i.e., the posterior part of the filiform papillae of tongue epithelium; Rentrop *et al.*, 1986); the epithelium overlying the plica palatinae (unpublished results); and in differentiating cells of the forestomach epithelium in which MK1 and MK10 are uniformly coexpressed with MK4 and MK13 (Schweizer *et al.*, 1988).

At the protein level, the structural specificity of the penultimate carboxy-terminal region of each mouse epidermal keratin (Fig. 5) has been exploited to compensate for the lack of monoclonal keratin antibodies. Conventional, however, monospecific antisera have been raised against synthetic peptides corresponding to these carboxy-terminal amino acid sequences (Roop *et al.*, 1984; Roop *et al.*, 1985). Besides confirming the differential expression of basal and suprabasal epidermal keratins by immunohistochemical studies, these antisera are especially helpful in investigating the mechanism by which the stratum corneum equivalents of MK1 and MK10 are generated. Studies by Roop *et al.* (1984) showed that an antiserum elicited against a carboxy-terminal peptide of MK1 (Fig. 5) stained all living suprabasal cells but not the stratum corneum of mouse epidermis. Since intact MK1 molecules are absent from cornified cells, the failure of the antiserum to stain this part of the epidermis cannot be due to masking of the determinant(s) recognized by this antiserum. The results instead indicate that the carboxy-terminal sequences of MK1 are removed piecemeal at the transition from living to dead cell layers, thereby giving rise to the 64-kDa and 62-kDa secondary proteins (Fig. 2a,b). This interpretation is supported by the finding that, in contrast to the MK1 precursor, the 64-kDa and 62-kDa proteins are not detected by the MK1 antiserum in Western blots (Fischer *et al.*, 1987). Since the carboxy-terminal domains of keratins are generally characterized by an accumulation of basic amino acid residues (see Fig. 5), the loss of these sequences accounts for the increasingly acidic nature of the 64-kDa and 62-kDa keratins and explains the typical staircase pattern

←

FIGURE 5 Alignment of the carboxy-terminal amino acid sequences of murine and human epidermal keratins and of differentiation-specific keratins of internal stratified epithelia. The amino acids are given by the one letter code. Black dots denote basic amino acid residues (arginine, R; histidine, H; lysine, K). Positionally preserved amino acids are indicated by a vertical bar. Sequences were taken from the following references. The MK5 sequence has not yet been published. MK16 is probably not expressed in mouse epidermis.

HK5 (Eckert and Rorke, 1988)
MK14 (Knapp *et al.*, 1987); HK14 (Marchuk *et al.*, 1985)
murine 60 kDa keratin, MK6 (Steinert *et al.*, 1984)
HK6 (Tyner *et al.*, 1985); HK16 (Rosenberg *et al.*, 1988)
MK17 (Knapp *et al.*, 1987); HK17 (Kartŏsova *et al.*, 1987)
MK1 (Steinert *et al.*, 1985); HK1 (Johnson *et al.*, 1985)
MK10 (Krieg *et al.*, 1985); HK10 (Zhou *et al.*, 1988)
MK4 (Knapp *et al.*, 1986a); HK4 (Leube *et al.*, 1988)
MK13 (Knapp *et al.*, 1986a); HK13 (Kuruc *et al.*, 1989)

formed by MK1 and its stratum corneum variants in NEPHGE gels (Fig. 2a,b). In contrast, an antiserum against the carboxy-terminal peptide of MK10 (Fig. 5) stained both living and dead suprabasal epidermal cells (Roop *et al.*, 1984). These results confirm that, unlike MK1, MK10 is not completely degraded in the stratum corneum. This also indicates that the secondarily generated MK10 variant, which exhibits a larger molecular weight than its precursor (Schweizer and Winter 1982b; Bowden *et al.*, 1984), has probably retained an intact carboxy terminus. However, post-translational cleavage of carboxy-terminal sequences of type I differentiation-specific keratins has been demonstrated with an antiserum against the carboxy-terminal peptide of the type I differentiation-specific keratin of internal mouse epithelia MK13 (Nischt *et al.*, 1988). Therefore, Western blot studies using isoelectrically (rather than NEPHGE) focused keratins of total mouse epidermis and of stratum corneum with the specific MK10 antiserum should help to further unravel the complex processing of this keratin.

For obvious reasons (and within certain limitations) mouse keratin specific cDNA probes and antisera are also important tools in the growing field of transgenic keratin expression. As previously mentioned, sequence comparison between human and murine epidermal keratins have shown that orthologous basal keratins are highly conserved (Figs. 3 and 5). Consequently, 3′-derived cDNA probes and antisera against carboxy-terminal peptides of these keratins tend to crossreact between the two species. Therefore, studies on the transgenic expression of HK14 in mice required the in-frame insertion of sequences coding for the antigenic portion of a nonepidermal protein into the HK14 gene construct in order to discriminate between the expression of the two keratins (Vassar *et al.*, 1989). In contrast, the sequences of orthologous differentiation-specific epidermal keratins K1 and K10 of mouse and humans are distinctly less conserved. Figure 5 shows that only 13 out of 20 amino acid residues are positionally preserved in the carboxy terminus of HK1 and MK1. This difference has, however, proved sufficient to confer species specificity to antisera elicited against the corresponding murine and human sequences, thus enabling their use in studies on transgenic HK1 expression in mice (Roop *et al.*, 1989). Since positional divergences in carboxy termini are even more pronounced for HK10 and MK10 and are particularly striking for the differentiation-specific keratin pair K4 and K13 of internal stratified epithelia (Fig. 5), it can be predicted that the respective antisera could also be used for studies on transgenic expression. The following sections show that mouse keratin specific cDNA probes and antisera are also indispensable tools for the elucidation of keratin patterns that deviate from the basic epidermal keratin pattern.

Body Site-Specific Variations of the Basic Epidermal Keratin Pattern

As discussed earlier, distinct skin sites of the adult mouse exhibit a pronounced propensity toward expression of local keratin patterns that, in addition to the keratins of the basic pattern (Fig. 2a,b), contain members of the mouse keratin multigene family. At present, this laboratory has identified one topologically restricted keratin that seems to be specific for the mouse (or related small rodents); one keratin that has an analog in human epidermis; and a keratin pair whose orthologous relationship to human keratins is presently unclear.

Murine Type II 70-kDa Keratin

One-dimensional resolution of adult mouse keratins obtained from the epidermis of different skin sites reveals a large 70-kDa keratin specific to ear, tail, and foot sole epidermis (Schweizer and Winter, 1982a; Schweizer *et al.*, 1987; Rentrop *et al.*, 1987). Upon two-dimensional analysis, this keratin, which consistently represents one of the most prominent of the keratin patterns, migrates within the pH range of type II keratins and analogously typically resolves into multiple isoelectric variants (Fig. 6b,c,d). Amazingly, in the adult animal, the three skin sites in which the 70-kDa keratin is expressed differ strikingly in their morphological appearance. Whereas ear epidermis is extremely thin (Fig. 7a), foot sole epidermis (especially that covering the footpads) represents an extremely thick epithelium (Fig. 7c,d). The most unusual morphology is encountered in mouse tail epidermis, which consists of regularly alternating areas of parakeratotic scale and orthokeratotic interscale regions (Fig. 7b and see also Fig. 12a). The particular morphology of each of the above-mentioned skin sites essentially develops during the first week after birth from the uniformly thick neonatal epidermis. This is also the period during which the expression of the 70-kDa keratin is induced. Whereas in tail epidermis the 70-kDa keratin can already be electrophoretically detected one day postnatum, its postnatal induction lasts about five days in foot sole epidermis and up to ten days in ear epidermis (Rentrop *et al.*, 1987).

Screening of a cDNA library constructed with polyA$^+$ RNA from adult mouse tail epidermis yields a cDNA clone, pke70, clearly encoding a type II keratin whose non-α-helical domains show no sequence homology with any of the known murine type II keratins. The use of a specific 3′ fragment of pke70 in Northern blots with epidermal mRNAs of various skin sites and from newborn mouse epidermis reveals a 2.8-kb mRNA species occurring only in adult mouse ear, tail, and footsole epidermis, indicating that the clone encodes the 70-kDa keratin (Rentrop *et al.*, 1987). *In situ* hybridization with the 3′ fragment indicates that the 70-kDa keratin mRNA is present in suprabasal cells of all three skin sites (Rentrop *et al.*, 1987). They exhibit, however, tissue-specific particularities. Whereas 70-kDa keratin transcripts are ubiquitously expressed in differentiating cells of ear epidermis (Fig. 7a), their occurrence in tail epidermis is strictly confined to the differentiating layers of the orthokeratotic interscale regions (Fig. 7b). In foot sole epidermis, the 70-kDa keratin is found in the entire foot sole epidermis

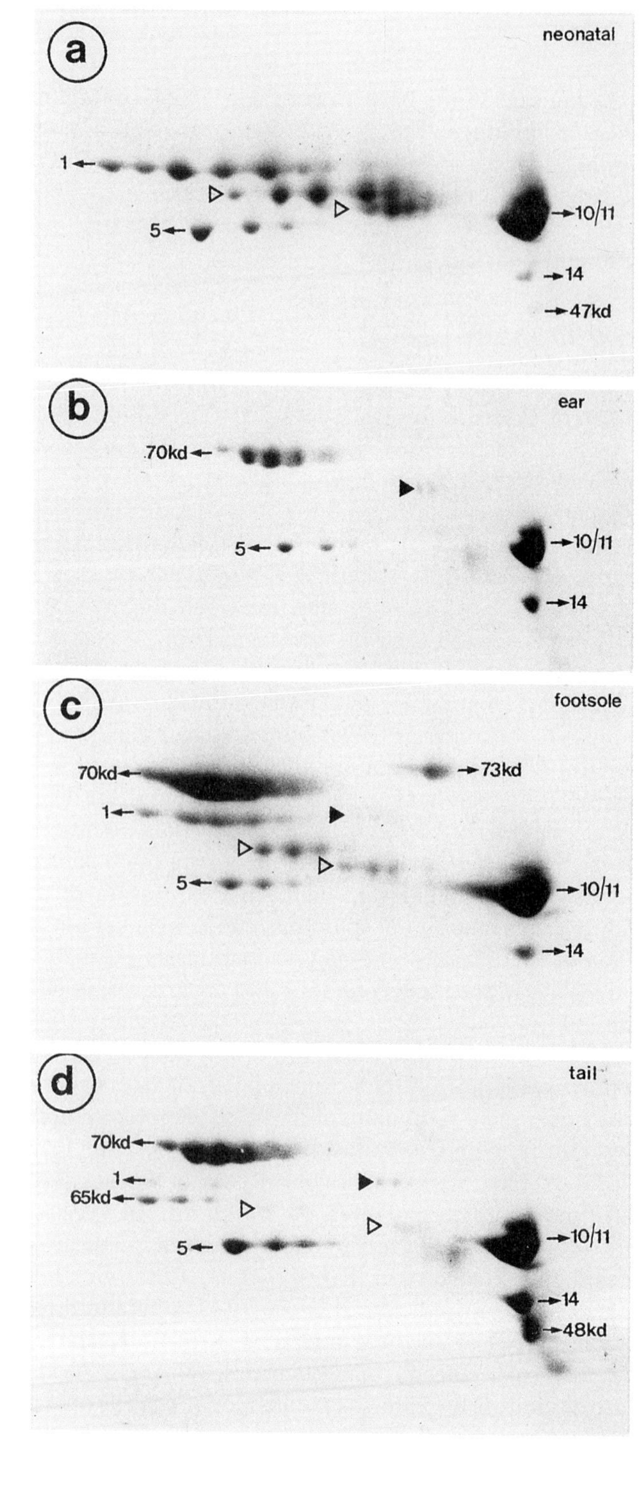

a
neonatal
1
5
10/11
14
47kd
b
ear
70kd
5
10/11
14
c
footsole
70kd
73kd
1
5
10/11
14
d
tail
70kd
1
65kd
5
10/11
14
48kd

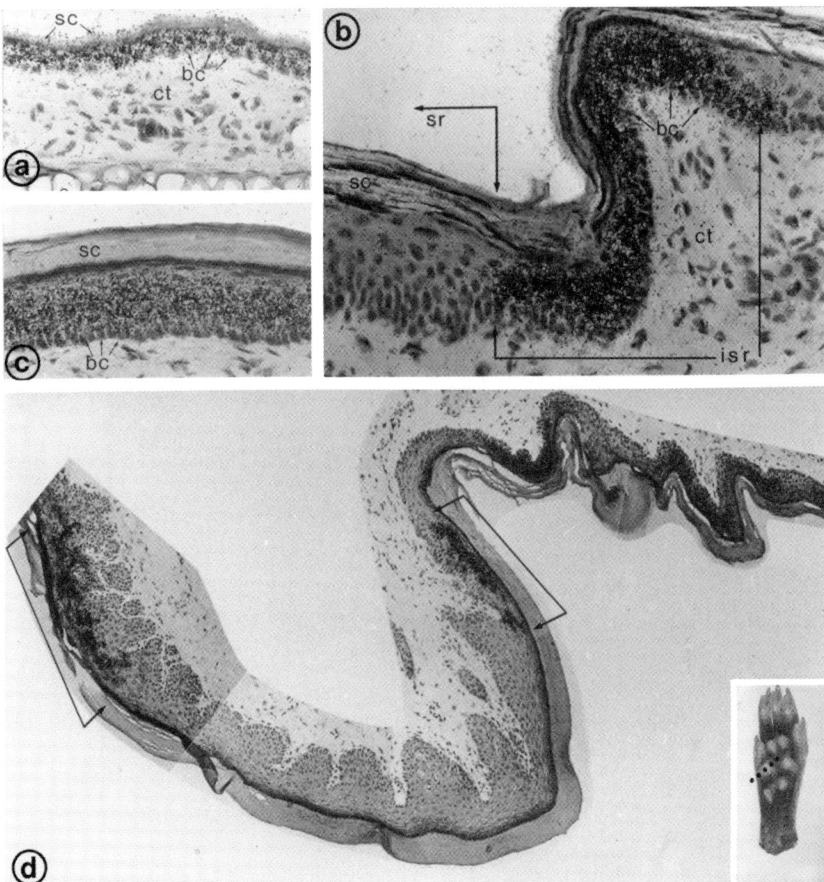

FIGURE 7 *In situ* hybridization of a ^{35}S-labeled specific 3′ probe of clone pke70 to frozen sections of (a) ear skin; (b) tail skin; (c) foot sole skin; and (d) footpad skin (the direction of this section through the tissue is indicated by a dotted line in the inset). bc, basal cells; sc, stratum corneum; isr, interscale region; sr, scale region; ct, connective tissue. Bar = 100 μm. (With permission from Rentrop *et al.*, 1987.)

←

FIGURE 6 Two-dimensional keratin patterns of (a) neonatal mouse epidermis; (b) adult mouse ear epidermis; (c) foot sole epidermis; and (d) tail epidermis. Open arrowheads indicate the 64-kDa and-62 kDa stratum corneum equivalents of MK1. Closed arrowheads denote the 67.5-kDa stratum corneum equivalent of the 70-kDa keratin. Proteins were resolved by NEP-HGE in the first dimension and by 9% SDS–PAGE in the second dimension.

proper (Fig. 7c). However, it is also demonstrable in the transitional epidermis at the base of the footpad. The epidermis of the lateral walls and in particular the apical region of the footpads are virtually devoid of 70-kDa keratin transcripts (Fig. 7d).

There is evidence that the suprabasally expressed type II 70-kDa keratin is subject to secondary modification in the stratum corneum of the respective skin sites similar to that of MK1. It is not yet clear whether a slightly more acidic protein variant seen occasionally below the 70-kDa keratin in well-resolved two-dimensional gels (Fig. 6b) represents such a secondarily derived protein. A smaller (but distinctly more acidic) weak protein (indicated by a closed arrowhead in Fig. 6b,c,d) could clearly be detected in the stratum corneum of foot sole epidermis (unpublished results). This protein exhibits almost the same molecular weight as MK1 (Fig. 6b,c,d) and is important because in one-dimensional gels of ear epidermis it simulates the presence of MK1, whereas two dimensionally resolved gels clearly show that MK1 is absent or present only in trace amounts in mouse ear epidermis (Fig. 6b).

Apart from its postnatally acquired body site specific expression, the 70-kDa keratin exhibits further remarkable properties. First, notwithstanding the presence of a specific 2.8-kb mRNA species in adult mouse ear, tail, and foot sole epidermis, *in vitro* translation of total polyA$^+$ RNA from the three skin sites does not lead to the expression of the 70-kDa keratin (see for example Fig. 8a,b). This is intriguing insofar as incubation of skin samples from ear, tail, and foot sole in a medium containing ^{35}S-methionine clearly results in a labeling of the 70-kDa keratin (Schweizer *et al.*, 1987). At present, the reason why the 70-kDa keratin mRNA is refractory to *in vitro* translation remains unclear. Second, the induction of a sustained hyperplasia in ear and tail epidermis by repetitive treatment with the phorbol ester TPA or retinoic acid leads to the suppression of 70-kDa keratin synthesis (Schweizer *et al.*, 1987. Thus, unlike keratins MK6 and MK17 the 70-kDa keratin is down-regulated in hyperproliferative mouse epidermis (Rentrop *et al.*, 1987; Schweizer *et al.*, 1987). Third, preliminary data from Southern blot analysis of murine and human genomic DNA indicate that the gene for the 70-kDa keratin is present only as a single copy in the mouse genome, whereas it seems to be absent from the human genome. Since a type II keratin of identical size and charge properties is also found in the respective skin sites of the rat and the multimammate mouse species *Mastomys natalensis* (Schweizer *et al.*, 1989), the 70-kDa keratin appears to be a characteristic member of the keratin multigene family of small rodents. At present, the functional significance of this large keratin is unknown. It is, however, noteworthy that the 70-kDa keratin occurs only in skin sites that contain comparatively little hair or no hair at all. It is also remarkable that in the three skin sites, the 70-kDa keratin is not expressed concomitantly with a new type I keratin. This implies that for filament formation, the 70-

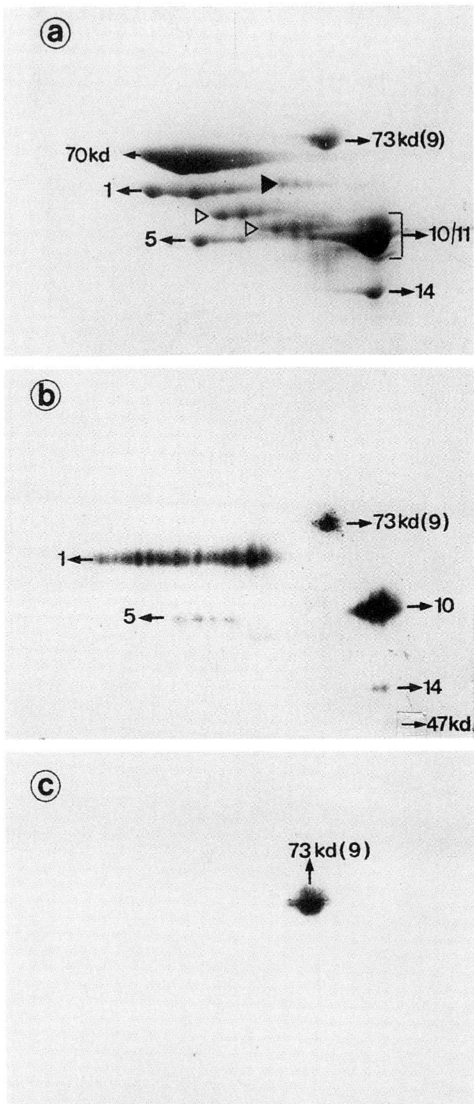

FIGURE 8 (a) Two-dimensional keratin pattern of adult mouse foot sole epidermis. (b) *In vitro* translation products of polyA + RNA of adult mouse foot sole epidermis. (c) Western blot of *a* with a specific antibody against HK9 (Knapp *et al.*, 1986b). Keratins were resolved by NEPHGE in the first dimension and by 8.5% SDS–PAGE in the second dimension. For the designation of open and closed arrowheads, see legend to Figure 6. (With permission from Schweizer *et al.*, 1989.)

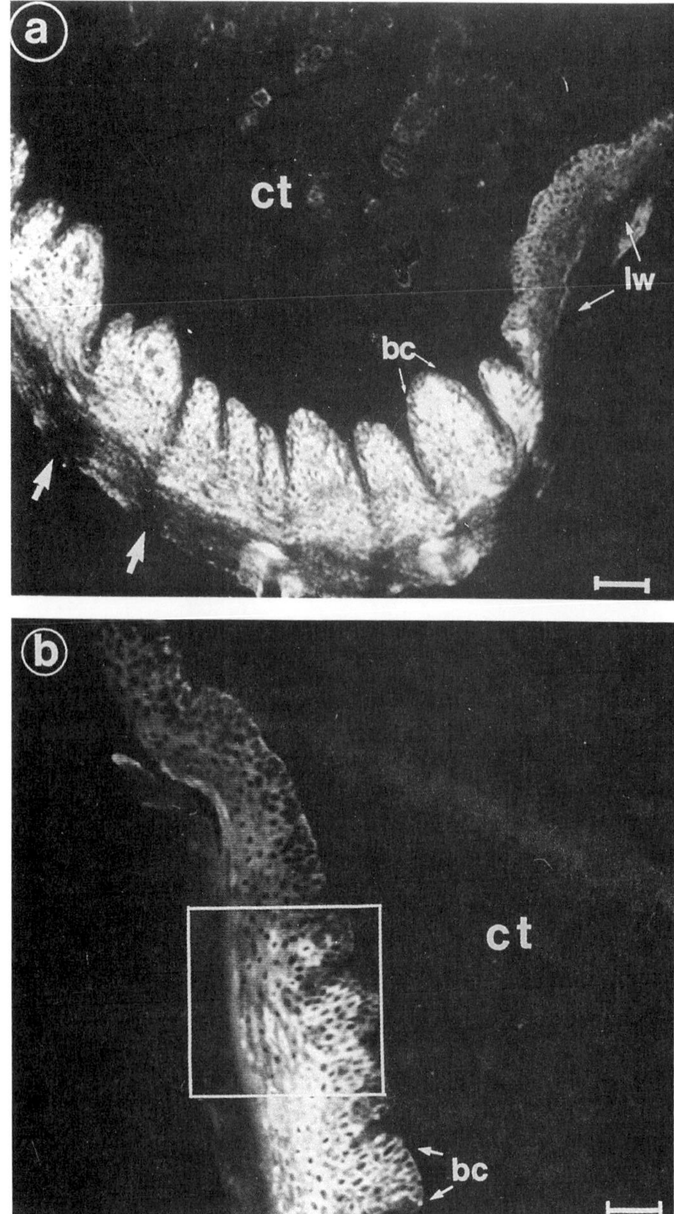

FIGURE 9 Immunolocalization of MK9 in frozen sections of adult (a–c) and newborn (d) mouse footsole epidermis using a specific antibody to HK9 (Knapp *et al.*, 1986b). Note the strong suprabasal staining in the apical epidermis of the footpads and the absence of staining in the small cell columns between the MK9-positive rete pegs (*arrows*) as well as the lateral walls of the footpads seen in *a*. The gradual disappearance of suprabasal MK9-positive cells at the

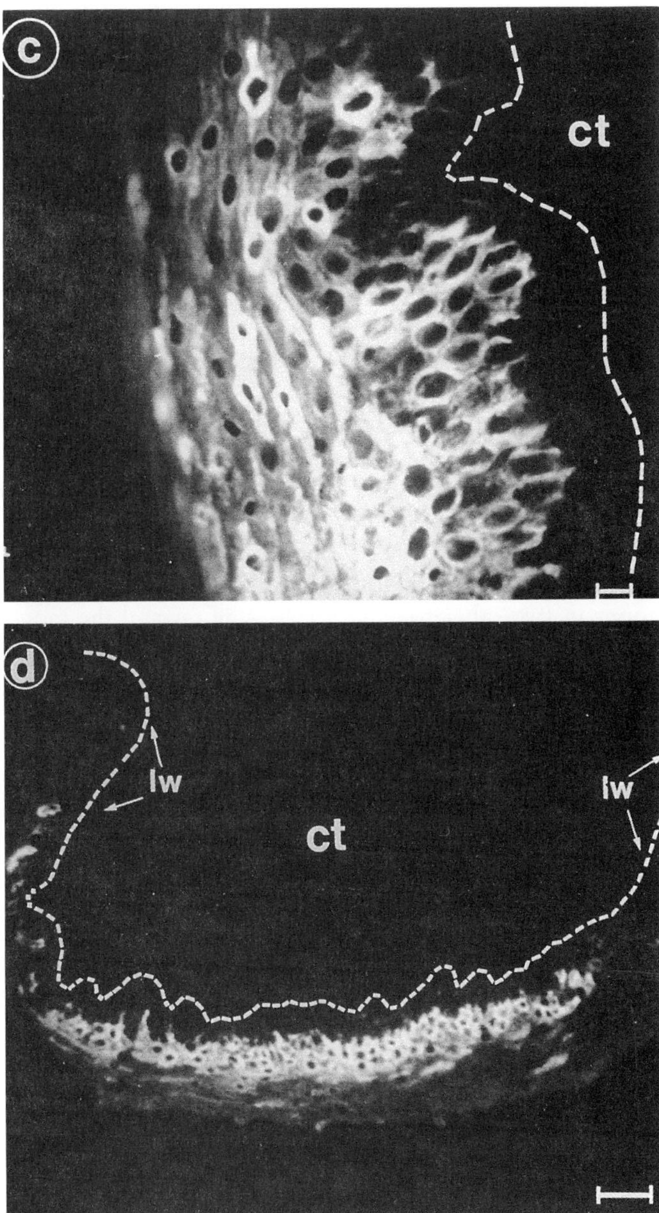

transition of the heavily undulated apical epidermis and the flat epidermis of the lateral walls of the footpads is indicated in *b* and *c*. The latter is a higher magnification of the boxed area in *b*. (d) A section of the developing footpad region of a 3 day-old mouse. bc, basal cells; lw, lateral walls; ct, connective tissue. Bars = 100 μm in *a;* 50 μm in *b,d;* 10 μm in *c*. (With permission from Schweizer *et al.*, 1989.)

kDa keratin must recruit a type I partner from the existing pool of acidic keratins. For obvious reasons this partner seems to be MK10.

Murine Type I MK9

The topological heterogeneity of epidermal keratin expression in humans is distinctly less pronounced than in mouse and is apparently restricted to the palmar and plantar epidermis. These skin sites have been shown to express high levels of an mRNA-encoded type I 64-kDa keratin which has been classified in the catalog of human keratins as HK9 (Moll *et al.*, 1982; Knapp *et al.*, 1986b). A keratin of almost identical size and charge properties has also been identified as a major keratin of bovine heel pad epidermis (Knapp *et al.*, 1986b).

Two-dimensional resolution of keratins from adult mouse fore and hind foot sole epidermis reveals an unusually large type I keratin of 73 kDa (Fig. 8a) that is absent from any other body site (Fig. 6a–d). This keratin is encoded by an mRNA of its own (Fig. 8b) and reacts specifically with an antibody against the human and bovine K9 in Western blots (Fig. 8c), thus indicating that the 73-kDa keratin represents the murine analog of HK9. Investigations in two other rodent species—the rat and the mouse species *Mastomys natalensis*—reveals K9 orthologs at 70.5 kDa and 68.5 kDa, respectively (Schweizer *et al.*, 1989). Therefore, the size of this keratin varies considerably among species, with MK9 representing the largest member of the mammalian keratin family reported so far (Schweizer *et al.*, 1989).

Indirect immunofluorescence studies with the antibody against HK9 showed positive staining only in suprabasal footpad epidermal cells, whereas the foot sole epidermis proper remained unstained Fig. 9a,b,c). The most intense staining was consistently noted in the apical epidermis of the footpads whose junction with the dermis is characterized by a pronounced formation of rete pegs (Fig. 9a). Vertical sections cut exactly through the center of a footpad show that the MK9-positive cell areas are separated from each other by small columns of MK9-negative cells ascending straight above the tips of the dermal papillae (Fig. 9a). Concomitant with the cessation of dermal–epidermal folding toward the lateral walls of the footpads, the number of MK9-positive cells gradually decreases and is no longer visible at the base of the footpads with their flat dermal–epidermal junction (Fig. 9b,c). It should be recalled that exactly these MK9-negative areas represent sites in which the type II 70-kDa keratin is expressed (Fig. 7d). The close relationship between the formation of rete pegs and MK9 expression is also seen in foot sole epidermis of newborn mice in which the postnatal evolution of a lobulated epidermis in the apical region of the developing footpads is accompanied by the gradual appearance of MK9-positive cells in the suprabasal cell layers (Fig. 9d). Moreover, in human palmar and plantar epidermis, which lacks anatomical equivalents of mouse footpads and which exhibits a continuous heavily lobulated dermal–epidermal junction, HK9 is almost homogenously expressed throughout the differentiated cell layers. From this it has been concluded that the expression of K9 is functionally related to skin sites firmly anchored in the underlying dermis and particularly destined to support body weight and to resist to mechanical wear and tear (Knapp *et al.*, 1986b, Schweizer *et al.*, 1989). The complex keratin expression in mouse foot sole epidermis is schematically illustrated in Figure 10. It implies that MK9 must compete with MK10 to form filaments with MK1 in the apex of the footpads, whereas, for the same reasons, the 70-kDa keratin and MK1 must compete for MK10 at the base of the footpads and in the foot sole epidermis proper.

At present, neither the human nor the murine K9 have been cloned. In view of the impressive size variability of the orthologous K9 species in SDS–PAGE with a maximal divergence of 9 kDa between HK9 and MK9 (Schweizer *et al.*, 1989), the elucidation of the primary structure would allow investigation of the evolution of the respective amino- and carboxy-terminal domains responsible for the observed size differences. Evidence for a possibly unusual amino acid composition of the head and tail portions of orthologous K9 species has recently been provided by investigations of the electrophoretic mobility of HK9 in two different gel systems. Whereas the mobility of both HK1 and HK10 remains unchanged relative to that of size markers, HK9 was shifted from 64 kDa in the Laemmli gel system to 73 kDa in the modified gel system (Egelrud and Lundström, 1989).

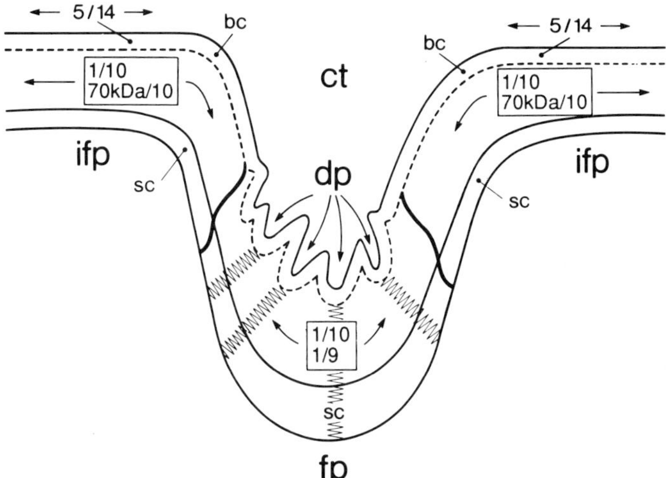

FIGURE 10 Schematic presentation of keratin expression in adult mouse footpad epidermis and the adjacent inter-footpad epidermis. The zigzag lines in the apical part of the footpad epidermis symbolize the narrow MK9 negative cell colums between the rete pegs. bc, basal cells; sc, stratum corneum; ifp, inter-foot pad; fp, foot pad; dp, dermal papillae; ct, connective tissue. (See also Fig. 9a; with permission from Schweizer *et al.*, 1989.)

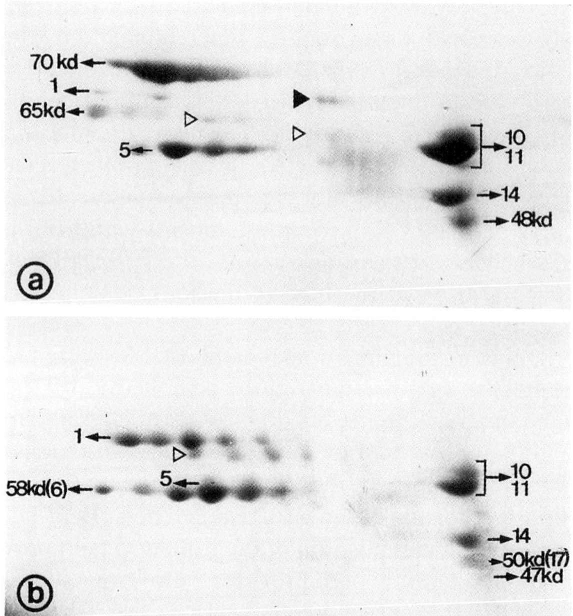

FIGURE 11 Keratin pattern of normal adult mouse tail epidermis (a) and tail epidermis treated daily for two weeks with 30 μg retinoic acid in 100 μl acetone (b). Keratins were resolved by NEPHGE in the first dimension and by 9% SDS–PAGE in the second dimension. For the designation of the open and closed arrowheads, see legend to Figure 6.

Murine Type II 65-kDa and Type I 48-kDa Keratins

The analysis of two-dimensionally resolved keratin patterns of different skin sites in the adult mouse reveals a type II 65-kDa keratin and a type I 48-kDa keratin as the major constituents of tail epidermis keratin pattern (Schweizer *et al.*, 1987 and Fig. 11a, and 6d). These keratins are virtually absent from any other skin site (Fig. 6a–d). The 65-kDa keratin can successfully be translated *in vitro* from polyA$^+$ RNA of adult mouse tail epidermis (Rentrop *et al.*, 1987). In a manner similar to the 70-kDa keratin, the synthesis of the 65-kDa keratin also occurs during the first postnatal week (i.e., the time during which the orthoparakeratotic epidermal patterning of the adult tail epidermis is established; Schweizer and Marks, 1977; Schweizer *et al.*, 1987). Since it can be shown that the 70-kDa keratin is selectively expressed in the orthokeratotic interscale region (Rentrop *et al.*, 1987 and Fig. 7b), it is tempting to speculate that the 65-kDa keratin and the 48-kDa keratin are associated with the parakeratotic scale epidermis unique to this skin site. This view is supported by the observation that unlike the orthokeratosis associated keratins MK1 and the 70-kDa keratin, the 65-kDa keratin is obviously not posttranslationally processed (Schweizer *et al.*, 1987). More importantly, the expression of the 65-kDa keratin and the 48-kDa keratin could completely be suppressed after orthokeratotic conversion of the parakeratotic scale epidermis by a two-week treatment with physiological doses of retinoic acid (Fig. 11b and Fig. 12a,b). This laboratory has identified cDNA clones that encode the 65-kDa and 48-kDa keratins. *In situ* hybridization to frozen sections of tail skin using specific 3′ fragments of the clones confirmed the expression of the keratins in suprabasal cells of the parakeratotic tail scale epidermis. The keratins are remarkable in that their carboxy-terminal domains contain high amounts of proline and cysteine residues that are, however, distinctly less frequent as in the carboxy-terminal domains of type II wool keratin 7c (Sparrow *et al.*, 1989) and of two murine type I hair keratins (Bertolino *et al.*, 1988; Bertolino *et al.*, 1989). The relationship of the 65-kDa keratin and the 48-kDa keratin to human keratins remains to be established.

Murine Hyperproliferation-Associated Keratins MK6 and MK17

Keratin analysis in wounded human epidermis or in epidermis affected by various diseases (including neoplasias) associated with cellular hyperproliferation leads to the occurrence of a new keratin pair that consists of a Type II 56-kDa keratin and a type I 48-kDa keratin (Moll *et al.*, 1982; Weiss *et al.*, 1984). These keratins are designated HK6 and HK16, respectively, in the catalog of human keratins (Moll *et al.*, 1982). As a rule, HK16 is coexpressed with another type I keratin of 46 kDa termed HK17 (Moll *et al.*, 1982). This keratin is difficult to resolve from HK16 by one- or two-dimensional gel electrophoresis. It should be emphasized that these keratins are not specific for hyperproliferative epidermis. They are also found in various nonepidermal keratinocytes undergoing hyperproliferation and are thought to represent markers for hyperproliferative keratinocytes in general (Weiss *et al.*, 1984).

Keratin analysis in hyperproliferative mouse epidermis leads to the detection of only two new keratins (Fig. 11b). The molecular weight of the new type II keratin has been determined to 60–58 kDa and that of its type I partner to 52–50 kDa (Winter *et al.*, 1980; Roop *et al.*, 1983; Breitkreutz *et al.*, 1984; Molloy and Laskin, 1988). Thus the diverging SDS–PAGE size estimates for both proteins overlap with the likewise varying molecular weight values electrophoretically determined for MK5 and MK14 (Fig. 2a,b).

Although it appears self evident that the hyperproliferation-associated murine type II keratin represents the analog of HK6, the identification of the corresponding orthologous cDNA clones remains a matter of conjecture. Investigations by E. Fuchs have shown that two distinct functional genes,

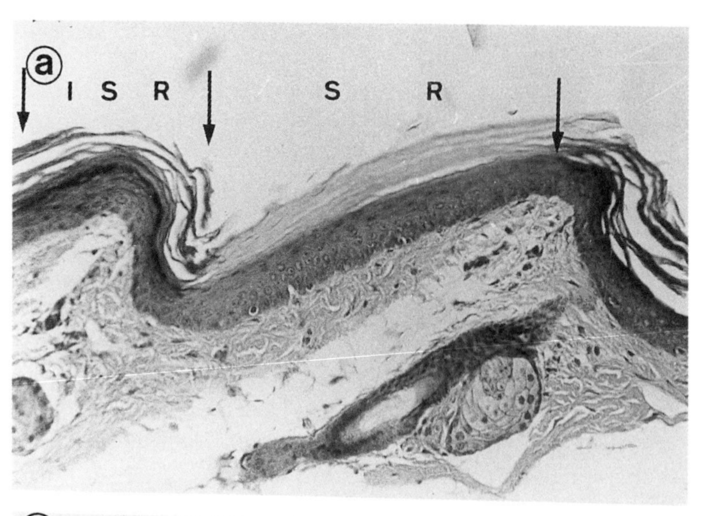
a
I S R
S R

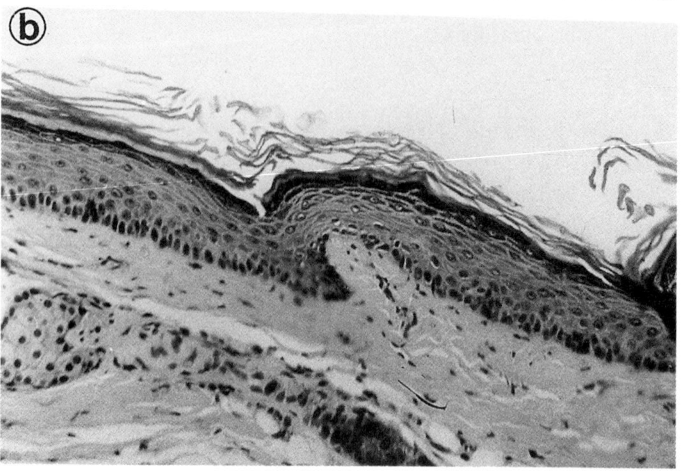
b

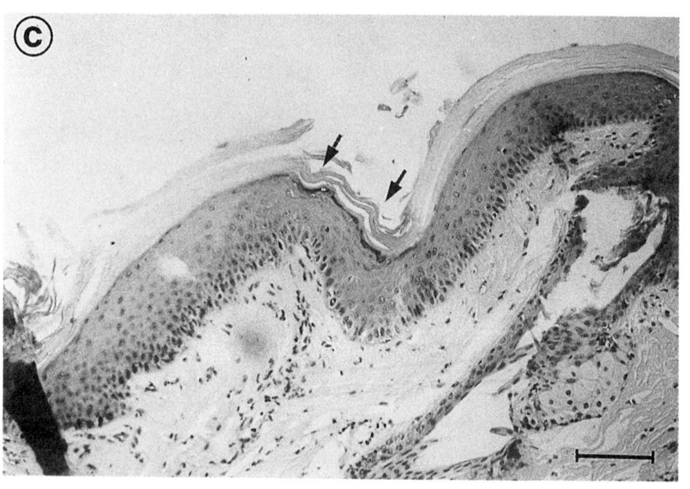
c

K6a and K6b, encode two highly homologous HK6 proteins (Hanakoglu and Fuchs, 1983; Tyner *et al.*, 1985). Both consist of 562 amino acid residues, differing only in the positions of three uncharged amino acids, and exhibit a predicted molecular weight of 60 kDa (Tyner *et al.*, 1985). On the other hand, Steinert *et al.* (1984) have isolated a murine cDNA clone that encodes a type II keratin of 552 amino acids and also has a predicted molecular weight of 60 kDa. The coding region of the murine clone shows an 85% sequence homology with that of HK6. However, on the basis of hybrid selection analysis, the authors suggest that the murine 60-kDa keratin represents the analog of HK5 (Steinert *et al.*, 1984) to which its sequence is also highly related (ca. 65%) (Galup and Darmon, 1988). However, it has been subsequently shown that an antibody raised against the synthetic peptide corresponding to the last 15 amino acids of the murine 60-kDa keratin (Fig. 5) does not stain normal mouse epidermis but reacts strongly with hyperproliferative epidermis. In addition, the amino acid alignments in Figure 5 indicate that the carboxy-terminal sequence of the mouse 60-kDa keratin is more closely related to HK6 than HK5, suggesting that the murine 60-kDa keratin represents the orthologous keratin of HK6. Supporting this assumption is the subsequent finding that the protein encoded by a mouse epidermal cDNA clone (mal 2), which was isolated from a polyA$^+$ RNA-selected cDNA library from TPA-treated hyperplastic mouse epidermis that was normalized by a library of control epidermis, is structurally identical with the 60-kDa keratin described by Steiner's group (Finch *et al.*, 1991).

In order to complete the orthologous relationship of murine and human hyperproliferation-associated keratins, it became necessary to determine whether the murine type I 52–50 kDa keratin corresponds to HK16 or HK17. Attempts to solve this question took a confusing course. In 1987, our laboratory isolated a partial murine cDNA clone pkSCC50 which, in hybrid selection experiments, selected for a 50-kDa keratin polyA$^+$ RNA from hyperproliferative mouse epidermis (Knapp *et al.*, 1987). Originally pkSCC50 was thought to code for the murine analog of MK16, since the murine clone showed no sequence homology with a postulated HK17 cDNA (RayChaudhury *et al.*, 1986). Subsequently, however, the postulated human clone was corrected to code for HK16, since its specific 3′ fragment did not hybridize in Northern blots with polyA$^+$ RNA from HeLa cells (Rosenberg *et al.*, 1988), which were known to synthesize only HK17

FIGURE 12 Morphological appearance of normal tail epidermis (a) and tail epidermis that has received applications of 30 μg retinoic acid in 100 μl acetone for two weeks (b) or that has been treated every two days with 20 μg TPA in 100 μl acetone for two weeks (c). Note the formation of an orthokeratinizing epidermis in the scale region in *b*, and the preservation of the ortho–parakeratinizing epidermal pattern (*arrows*) in *c*. ISR, interscale region; SR, scale region. Bar = 200 μm.

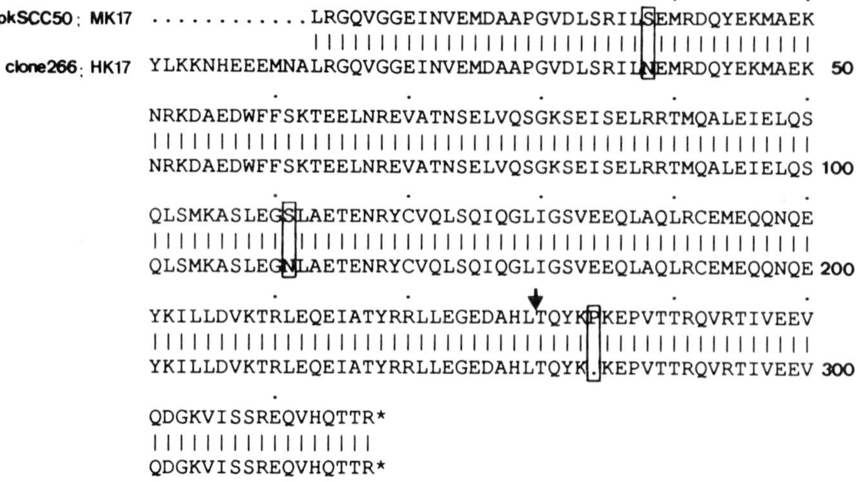

FIGURE 13 Amino acid sequence comparison of MK17 [clone pkSCC50 (Knapp *et al.*, 1987)] and HK17 [clone 266 (Kartasova *et al.*, 1987)]. Amino acids are designated by the one letter code. The arrow indicates the end of the α-helical domains. Two S⟶N changes in the α-helical domain and a proline residue that is absent from the carboxy terminus of HK17 are boxed.

(Franke *et al.*, 1981). Concomitantly, Kartasova *et al.* (1987) reported a new type I human keratin clone 266 isolated from cultured human keratinocytes; which had no sequence homology with any of the known human type I keratins. Surprisingly, the keratin encoded by the human clone 266 exhibited a 98% sequence homology with that derived from our murine clone pkSCC50 (Fig. 13). At the nucleotide level, this high homology extended to the 3′ noncoding region (not shown). It appears, therefore, that both clones are orthologs and code for human and murine K17.

According to this assignment, the murine hyperproliferation-associated keratin pair would consist of MK6 of 60 kDa and MK17 whose actual molecular mass cannot yet be predicted. In view of the known difficulties in discriminating electrophoretically between HK16 and HK17, it is necessary to ascertain whether an analog of HK16 is expressed in low amounts in hyperproliferative mouse epidermis. In view of the evidence that basal- and hyperproliferation-associated type I keratins K14 and K17 are extremely conserved in mouse and humans (Figs. 3, 5, and 13) and assuming that this is also true for a presumptive MK16, Northern blots have been performed using mRNA from mouse and human epidermis in various hyperproliferative conditions. These blots were probed with a specific 3′ fragment of HK16. Whereas the probe detected a 1.6-kb mRNA species in human epidermis (Rosenberg *et al.*, 1988), no signal was seen in mouse hyperproliferative epidermis. Likewise, the HK16 probe did not react in Southern blots with mouse genomic DNA, whereas a similar probe derived from the MK17 clone reacted with both human and murine genomic DNA (unpublished results). These preliminary data suggest that K16 is not part of the murine keratin multigene family but rather represents an evolutionary acquisition of humans or higher mammals.

Apart from their *de novo* synthesis in hyperproliferative mammalian epidermis, the expression of K6 and K16/17 is also a consistent feature of cultured epidermal keratinocytes and hence they are frequently also referred to as "culture-type keratins" (Weiss *et al.,* 1984). Since the differentiation-specific keratins K1 and K10 are usually not expressed in routine cultures of submerged epidermal keratinocytes, the keratin pattern of mouse epidermal cells *in vitro* consists of the basal keratins MK5 and MK14 including the 47 kDa keratin and the keratin pair MK6 and MK17 (Fig. 14a). It should be emphasized that no culture system of human or murine normal epidermal keratinocytes has been developed in which the expression of the culture-type keratins is suppressed. More importantly, these keratins are also rapidly induced when freshly excised, but otherwise intact, skin samples are placed in culture medium (Tyner and Fuchs, 1986). This suggests that any removal of epidermal cells from their normal environment irrevocably entails the synthesis of these keratins. The notorious presence of the hyperproliferation-associated keratins in any *in vitro* model of normal epidermal cells should be kept in mind when those systems are used as living skin equivalents for the testing of pharmacokinetically active substances or factors as the substances to be tested may be meeting a situation *in vitro* that needs to be created first *in vivo*.

The conspicuous association of the hyperproliferative keratins with both hyperproliferative epidermis *in vivo* and cultured epidermal keratinocytes *in vitro* raises the question of which cell population synthesizes these keratins. Using a monospecific antibody to MK6 for immunohistochemical studies and a specific 3′ fragment of MK17 for *in situ* hybridization, it can be shown in hyperplastic mouse epidermis that MK6 and MK17 are synthesized in differentiating cell layers (Fisher *et al.*, 1987; Knapp *et al.*, 1987). The abolition, however, of a clear cut basal–suprabasal demarcation and the creation of an enlarged basilar compartment in hyperplastic epidermis have made it difficult to precisely define the site of onset of synthesis (Fig. 15).

In addition, in routinely cultured mouse epidermal keratinocytes, MK6 and MK17 are synthesized preferentially if not exclusively in those cells undergoing *in vitro* differentiation. This can be demonstrated by culturing the cells in low Ca^{2+} medium (0.05 m*M*) which is known to favor cellular proliferation rather than differentiation (Hennings *et al.*, 1980). Raising the Ca^{2+} concentration in those cultures to 1.2 m*M* Ca^{2+} (high Ca^{2+} conditions) leads to enhanced cell differentiation concomitant with a drastic increase in the amounts of MK6 and MK 17 (Fig. 14b, lanes 1,2). The associa-

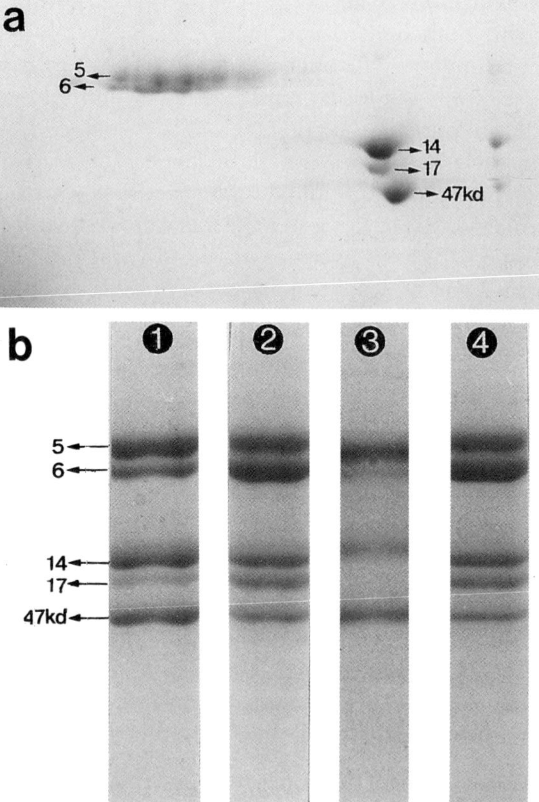

FIGURE 14 (a) Keratin pattern of cultured primary epidermal keratinocytes of newborn mouse epidermis. Keratins were resolved by NEPHGE in the first dimension and by 9% SDS–PAGE in the second dimension. (b) One-dimensional keratin patterns of the murine epidermal cell line SP1 (Strickland *et al.*, 1988) cultured under low Ca^{2+} conditions (lane 1); high Ca^{2+} conditions (lane 2); high Ca^{2+} conditions followed by a 30-hr culture period in low Ca^{2+} medium, which allows a separation into adhering, basal type cells (lane 3); and shedded differentiated cells (lane 4). Note the increase in intensity of the MK6 and MK17 bands after the high Ca^{2+} switch (lanes 1 and 2) and the association of MK6 and MK17 with differentiated cells (lanes 3 and 4). For further details see Sutter *et al.*, 1991.

tion of MK6 and MK17 expression with differentiated cells is even better visible when high Ca^{2+} cultures are returned to a low Ca^{2+} medium. Under these conditions, which are known as "Ca^{2+} stripping" (Jensen and Bolund, 1988; Read and Watt, 1988), differentiated cells are rapidly shed into the culture medium and can easily be separated from the adhering proliferating cells (Fig. 14b, lanes 3 and 4). It should be emphasized that, for practical reasons, the experiments shown in Figure 14b were conducted in an immor-

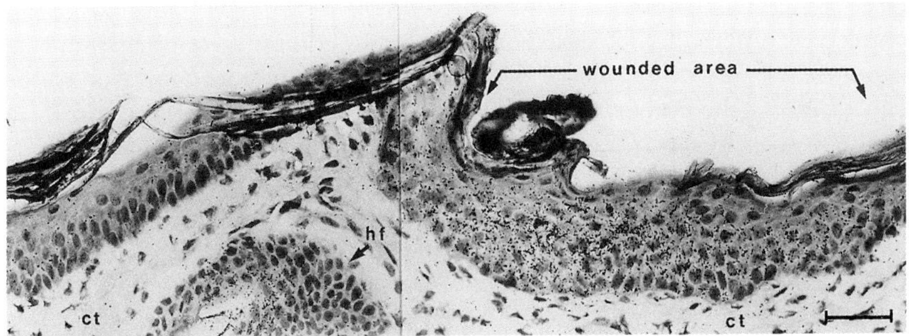

FIGURE 15 *In situ* hybridization of a ^{35}S-labeled specific 3′ probe derived from MK17 cDNA clone pkSCC50 (Knapp *et al.*, 1987) to a frozen section of normal tail skin showing (at right) a wounded and hyperplastic area probably caused by fighting. Note the presence of label mainly in suprabasal cells of the wounded epidermis and the absence of hybridization signals in the healthy epidermis at left. Note also that a distinct portion of hair follicle cells (*arrow*) is labeled by the MK17-specific DNA probe. hf, hair follicle; ct, connective tissue. Bar = 100 μm. (With permission from Knapp *et al.*, 1987.)

talized murine epidermal cell line SP1 (Strickland *et al.*, 1988). Differentiation-associated expression of MK6 and MK17 can, however, also be demonstrated in cultured primary mouse keratinocytes (Yuspa *et al.*, 1989).

The induction of the hyperproliferative keratins *in vivo* occurs very rapidly after the hyperproliferative stimulus. In mouse epidermis, MK6 can already be demonstrated 4 hr after treatment with TPA (D. R. Roop, personal communication) and organ cultures of human skin samples allow the detection of the hyperproliferation-associated keratins after 1 hr (Tyner and Fuchs, 1986). The rapid induction of these keratins has led E. Fuchs and coworkers to propose that the corresponding mRNAs may be present per se in normal epidermis, and are released from a translational block only under conditions of hyperproliferation (Tyner and Fuchs, 1986). This conclusion, obtained from experiments in surgically removed human skin, could not be confirmed in normal mouse epidermis where MK17 transcripts were undetectable by *in situ* hybridization (Knapp *et al.*, 1987; and Fig. 15).

The functional significance of K6, K16, and K17 in epidermis and in cultured epidermal cells remains mysterious. In body epidermis, their synthesis is undoubtedly related to a state of cellular hyperproliferation (Weiss *et al.*, 1984; Stoler *et al.*, 1988), although a variety of other stratified epithelia (i.e., the outer root sheath of hair follicles; Stark *et al.*, 1987; see also Fig. 15), the posterior part of the filiform tongue papillae (Rentrop *et al.*, 1986), and also human plantar epidermis (Moll *et al.*, 1982; Moll *et al.*, 1984) express these keratins normally. In cultured epidermal cells, a relationship to hyperproliferation is difficult to assess. It is, however, noteworthy that nearly normal skin equivalents that exhibit a mitotic index comparable to that of *in vivo* epidermis, still express these keratins. Interestingly, cultured rabbit corneal keratinocytes express K6, K16, and K17 only transiently during exponential growth and cease their synthesis at confluence in favor of the expression of the cornea-type differentiation-specific keratins K3 and K12. Thus, in these respects, corneal keratinocytes behave fundamentally differently from epidermal keratinocytes (Schermer *et al.*, 1989). At present the most reasonable explanation for the synthesis of K6, K16, and K17 in hyperproliferative epidermis and cultured epidermal cells seems to be that these keratins are expressed as an alternative suprabasal keratin pair under conditions that are nonpermissive for an undisturbed expression of the normal differentiation-specific epidermal keratin pair (Schermer *et al.*, 1989).

Keratin Expression in Mouse Epidermal Tumors

The most frequently used model for mouse skin carcinogenesis is the so-called two-stage carcinogenesis protocol. Basically, this protocol consists of tumor initiation of epidermal cells by a single topical application of a subtumorigenic dose of a chemical carcinogen (e.g., DMBA) followed by repetitive long-term tumor promotion with a nontumorigenic hyperplasiogen (e.g., TPA). (For review, see Hecker *et al.*, 1982; Slaga, 1984.) This treatment leads to the formation of multiple papillomas (Fig. 16a) and, upon continued promotion, to the development of squamous cell carcinomas (Fig. 16b), usually occurring within the site of preexisting benign lesions. This stepwise formation of both benign and malignant epidermal tumors is, therefore, ideally suited to investigate alterations in keratin expression during mouse skin carcinogenesis.

Since, a priori, papillomas represent a special type of epidermal hyperplasia, it would be expected that their keratin patterns would differ from those of normal epidermis by the presence of MK6 and MK17 (Fig. 17a,c–e). Moreover, even one-dimensionally resolved papilloma keratins reflect the perturbation of normal differentiation by the loss of the ordered generation of the 64-kDa and 62-kDa stratum corneum equivalents of MK1 (Fig. 17a,c–e). A further typical feature of papilloma keratin patterns is the visible decrease of MK1 and MK10 expression (Fig. 17a,c–e; the latter being only discerned well in two-dimensionally resolved gels). The progressive down-regulation of differentiation-specific keratin synthesis during tumor promotion has most convincingly been demonstrated by *in situ* hybridization studies that revealed a dramatic decrease in the number of MK1 and MK10 transcripts during malignant progression (Roop *et al.*, 1988). In fact, once a papilloma has been converted into a squamous cell carcinoma, MK1 and MK10 are no longer expressed (Winter *et al.*, 1980; Nelson and Slaga, 1982; Toftgard *et al.*, 1985) so that the keratin pattern of the malignant phenotype of DMBA/TPA-induced mouse skin tumors consists of

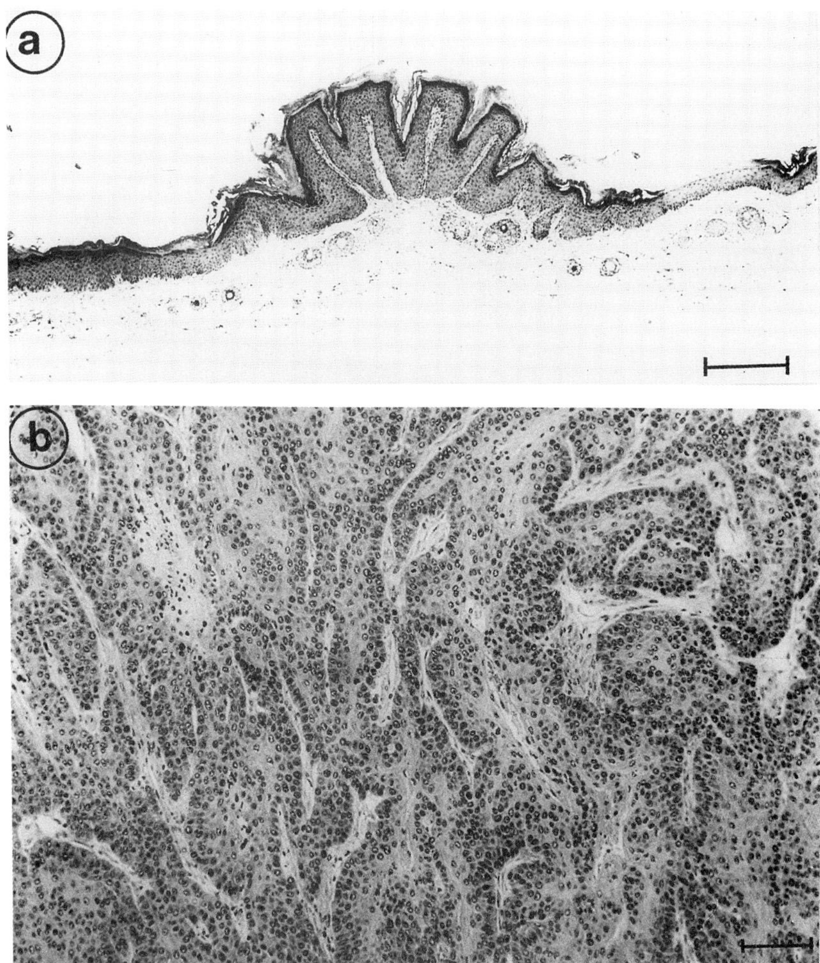

FIGURE 16 Morphological appearance of a DMBA/TPA-induced early mouse skin papilloma (a) and a squamous cell carcinoma (b). Bar = 250 μm in *a* and 25 μm in *b*.

MK5/MK14, MK6/MK17, and the 47-kDa keratin whose intensity is considerably increased relative to that observed in papillomas (Fig. 17f,g). Also carcinomas frequently, but not consistently, express MK19 at varying levels (Fig. 17f,g).

These observations, which have been verified independently in three different laboratories (Winter *et al.*, 1980; Nelson and Slage, 1982; Toftgard, *et al.*, 1985), lead to the proposal that keratin analysis may be a reliable tool to discriminate between benign and malignant skin tumors (Winter *et al.*, 1980). It must be emphasized, however, that the suppression

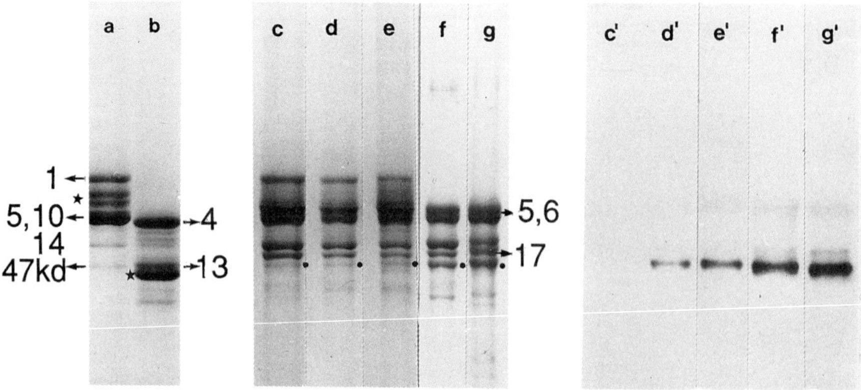

FIGURE 17 One-dimensional keratin patterns of newborn mouse epidermis (lane a); adult mouse tongue epithelium (lane b); three different DMBA/TPA-induced papillomas (lanes c,d,e) and two DMBA/TPA-induced squamous cell carcinomas (lanes f,g). The asterisks indicate the 64-kDa and 62-kDa stratum corneum equivalents of MK1 (lane a) and the 46-kDa stratum corneum equivalent of MK13 (lane b; Nischt *et al.*, 1988). The black dots in *c–g* denote a 47-kDa protein in papillomas and carcinomas that has the same mobility as the faint epidermal 47-kDa keratin (lane a) and MK13 of internal stratified epithelia (lane b). Lanes c′–g′ represent immunoblots of the papilloma and carcinoma keratins of lanes *c–g* with an MK13 monospecific antiserum (Nischt *et al.*, 1988). 10% SDS–PAGE. (With permission from Nischt *et al.*, 1988.)

of MK1 and MK10 in DMBA/TPA-induced mouse skin carcinomas cannot be generalized. Although human skin carcinomas in general exhibit drastically reduced levels of the differentiation-specific keratins, a certain percentage have minor but significant amounts of HK1 and HK10 (Moll *et al.*, 1982; Nelson *et al.*, 1984; Moll *et al.*, 1984; Huszar, *et al.*, 1986). Therefore, the diagnosis of a human skin tumor as a benign or malignant lesion solely on the presence or absence of HK1 and HK10 would be unwarranted. The loss of MK1 and MK10 expression in the course of DMBA/TPA-mediated mouse skin carcinogenesis could be explained by the formation of a dedifferentiated type of squamous cell carcinoma. This view does not appear tenable, since squamous cell carcinomas produced by the two-stage protocol clearly belong to the moderately differentiated type of these tumors (Fig. 16b). Thus, the reasons for the conspicuous suppression of MK1 and MK10 synthesis in DMBA/TPA-induced malignant skin tumors must lie elsewhere. The following sections provide evidence that DMBA/TPA-mediated mouse skin carcinogenesis involves particular alterations that may ultimately be related to the suppression of MK1 and MK10 in malignant tumors.

Aberrant Expression of MK13 in DMBA/TPA-Induced Mouse Skin Tumors

In the mouse, the type I, 47-kDa keratin MK13 is normally expressed together with its 57-kDa type II partner MK4 in the differentiating cell layers of internal stratified epithelia that line the oral cavity and the upper digestive tract (Knapp *et al.,* 1986a; and Fig. 17, lane b). MK 13 is not expressed in normal mouse epidermis with the exception of embryonic epidermis (day 15–16 of gestation), which expresses MK4 and MK13 in the overlaying periderm (Nischt *et al.,* 1988). This keratin is also absent from hyperproliferative epidermis produced by wounding or short- and long-term treatment with TPA. Likewise, MK13 cannot be detected in cultured primary keratinocytes under low and high CA^{2+} conditions (Nischt *et al.,* 1988). However, it can be demonstrated that MK13 is aberrantly expressed without MK4 in DMBA/TPA-induced mouse skin tumors (Nischt *et al.,* 1988). Typically, the expression of MK13 in papillomas is variable. Western blots of keratins from individual papillomas collected early during TPA promotion using a monospecific antibody against the carboxy-terminal sequence of MK13 (Fig. 5) reveal tumors with either weak or strong MK13 expression. However, these blots also reveal tumors that do not express the keratin (Fig. 17, lanes c′–e′). Analysis of papillomas at later times of promotion show that the percentage of MK13-positive tumors steadily increase so that after 20–25 weeks of promotion virtually all papillomas express the keratin. However, differences in the extent of its expression are maintained (Nischt *et al.,* 1988; Gimenez-Conti *et al.,* 1990; and unpublished results). In contrast to papillomas, squamous cell carcinomas express MK13 constitutively and in high amounts (Fig. 17, lanes f′,g′). Both indirect immunofluorescence studies with the MK13 specific antibody and *in situ* hybridization with a specific 3′ fragment of MK13 cDNA (Knapp *et al.,* 1986a) indicate that, as in normal internal stratified epithelia, MK13 expression in DMBA/TPA-induced papillomas occurs only in the differentiating cell layers of the lesions. Sections of papillomas staining weakly positive for MK13 in Western blots reveal MK13-expressing cells that are either isolated or form small foci. In contrast, papillomas staining strongly positive for MK13 in Western blots, contain suprabasal MK13-expressing cell foci that are considerably extended in size and occasionally occupy nearly all of the suprabasal papilloma epithelium (Nischt *et al.,* 1988; Gimenez-Conti *et al.,* 1990). Compartmentalized expression of MK13 is also observed in sections of squamous cell carcinomas. Again, MK13 transcripts occurred only in those areas with signs of rudimentary differentiation (Nischt *et al.,* 1988).

In view of particular expression characteristics, MK13 can therefore be regarded as a marker for malignant progression of mouse skin tumors. Hence, the elucidation of the molecular mechanisms responsible for the transformation-dependent aberrant expression of the keratin would contribute to the understanding of DMBA/TPA-mediated mouse skin carcinogenesis.

Molecular Mechanisms Involved in Aberrant MK13 Expression

In order to unravel regulatory DNA sequence elements involved in the tissue-specific expression of MK13, the 5′ flanking region of the MK13 gene has been analyzed with regard to potential CpG methylation sites of potential regulatory function. Amazingly, within a 3.2-kbp stretch of the 5′ flanking region of the gene, those elements occur only sporadically and distally far from the transcriptional site of origin. Only one CpG site, termed "M1," was located within an MspI/HpaII recognition site C C G G and was thus available for differential methylation analysis of genomic DNA of MK13 expressing and nonexpressing epithelia (Winter *et al.*, 1990). The study clearly indicates that in MK13 nonexpressing normal tissues or cells, a diagnostic 3.7-kbp fragment containing the M1 site is not further cleaved by HpaII (Fig. 18, lanes a–d). In contrast, a 2.6-kbp fragment is generated by HpaII DNA digests of MK13-expressing internal epithelia (Fig. 18, lanes e–g). In addition, genomic DNA from MK13-expressing internal epithelia exhibit a DNaseI hypersensitive region around the M1 site that is not found in DNA of MK13 nonexpressing epidermis (Winter *et al.*, 1990). These results clearly indicate that an unmethylated and DNaseI hypersensitive M1 site correlates with the tissue-specific expression of the MK13 gene in internal stratified epithelia. It is, therefore, of interest to know whether, in DMBA/TPA-induced mouse epidermal tumors with aberrant MK13 expression, the methylated M1 site of the normal target tissue is altered. It can be shown that, in strict proportion to the variable extent of suprabasal MK13 protein expression in papillomas, DNA copies with an unmethylated M1 site are generated at the expense of methylated DNA copies (Fig. 18, lanes h,i), whereas as with normal internal stratified epithelia, carcinomas with constitutive MK13 expression contain only unmethylated DNA copies (Fig. 18, lanes j,k). Thus, these investigations show that the unmethylated M1 site in the 5′ flanking region of the MK13 gene, which is potentially involved in the regulation of the normal tissue-specific expression of MK13, is newly created along with the aberrant MK13 expression in DMBA/TPA-induced mouse skin tumors (Winter *et al.*, 1990).

Another approach aimed at gaining more insight into possible mechanisms governing aberrant MK13 expression in DMBA/TPA-induced mouse skin tumors has originated from the observation that HK13 is not expressed in human skin tumors (Kuruc *et al.*, 1989) and that, in the mouse model, the DMBA initiation step represents the decisive event responsible for the subsequent expression of MK13 in TPA-promoted tumors (Nischt *et al.*, 1988). In addition, aberrant MK13 expression has also been observed in papillomas

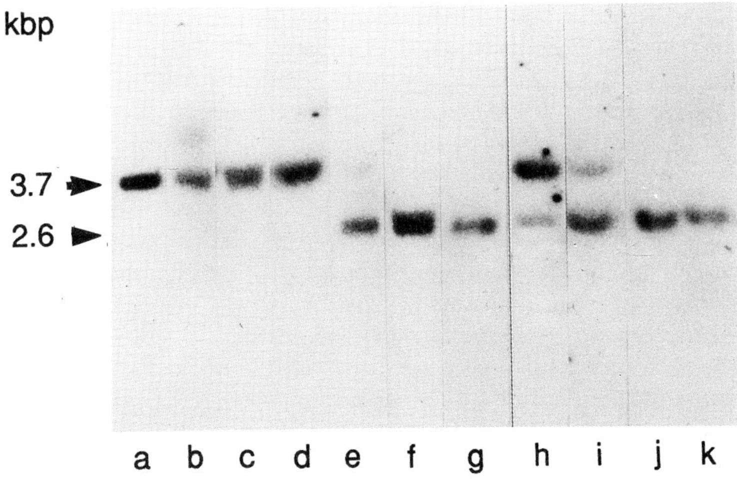

FIGURE 18 Determination of the methylation state of the CpG methylation site M1 in the 5′-flanking region of the MK13 gene in MK13 nonexpressing (lanes a–d) and expressing normal epithelia and cells (lanes e–g) and in DMBA/TPA-induced papillomas (lanes h,i) and carcinomas (lanes j,k). The 3.7-kbp fragment indicates a methylated M1 site; the 2.6-kbp fragment denotes an unmethylated M1 site. MK13-nonexpressing epithelia: footsole epidermis (a); neonatal epidermis (b); primary epidermal keratinocytes (c); tail epidermis (d). MK13-expressing epithelia: forestomach epithelium (e); primary forestomach keratinocytes (f); tongue epithelium (g). Tumors: papilloma weakly positive for MK13 in Western blots (h); papilloma strongly positive for MK13 in Western blots (i); transplantable SCC (j); *in situ* SCC (k). (Modified from Winter *et al.*, 1990.)

that are initiated by MNNG or urethane instead of DMBA (Schweizer *et al.*, unpublished results). It is now generally accepted that a common feature of these chemical carcinogens is the early activation of the cellular Ha-*ras* gene by point mutation of distinct codons (Balmain and Brown, 1988). Since DMBA initiation of mouse skin leads to point mutations in codon 61 of the Ha-*ras* gene in virtually all papillomas (Brown *et al.*, 1990, and Schweizer *et al.*, unpublished results), it has been hypothesized that the activation of the gene might be important for the aberrant expression of MK13 (Sutter *et al.*, 1992). Investigation of this benefited from the observation that infection of primary mouse keratinocytes with a replication-defective retrovirus containing the *v*-Ha *ras* gene and subsequent grafting of the infected cells to nude mice leads to the rapid formation of benign skin tumors that are morphologically indistinguishable from their chemically induced analogs (Roop *et al.*, 1986). Analysis of *v*-Ha-*ras*-induced papillomas for MK13 expression reveal that all of them displayed more or less extended suprabasal foci of MK13 expressing cells. Moreover, also these tumors exhibit the generation of DNA copies with an unmethylated M1 site that correlates with the extent of suprabasal MK13 protein expression (Sutter *et al.*, 1991). Thus, all features of DMBA/TPA-induced papillomas were shared by *v*-Ha-*ras*-induced papillomas, indicating that the aberrant synthesis of MK13 in mouse skin papillomas is primarily a consequence of Ha-*ras* oncogene activation. However, in view of the current belief that papillomas arise from the clonal expansion of an initiated Ha-*ras* mutated cell (for review, see Boyd and Barrett, 1990), the focal and variable MK13 expression in papillomas strongly suggests that besides the activation of the Ha-*ras* gene as a primary event, additional events—again possibly concerning the Ha-*ras* locus itself—must take place in a subpopulation of cells before MK13 can be aberrantly expressed (Sutter *et al.*, 1991).

The profound influence of the activated Ha-*ras* gene on the normal differentiation program of epidermal keratinocytes is well documented. It has been shown that the introduction of the *v*-Ha-*ras* gene into primary mouse keratinocytes not only produces reduction of the Ca^{2+}-mediated induction of epidermal differentiation markers but was also a prerequisite for the Ca^{2+}-dependent abnormal expression of the simple keratins MK8 and MK18 (Yuspa *et al.*, 1985; Cheng *et al.*, 1990) that are normally not expressed in epidermal cells. In concert with these data, the aberrant expression of MK13 in papillomas emphasizes the complexity of the keratinocyte phenotype produced by Ha-*ras* oncogene activation. Furthermore, investigations by Gimenez-Conti *et al.* (1990) have unequivocally shown that the suprabasal cell population that expresses MK13 inside papillomas does not synthesize MK1 and MK10, whereas these keratins are expressed in all suprabasal cells adjacent to the MK13-expressing foci. Thus, it appears that the gain of aberrant MK13 expression by a papilloma cell is accompanied by the loss of ability to synthesize a normal differentiation-specific keratin pair. The steady increase of the MK13-expressing cell population in papillomas during promotion, which ultimately represents the unique cell population of the malignant phenotype, may explain why, in contrast to human epidermal carcinomas, DMBA/TPA-induced malignant mouse skin tumors exhibit in conjunction with aberrant MK13 expression an apparantly Ha-*ras* oncogene-mediated suppression of MK1 and MK10 synthesis. It is clear that further experiments are needed to verify this hypothesis.

Concluding Remarks

We began with the discussion of the basic pattern of mouse epidermal keratin expression and continued with the presentation of keratins that are only expressed under specific circumstances. These special keratin members allow an expansion of the basic epidermal keratin pattern and Figure 19 is a schematic representation of this catalog of potential mouse epidermal keratins. No claim is made, however, that this catalog is comprehensive. Further

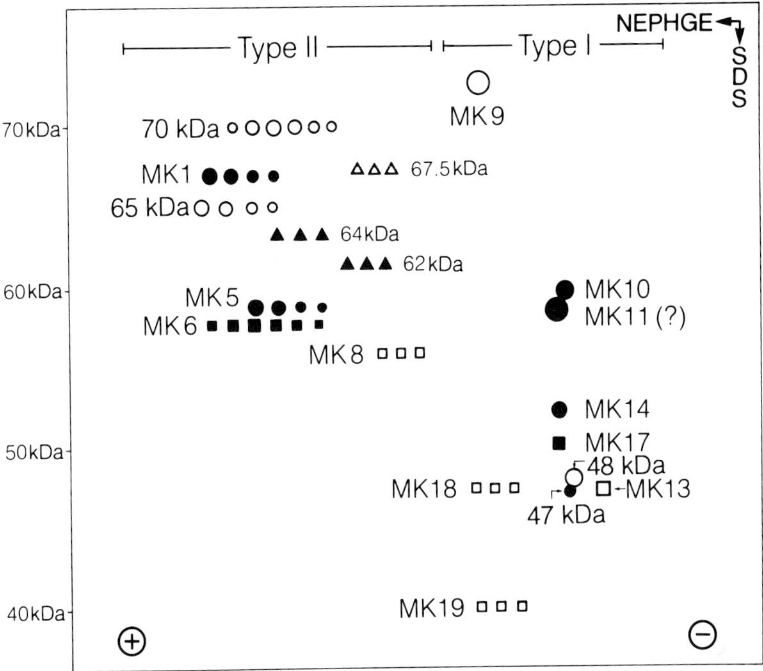

FIGURE 19 Schematic presentation of the mouse epidermal keratin repertoire. Full circles, basic epidermal keratin pattern [MK1, MK10, MK11 (?); MK5, MK14, 47kD keratin]; full triangles, MK1-posttranslationally derived keratins (64-kDa, 62-kDa keratins); open circles, body-site-specific keratins (70-kDa, 65-kDa and 48-kDa keratins; MK9); open triangles, 70-kDa keratin posttranslationally derived keratin (67.5-kDa keratin); full squares, hyperproliferation-associated or culture-type keratins (MK6, MK17); open squares, transformation-associated keratins (MK8, MK18, MK13, MK19).

studies may well reveal the existence of additional new keratins or keratins that, just as MK13, MK8, and MK18, are expressed aberrantly under distinct circumstances. Moreover, the existence of MK11 and MK16 as genuine translation products as well as the possible orthologous relationship of the type II 65-kDa and the type I 48-kDa parakeratotic tail epidermis keratin pair and of the type I 47-kDa basal cell keratin to human keratins needs to be clarified. In general, however, this catalog reflects the current status of our knowledge concerning the repertoire of keratins in normal, hyperproliferative, and transformed mouse epidermis.

Addendum

After submission of this chapter, several new results in the area of murine epidermal keratins have appeared that merit discussion:

1. Murine 70 kD keratin. Contrary to the assumption that the type II 70 kDa keratin might be specific for small rodents, recent results strongly indicate that this keratin represents the murine ortholog of the human epidermal keratin HK2 whose sequence has recently been elucidated (Collin, 1992; Collin *et al.*, 1992). This evidence is currently being followed up in our laboratory.
2. Murine type II 65 kD and type I 48 kDa keratins. This laboratory has recently characterized these keratins in more detail (Tobiasch *et al.*, 1992). Their suprabasal expression in normal mouse epithelia is indeed associated with a particular type of parakeratosis that is confined to tail scale epidermis and the posterior unit of the lingual filiform papillae. They are, however, also coexpressed with "hard" α-keratins in cortex cells of the hair follicle and in differentiating cells of the central unit of the lingual filiform papillae. Comparative structural analyses and phylogenetic investigations have led to the proposal to call these keratins "hair related" keratins, i.e., MHRb-1 (65-kDa keratin) and MHRa-1 (48 kDa keratin). There is evidence that two human keratins, designated Hbx and Hax by Heid *et al.* 1988a,b, represent the human orthologs of the murine "hair related" keratins (Tobiasch *et al.*, 1992).
3. Murine hyperproliferation associated keratins MK6 and MK17. Recently, the sequence of a human type I keratin expressed in Hela cells has been published (Flohr *et al.*, 1992). Sequence comparisons with HK14, HK15, and "HK17" suggested that this keratin represents a new member of the human keratin family. However, as described in this section, above, the "HK17" sequence used for comparison actually codes for HK16. Thus, the clone presented by Flohr *et al.* describes the full length sequence of HK17.

Acknowledgments

I am grateful to Hermi Winter for critically reading the manuscript and to Michael Rogers for stylistic help.

References

Balmain, A. and Brown, K. (1988). Oncogene activation in chemical carcinogenesis. *Adv. Cancer Res.* **51**, 147–182.

Banks-Schlegel, S. P. (1982). Keratin alterations during embryonic epidermal differentiation: a presage of adult epidermal maturation. *J. Cell Biol.* **93**, 551–559.

Baumberger, J. C. (1942). Methods for the separation of epidermis from dermis and some physiological and chemical properties of isolated epidermis. *J. Natl. Cancer Inst.* **2**, 413–423.

Bertolino, A. P., Checkla, D. M., Notterman, R., Sklaver, I., Schiff, T. A., Freedberg, I. M., and DiDona, G. J. (1988). Cloning and characterization of a mouse Type I hair keratin cDNA. *J. Invest. Dermatol.* **91,** 541–546.

Bertolino, A. P., Checkla, D. M., Heitner, S., Freedberg, I. M., and Yu, D. W. (1989). Differential expression of Type I hair keratins. *J. Invest. Dermatol.* **94,** 297–303.

Bowden, P. E., Quinlan, R. A., Breitkreutz, D., and Fusenig, N. E. (1984). Proteolytic modification of acidic and basic keratins during terminal differentiation of mouse and human epidermis. *Eur. J. Biochem.* **142,** 29–36.

Boyd, J. A. and Barrett, J. C. (1990). Genetic and cellular basis of multistep carcinogenesis. *Pharm. Ther.* **46,** 469–486.

Breitkreutz, D., Bowden, P. R., Quinlan, R., Franke, W. W., and Fusenig, N. E. (1983). Precursor–product relationship between prekeratin and keratin in mouse and human epidermis. *J. Invest. Dermatol.* **80,** 334.

Breitkreutz, D., Bohnert, A., Herzmann, E., Bowden, P. E., Boukamp, P., and Fusenig, N. E. (1984). Differentiation-specific functions in cultures and transplanted mouse keratinocytes. *Differentiation* **26,** 154–169.

Brown, K., Buchmann, A., and Balmain, A. (1990). Carcinogen-induced mutations in the mouse *c-Ha-ras* gene provide further evidence of multiple pathways for tumor progression. *Proc. Natl. Acad. Sci. (U.S.A.)* **87,** 538–542.

Cheng, C. K., Kilkenny, A. E., Roop, D. R., and Yuspa, S. H. (1990). The *v-ras* oncogene inhibits the expression of differentiation markers and facilitates expression of cytokeratins 8 and 18 in mouse keratinocytes. *Mol. Carcinog.* **3,** 363–373.

Collin, C. (1992). Deux cytokératins 2, spécifiques de deux programmes de kératinisation épitheliale. Ph.D. Thesis, Université Paris VII.

Collin, C., Moll, I., Kubicka, S., Ouhayoun, J. P., and Franke, W. W. (1992). Characterization of human cytokeratin 2, an epidermal cytoskeletal protein synthesized late during differentiation. *Exp. Cell Res.* **202,** 132–141.

Cowdry, E. V. (1952). In "Laboratory Techniques in Biology and Medicine, 3 ed. Williams & Wilkins Co., Baltimore.

Darmon, M. Y.;, Sémat, A., Darmon, M. C., and Vasseur, M. (1987). Sequence of a cDNA-encoding human keratin No10 selected according to structural homologies of keratins and their tissue-specific expression. *Mol. Biol. Rep.* **12,** 277–283.

Eckert, R. L., and Rorke, E. A. (1988). The sequence of the human epidermal 58-kDa (#5) Type II keratin reveals an absence of 5′ upstream sequence conservation between coexpressed epidermal keratins. *DNA* **7,** 337–345.

Egelrud, T., and Lundström, A. (1989). Stepwise modification of keratin polypeptides during keratinization in palmar–plantar epidermis. *Acta Derm. Venereol. (Stockholm)* **69,** 105–110.

Eichner, R., Bonitz, P., and Sun, T.-T. (1984). Classification of epidermal keratins according to their immunoreactivity, isoelectric points, and mode of expression. *J. Cell Biol.* **98,** 1388–1396.

Finch, J., Andrews, K., Krieg, P., Fürstenberger, G., Slaga, T., Ootsuyama, A., Tanooka, H. and Bowden, T. J. (1991). Identification of a cloned sequence activated during multistage carcinogenesis in mouse skin. *Carcinogenesis* **12,** 1519–1522.

Fisher, C., Jones, A., and Roop, D. R. (1987). Abnormal expression and processing of keratins in pupoid fetus (*pf/pf*) and repeated epilation (*Er/Er*) mutant mice. *J. Cell Biol.* **105,** 1807–1819.

Flohr, T., Buwitt, U., Bonnehold, B., Decker, T., and Boettger, E. C. (1992). Interferon γ regulates the expression of a novel keratin class I gene. *Europ. J. Immun.* 22, 975–979.

Franke, W. W., Schiller, D. L., Moll, R., Winter, S., Schmid, E., Engelbrecht, I., Denk, H., Krepler, R., and Platzer, B. (1981). Diversity of cytokeratins: differentiation-specific expression of cytokeratin polypeptide in epithelial cells and tissues. *J. Mol. Biol.* **153,** 933–959.

Franke, W. W., Schmid, E., Schiller, D. L., Winter, S., Jarasch, E. D., Moll, R., Denk, H., Jackson, B. W., and Illmensee, K. (1982). Differentiation-related patterns of expression of proteins of intermediate-sized filaments in tissues and cultured cells. *Cold Spring Harbor Symposia on Quantitative Biology, Vol. XLVI,* pp. 431–453.

Fuchs, E. (1988). Keratins as biochemical markers of epithelial differentiation. *TIG* **4,** 277–281.

Fuchs, E., and Green, H. (1980). Changes in keratin gene expression during terminal differentiation. *Cell* **19,** 1033–1042.

Galup, C., and Darmon, M. Y. (1988). Isolation and characterization of a cDNA clone coding for human epidermal keratin K5. Sequence of the carboxy-terminal half of this keratin. *J. Invest. Dermatol.* **91,** 39–42.

Garrels, J. I. and Gibson, W. (1976). Identification and characterization of multiple forms of actin. *Cell* **9,** 793–805.

Gimenez-Conti, I., Aldaz, G. M., Bianchi, A. B., Roop, D. R., and Conti, C. J. (1990). Early expression of Type I K13 keratin in the progression of mouse skin papillomas. *Carcinogenesis* **11,** 1995–1999.

Hanakoglu, I., and Fuchs, E. (1983). The cDNA sequence of a type II cytoskeletal keratin reveals constant and variable structural domains among keratins. *Cell* **33,** 915–924.

Hanson, J. (1947). The histogenesis of the epidermis in the rat and mouse. *J. Anat.* **81,** 174–197.

E. Hecker, N. E. Fusenig, W. Kunz, F. Marks, and H. W. Thielmann (eds.). (1982). In "Carcinogenesis—a Comprehensive Survey. Vol. 7." New York, Raven Press.

Heid, H. W., Moll, I., Franke, W. W. (1988a). Patterns of expression of trichocytic and epithelial cytokeratins in mammalian tissues. I. Human and bovine hair follicles. *Differentiation* **37,** 137–157.

Heid, H. W., Moll, I., and Franke, W. W. (1988b). Patterns of expression of trichocytic and epithelial cytokeratines in mammalian tissues. II. Concomitant and mutually exclusive synthesis of trichocytic and epithelial cytokeratins in diverse human and bovine tissues (hair follicle, nail bed and matrix, lingual papillae, thymic reticulum). *Differentiation* **37,** 215–230.

Hennings, H., Michael, D., Cheng, C. K., Steinert, P. M., Holbrook, K. A., and Yuspa, S. H. (1980). Calcium regulation of growth and differentiation of mouse epidermal cells in culture. *Cell* **19,** 245–254.

Huszar, M., Gigi-Leitner, O., Moll, R., Franke, W. W., and Geiger, B. (1986). Monoclonal antibodies to various acidic (Type I) cytokeratins of stratified epithelia. *Differentiation* **31,** 141–153.

Jensen, P. K. A. and Bolund, L. (1988). Low Ca^{2+} stripping of differentiated cell layers in human epidermal cultures: An *in vitro* model of epidermal regeneration. *Exp. Cell Res.* **175,** 63–73.

Johnson, L. D., Idler, W. W., Zhou, X. M., Roop, D. R., and Steinert, P. M. (1985). Structure of a gene for the human epidermal 67-kDa keratin. *Proc. Natl. Acad. Sci. U.S.A.* **82,** 1896–1900.

Kartasova, T., Cornelissen, B. J. C., Belt, P., and van de Putte, P. (1987). Effects of UV, 4-NQO and TPA on gene expression in cultured human keratinocytes. *Nucleic Acids Res.* **15,** 5045–5056.

Knapp, B., Rentrop, M., Schweizer, J., and Winter, H. (1986a). Nonepidermal members of the keratin multigene family: cDNA sequences and *in situ* localization of the mRNAs. *Nucleic Acids Res.* **14,** 751–763.

Knapp, A. C., Franke, W. W., Heid, H., Hatzfeld, M., Jorcano, J. L., and Moll, R. (1986b). Cytokeratin No. 9, an epidermal Type I keratin characteristic of a special program of keratinocyte differentiation displaying body site specificity. *J. Cell Biol.* **103,** 657–667.

Knapp, B., Rentrop, M., Schweizer, J., and Winter, H. (1987). Three cDNA sequences of mouse Type I keratins: Cellular localization of the mRNA in normal and hyperproliferative tissues. *J. Biol. Chem.* **262,** 938–945.

Krieg, T. M., Schafer, M. P., Cheng, C. K., Filpula, D., Flaherty, P., Steinert, P. M. and Roop, D. R. (1985). Organization of a Type I keratin gene. Evidence for evolution of intermediate filaments from a common ancestral gene. *J. Biol. Chem.* **260,** 5867–5870.

Kuruc, N., Leube, R. E., Moll, I., Bader, B. L., and Franke, W. W. (1989). Synthesis of cytokeratin 13, a component characteristic of internal stratified epithelia, is not induced in human epidermal tumors. *Differentiation* **42,** 111–123.

Lersch, R., and Fuchs, E. (1988). Sequence and expression of a type II keratin, K5, in human epidermal cells. *Mol. Cell. Biol.* **8,** 486–493.

Lersch, R., Stellmach, V., Stocks, C., Guidice, C., and Fuchs, E. (1989). Isolation, sequence, and expression of a human keratin K5 gene: Transcriptional regulation of keratins and insights into pairwise control. *Mol. Cell. Biol.* **9,** 3685–3697.

Leube, R. E., Bader, B. L., Bosch, F. X., Zimbelmann, R., Achtstaetter, T., and Franke, W. W. (1988). Molecular characterization and expression of the stratification related cytokeratins 4 and 15. *J. Cell Biol.* **106,** 1249–1261.

Marchuk, D., McCrohon, S., and Fuchs, E. (1984). Remarkable conservation of structure among intermediate filament genes. *Cell* **39,** 491–498.

Marchuk, D., McCrohon, S., and Fuchs, E. (1985). Complete sequence of a gene encoding a human type I keratin: Sequences homologous to enhancer elements in the regulatory region of the gene. *Proc. Natl. Acad. Sci. U.S.A.* **82,** 1609–1613.

Marrs, J. M., and Vorhees, J. J. (1971). A method for bioassay of an epidermal chalone-like inhibitor. *J. Invest. Dermatol.* **56,** 174–181.

Mischke, D., and Wild, G. A. (1987). Polymorphic keratins in human epidermis *J. Invest. Dermatol.* **88,** 191–197.

Moll, R., Franke, W. W., Schiller, D. L., Geiger, B., and Krepler, R. (1982). The catalog of human cytokeratin polypeptides: pattern of expression of specific cytokeratins in normal epithelia, tumors, and cultured cells. *Cell* **31,** 11–24.

Moll, R., Moll, I., and Franke, W. W. (1984). Differences of expression of cytokeratin polypeptides in various epithelial skin tumors. *Arch. Dermatol. Res.* **270,** 349–363.

Molloy, C. J., and Laskin, J. D. (1988). Keratin polypeptide expression in mouse epidermis and cultured epidermal cells. *Differentiation* **37,** 86–97.

Nelson, K. G., and Slaga, T. J. (1982). Keratin modifications in epidermis, papillomas and carcinomas during two stage carcinogenesis in the Sencar mouse. *Cancer Res.* **42,** 4176–4181.

Nelson, W. G., Battifora, H., Santana, H., and Sun, T.-T. (1984). Specific keratins as molecular markers for neoplasms with a stratified origin. *Cancer Res.* **44,** 1600–1603.

Nischt, R., Rentrop, M., Winter, H., and Schweizer, J. (1987). Localization of a novel mRNA in keratinizing epithelia of the mouse: Evidence for the sequential activation of differentiation specific genes. *Epithelia* **1,** 165–177.

Nischt, R., Roop, D. R., Mehrel, T., Yuspa, S. H., Rentrop, M., Winter, H., and Schweizer, J. (1988). Aberrant expression during two-stage mouse skin carcinogenesis of a Type I 47 k-Da keratin, K13, normally associated with terminal differentiation of internal stratified epithelia. *Mol. Carcinog.* **1,** 96–108.

O'Farrell, P. H. (1975). High-resolution two-dimensional electrophoresis of proteins. *J. Biol. Chem.* **250,** 4007–4021.

Quellet, T., Jussier, M., Bélanger, C., Kessous, A., and Royal, A. (1986). Differential expression of keratin genes during mouse development. *Dev. Biol.* **113,** 282–287.

Quellet, T., Lussier, M., Babai, F., Lapointe, L., and Royal, A. (1990). Differential expression of the epidermal K1 and K10 keratin genes during mouse embryo development. *Biochem. Cell Biol.* **68,** 448–453.

RayChaudhury, A., Marchuk, D., Lindhurst, M., and Fuchs, E. (1986). Three tightly linked genes encoding human type I keratins: Conservation of sequence with 5′ untranslated leader and 5′ upstream regions of coexpressed keratin genes. *Mol. Cell. Biol.* **6,** 539–548.

Read, J., and Watt, F. M. (1988). A model for *in vitro* studies of epidermal homoeostasis:

Proliferation and involucrin synthesis by cultures human keratinocytes during recovery after stripping off the suprabasal layers. *J. Invest. Dermatol.* **90,** 739–743.

Rentrop, M., Knapp, B., Winter, H., and Schweizer, J. (1986). Differential localization of distinct keratin in RNA species in mouse tongue epithelium by *in situ* hybridization with specific cDNA probes. *J. Cell Biol.* **103,** 2583–2591.

Rentrop, M., Nischt, R., Knapp, B., Schweizer, J., and Winter, H. (1987). An unusual Type II 70 kilodalton keratin protein of mouse epidermis exhibiting postnatal body site specificity and sensitivity to hyperproliferation. *Differentiation* **34,** 189–200.

Rieger, M., and Franke, W. W. (1988). Identification of an orthologous mammalian cytokeratin gene. High degree of intron sequence conservation during evolution of human cytokeratin 10. *J. Mol. Biol.* **204,** 841–856.

Roop, D. R., Hawley-Nelson, P., Cheng, C. K., and Yuspa, S. H. (1983). Keratin gene expression in mouse epidermis and cultured epidermal cells. *Proc. Natl. Acad. Sci. U.S.A.* **80,** 716–720.

Roop, D. R., Cheng, C. K., Titterington, L., Meyers, C. A., Stanley, J. R., Steinert, P. M. and Yuspa, S. H. (1984). Synthetic peptides corresponding to keratin subunits elicit highly specific antibodies. *J. Biol. Chem.* **259,** 8037–8040.

Roop, D. R., Cheng, C. K., Toftgard, R., Stanley, J. R., Steinert, P. M., and Yuspa, S. H. (1985). The use of cDNA clones and monospecific antibodies as probes to monitor keratin gene expression. *Annals N.Y. Acad. Sci.* **455,** 426–435.

Roop, D. R., Lowy, D. R., Tambourin, P. E., Strickland, J., Harper, J. R., Balaschek, M., Spengler, E. F., and Yuspa, S. H. (1986). An activated Harvey *ras* oncogene produces benign tumors on mouse epidermal tissues. *Nature* **323,** 822–824.

Roop, D. R., Krieg, T. M., Mehrel, T., Cheng, C. K. and Yuspa, S. H. (1988). Transcriptional control of high molecular weight keratin gene expression in multistage mouse skin carcinogenesis. *Cancer Res.* **48,** 3245–3252.

Roop, D. R., Chung, S., Cheng, C. K., Steinert, P. M., Sinha, R., Yuspa, S. H., and Rosenthal, D. S. (1989). Epidermal differentiation and its modulation by retinoids. In "Pharmacology of Retinoids in the Skin." (U. Reichert, and B. Shroot, eds.) Karger, Basel, New York pp. 1–7.

Rosenberg, M., RayChaudhury, A., Shows, T. B., Le Beau, M. M., and Fuchs, E. (1988). A group of Type I keratin genes on human chromosome 17: characterization and expression. *Mol. Cell. Biol.* **8,** 722–736.

Schermer, A., Jester, J. V., Hardy, C., Milano, D., and Sun, T.-T. (1989). Transient synthesis of K6 and K16 keratins in regenerating rabbit corneal epithelium: keratin markers for an alternative pathway of keratinocyte differentiation. *Differentiation* **42,** 103–110.

Schweizer, J., and Marks, F. (1977). A developmental study of the distribution and frequency of Langerhans cells in relation to formation of patterning in mouse tail epidermis. *J. Invest. Dermatol.* **69,** 198–204.

Schweizer, J., and Goerttler, K. (1980). Synthesis *in vitro* of keratin polypeptides directed by mRNA isolated from newborn and adult mouse epidermis. *Euro. J. Biochem.* **112,** 243–249.

Schweizer, J., and Winter, H. (1982a). Changes in regional keratin polypeptide pattern during phorbol ester-mediated reversible and permanently sustained hyperplasia of mouse epidermis. *Cancer Res.* **42,** 1517–1529.

Schweizer, J., and Winter, H. (1982b). Keratin polypeptide analysis in fetal and in terminally differentiating newborn mouse epidermis. *Differentiation* **22,** 19–24.

Schweizer, J., Kinjo, M., Fürstenberger, G., and Winter, H. (1984). Sequential expression of RNA-encoded keratin sets in neonatal mouse epidermis: Basal cells with properties of terminally differentiating cells. *Cell* **37,** 159–177.

Schweizer, J., Fürstenberger, G., and Winter, H. (1987). Selective suppression of two postnatally acquired 70-kDa and 65-kDa keratin proteins during continuous treatment of adult mouse tail epidermis with vitamin A. *J. Invest. Dermatol.* **89,** 125–131.

Schweizer, J., Rentrop, M., Nischt, R., Kinjo, M., and Winter, H. (1988). The intermediate filament system of the keratinizing mouse forestomach epithelium: Coexpression of keratins of internal squamous epithelia and of epidermal keratins in differentiating cells. *Cell Tiss. Res.* **253,** 221–229.

Schweizer, J., Baust, J., and Winter, H. (1989). Identification of murine Type I keratin 9 (73-kDa) and its immunolocalization in neonatal and adult mouse foot sole epidermis. *Exp. Cell Res.* **184,** 193–206.

Skerrow, D., and Skerrow, C. J. (1983). Tonofilament differentiation in human epidermis. Isolation and polypeptide chain composition of keratinocytes subpopulations. *Exp. Cell Res.* **143,** 27–35.

Slaga, T. J., (ed.) (1984). "Mechanisms of Tumor Promotion, Vol. 1–4." CRC Press, Boca Raton, Florida.

Sparrow, L. G., Robinson, C. P., MacMahon, D. T. W., and Rubira, M. R. (1989). The amino acid sequence of component 7c, a Type II intermediate filament protein from wool. *Biochem. J.* **261,** 1015–1022.

Stark, H. J., Breitkreutz, D., Limat, A., Bowden, P. E., and Fusenig, N. E. (1987). Keratins of human hair follicle: "hyperproliferative" keratins consistently expressed in outer root sheath cells *in vivo* and *in vitro*. *Differentiation* **35,** 236–248.

Steinert, P. M., and Roop, D. R. (1988). Molecular and cellular biology of intermediate filaments. *Ann. Rev. Biochem.* **57,** 593–636.

Steinert, P. M., Rice, R. H., Roop, D. R., Trus, B. L. and Steven, A. C. (1983). Complete amino acid sequence of a mouse epidermal keratin subunit and implications for the structure of intermediate filaments. *Nature* **302,** 794–800.

Steinert, P. M., Parry, D. A., Racoosin, E. L., Idler, W. W., Steven, A. C., Trus, B. L., and Roop, D. R. (1984). The complete cDNA and deduced amino acid sequence of a Type II mouse epidermal keratin of 60,000 Da: Analysis of sequence differences between Type I and Type II keratins. *Proc. Natl. Acad. Sci. U.S.A.* **81,** 5709–5713.

Steinert, P. M., Parry, D. A. D., Idler, W. W., Johnson, L. D., Steven, A. C., and Roop, D. R. (1985). Amino acid sequences of mouse and human epidermal Type II keratins of M_r 67,000 provide a systematic basis for the structural and functional diversity of the end domains of keratin intermediate filaments. *J. Biol. Chem.* **260,** 7142–7149.

Stoler, A., Kopan, R., Duric, M., and Fuchs, E. (1988). Use of monospecific antisera and cRNA probes to localize the major changes in keratin expression during normal and abnormal epidermal differentiation. *J. Cell Biol.* **197,** 427–446.

Strickland, J. E., Greenhalgh, D. A., Koceva-Chyla, A., Hennings, H., Restrepo, C., Balaschek, M., and Yuspa, S. H. (1988). Development of murine epidermal cell lines which contain an activated Ha-*ras* oncogene and form papillomas in skin grafts on athymic nude mouse hosts. *Cancer Res.* **48,** 165–169.

Sun, T.-T., Eichner, R., Nelson, W. G., Tseng, S. C. G., Weiss, R. A., Jarvinen, M. and Woodcock-Mitchell, J. (1983). Keratin classes: molecular markers for different types of epithelial differentiation. *J. Invest. Dermatol.* **81,** 1093–1098.

Sun, T.-T., Eichner, R., Schermer, A., Cooper, D., Nelson, W. G., Weiss, R. A. (1984). Classification, expression, and possible mechanisms of evolution of mammalian epithelial keratins: A unifying model. In: "Cancer Cells, Vol. I." A. Levine, G. F. VandeWoude, W. E. Topp, and J. D. Watson (eds.) Cold Spring Harbor, N.Y., pp. 169–176.

Sutter, C., Nischt, R., Winter, H., and Schweizer, J. (1991). Aberrant expression of keratin K13 induced by Ca^{2+} and vitamin A in mouse epidermal cell lines. *Exp. Cell Res.* **195,** 183–193.

Sutter, C., Strickland, J. E., Welty, D. J., Yuspa, S. H., Winter, H., and Schweizer, J. (1991). *v-Ha-ras* induced mouse skin papillomas exhibit aberrant expression of keratin K13 as do their DMBA/TPA-induced analog. *Mol. Carcinog.* **4,** 467–476.

Tobiasch, E., Winter, H., and Schweizer, J. (1992). Structural features and sites of expression of a new murine 65kD and 48kD hair related keratin pair associated with a special type of parakeratotic epithelial differentiation. *Differentiation* **50,** 163–187.

Toftgard, R., Yuspa, S. H., and Roop, D. R. (1985). Keratin gene expression in mouse skin tumors and in mouse skin treated with TPA. *Cancer Res.* **45**, 5845–5840.

Tyner, A. L. and Fuchs, E. (1986). Evidence for posttranscriptional regulation of the keratins expressed during hyperproliferation and malignant transformation in human epidermis. *J. Cell Biol.* **193**, 1945–1955.

Tyner, A. L., Eichman, M. J., and Fuchs, E. (1985). The sequence of a Type II keratin gene expressed in human skin: Conservation of structure among all intermediate filament genes. *Proc. Natl. Acad. Sci. U.S.A.* **82**, 4583–4687.

Vassar, R., Rosenberg, M., Ross, S., Tyner, A. and Fuchs, E. (1989). Tissue-specific and differentiation-specific expression of a human K14 keratin gene in transgenic mice. *Proc. Natl. Acad. Sci. U.S.A.* **86**, 1563–1567.

Weiss, L. W., and Zelickson, A. S. (1975). Embryology of the epidermis: Ultrastructural aspects. Maturation and primary appearance of dendritic cells in the mouse with mammalian comparisons. *Acta Dermatovenerol.* (*Stockholm*) **55**, 431–442.

Weiss, R. A. R., Eichner, R., and Sun, T.-T. (1984). Monoclonal antibody analysis of keratin expression in epidermal dieseases: a 48- and 56 kDa keratin as markers for hyperproliferation keratinocytes. *J. Cell Biol.* **98**, 1397–1406.

Wild, G. A., and Mischke, D. (1986). Variation and frequency of cytokeratin polypeptide patterns in human squamous nonkeratinizing epithelium. *Exp. Cell Res.* **162**, 114–126.

Winter, H., Schweizer, J., and Goerttler, K. (1980). Keratins as markers of malignancy in mouse epidermal tumors. *Carcinogenesis* **1**, 391–398.

Winter, H., Rentrop, M., Nischt, R., and Schweizer, J. (1990). Tissue-specific expression of murine keratin K13 in internal stratified squamous epithelia and its aberrant expression during two-stage mouse skin carcinogenesis is associated with the methylation state of a distinct CpG site in the remote 5′ flanking region of the gene. *Differentiation* **43**, 105–114.

Yuspa, S. H., Kilkenny, A. E., Stanley, J., Lichti, U. (1985). Keratinocytes blocked in phorbol ester responsive early stages of terminal differentiation by sarcoma viruses. *Nature* **314**, 459–462.

Yuspa, S. H., Kilkenny, A. E., Steinert, P. M., and Roop, D. R. (1989). Expression of murine epidermal differentiation markers is tightly regulated by restricted extracellular calcium concentrations in vitro. *J. Cell Biol.* **109**, 1207–1217.

Zhou, X. M., Idler, W. W., Steven, A. C., Roop, D. R., and Steinert, P. M. (1988). The complete sequence of the human intermediate filament chaine keratin 10. Subdomainal divisions and model for folding of end domain sequences. *J. Biol. Chem.* **263**, 15584–15589.

3

Phenotypic Expression and Processing of Filaggrin in Epidermal Differentiation

Beverly A. Dale, Richard B. Presland, Philip Fleckman, Ephraim Kam, and Katheryn A. Resing

Introduction and Objectives

The epidermis is ideal for the study of the relationship of morphology, biochemistry, and function. Epidermal differentiation results in an organized tissue in which morphologically distinguishable cells are arranged in discrete layers of basal, spinous, granular, and cornified cells. Biochemical maturation corresponds to the morphological differentiation, proceeding from the

Molecular Biology of the Skin: The Keratinocyte

proliferative basal cells to the anucleate cells of the stratum corneum. The cells of the stratum corneum provide a mechanically resistant skin surface via their densely packed structural proteins within a toughened cell membrane. The structural components responsible for the mechanical properties include keratin intermediate filaments and filaggrin, as well as proteins of the cornified cell envelope (e.g., loricrin, involucrin, and others), all of which are synthesized in the underlying living cell layers and undergo biochemical and organizational changes during the transition of granular cells into those of the stratum corneum.

This chapter focuses on filaggrin, a cationic protein normally present in the stratum corneum of mammalian epidermis. Filaggrin associates with keratin intermediate filaments, resulting in the formation of densely organized aggregates *in vitro* and aiding in the dense packing of keratin filaments during epidermal differentiation *in vivo* (Dale *et al.*, 1978; Steinert *et al.*, 1981). With keratin filaments present throughout the epidermis, the premature expression and function of filaggrin could result in aberrant aggregation of keratin filaments and subsequent cell or tissue damage. The keratinocyte has evolved several means of preventing this potentially damaging situation. First, filaggrin is expressed in the epidermal granular cell layer late in the process of epidermal differentiation. Second, it is expressed as a functionally inactive, precursor form (profilaggrin), which does not aggregate keratin filaments effectively. Profilaggrin is large (MW = 350 to 1,000 kDa, depending on species), highly phosphorylated, and relatively insoluble. It has an unusual repeating structure that is reflected in the profilaggrin gene the expression of which appears to be regulated predominantly at the level of transcription. Third, profilaggrin is effectively sequestered by accumulation in keratohyalin in the granular cell layer. Posttranslational processing of profilaggrin to filaggrin by dephosphorylation and proteolysis is dependent on the action of specific enzymes, one or more of which may be activated during transition of the granular cell to a cornified cell. Thus, premature aggregation of keratin filaments is prevented. Furthermore, this pathway provides several potential points of control of the complex posttranslational events that are coordinated with late events of epidermal differentiation.

An overview of the filaggrin pathway is shown in Figure 1, illustrating profilaggrin expression, accumulation into keratohyalin, processing to filaggrin, aggregation with keratin filaments, and subsequent breakdown to free amino acids. These steps are being investigated by a variety of techniques *in vivo* and *in vitro*. Profilaggrin is a polymeric protein that consists of 10–12 repeating filaggrin units in the human, and at least 20 in rodents. Each repeat consists of a filaggrin domain plus a short peptide that serves as a "linker" between repeats and is proteolytically removed during processing to filaggrin. Filaggrin aggregates with keratin filaments and eventually undergoes modification and degradation to free amino acids that may function osmotically to retain moisture within the cells of the stratum corneum.

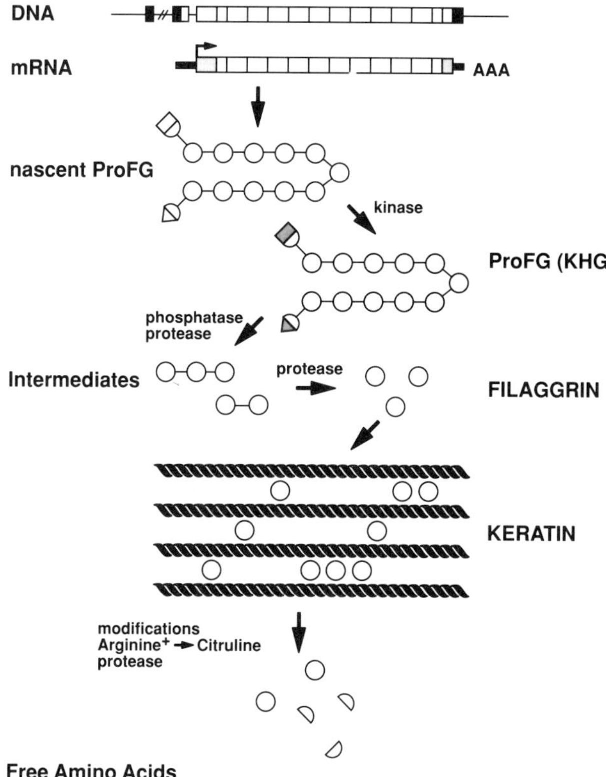

FIGURE 1 Diagram of the pathway of filaggrin synthesis, interaction with keratin filaments, and degradation as it occurs in the granular and cornified cells of mammalian epidermis.

The properties of filaggrin and profilaggrin and the functional interaction of filaggrin with keratin intermediate filaments have been the subject of several reviews (Dale *et al.,* 1990a; Dale *et al.,* 1985a). Therefore, this chapter will emphasize new studies on improved understanding of the regulation of profilaggrin/filaggrin expression from studies *in vitro,* the structure and expression of the profilaggrin gene, and recent work on enzymatic processing events.

Expression of Profilaggrin

Immunologic and Morphologic Studies

Immunologic labeling studies show that profilaggrin is first expressed in the granular layer of the epidermis during normal epidermal differentiation (Fig. 2A) and at the time of interfollicular keratinization during fetal development

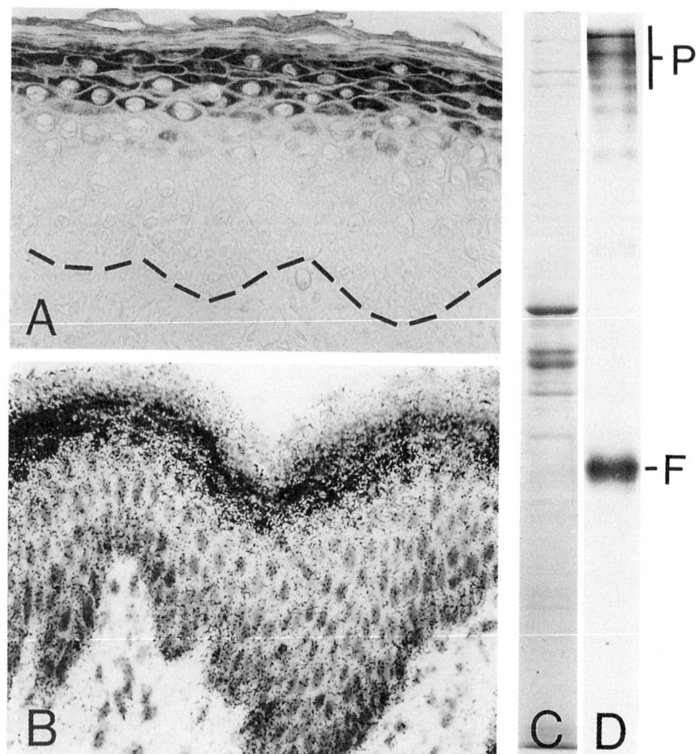

FIGURE 2 Localization of profilaggrin and filaggrin expression. (A) Detection of filaggrin immunoreactive protein in the granular layer of newborn epidermis. (B) *In situ* hybridization using a cRNA probe for human filaggrin showing mRNA expression in the granular layer. (C–D) Epidermal extracts separated by SDS–polyacrylamide gel electrophoresis and stained with Coomassie brilliant blue (C) for protein and blotted to nitrocellulose and stained with monoclonal antibody to human filaggrin (D). P, profilaggrin; F, filaggrin.

(Dale *et al.*, 1985b; Dale and Ling, 1979; Murozuka *et al.*, 1979). Profilaggrin mRNA is detectable by *in situ* hybridization in the granular layer in tissue (Fig. 2B; Fisher *et al.*, 1987; McKinley-Grant *et al.*, 1989). In cultured keratinocytes, profilaggrin is expressed only in confluent, well-stratified cells (Fleckman *et al.*, 1985). Profilaggrin mRNA is detectable 1–2 days after cells reach confluence in both human and rat keratinocytes, while profilaggrin protein can be visualized using antibodies on Western blots by 2–4 days after confluence (Fig. 3). Profilaggrin conversion to filaggrin occurs in a cell line derived from rat keratinocytes (Fig. 3D) but not in human keratinocytes grown in standard culture conditions adapted from Rheinwald and Green (1975) (Fleckman *et al.*, 1985; Fleckman *et al.*, 1987; Fig. 3C). Nuclear runoff studies demonstrate profilaggrin transcription increases from barely

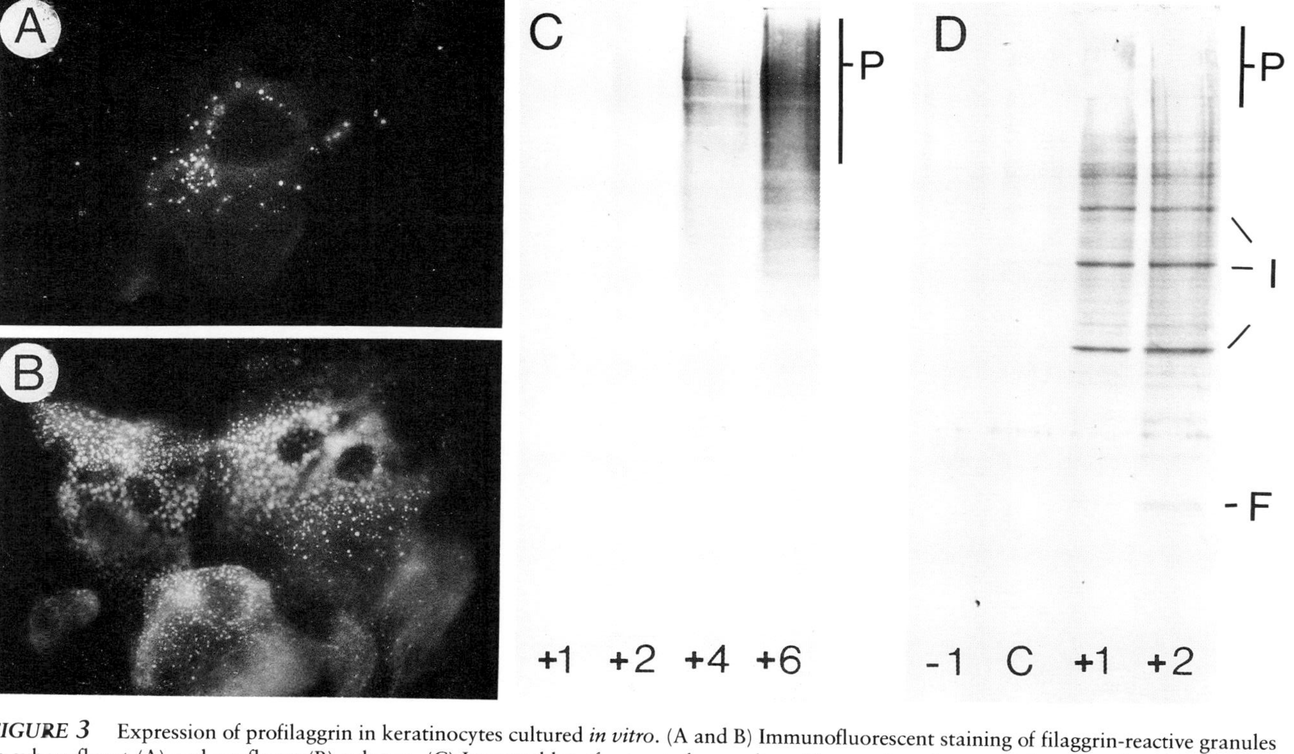

FIGURE 3 Expression of profilaggrin in keratinocytes cultured *in vitro*. (A and B) Immunofluorescent staining of filaggrin-reactive granules in subconfluent (A) and confluent (B) cultures. (C) Immunoblot of extracts from cultured human keratinocytes. The time after confluence is indicated in days. (D) Immunoblot of extracts from cultured rat keratinocytes. The time relative to confluence (C *at bottom of figure*) is indicated in days. Note that filaggrin immunoreactive staining is weak and variable prior to confluence, but that particulate staining in cells and immunoreactive profilaggrin (P) is detectable after confluence. Rat keratinocytes express profilaggrin (poorly blotted in this sample) and processing to intermediates (I) and filaggrin (F) occurs with time after confluence.

detectable levels in subconfluent keratinocytes to high levels soon after confluence (Nirinsuksiri and Fleckman, unpublished data). These studies suggest that profilaggrin expression is regulated at the level of transcription both *in vivo* and *in vitro*.

In each of these situations, the presence of profilaggrin correlates with the occurrence of keratohyalin granules (KHG) (Figs. 2A and 3B). Early solubility and radiolabeling studies (Ball *et al.*, 1978; Fukuyama and Epstein, 1975; Sibrack *et al.*, 1974) and later immunoelectron microscopic labeling studies demonstrate the presence of profilaggrin in the histidine and phosphorus-rich, irregularly shaped type of keratohyalin in human and rodents. Profilaggrin is not detected in the cysteine-rich "dense-granule" form of keratohyalin typically found in rodent epidermis (Jessen *et al.*, 1976; Manabe *et al.*, 1991); this is the localization of loricrin (Steven *et al.*, 1990). The morphology of human KHGs differs significantly from that of the rodent. In normal human epidermis, KHGs are initially seen in lower granular cells as small, irregular granules; these granules grow into larger, stellate masses in upper granular cells (reviewed in Holbrook, 1989) which can be immunolabeled with antibody to filaggrin (Fig. 4). In the lowest 3–4 cell layers of the stratum corneum, immunolabeling is dense and homogeneous throughout the cells, demonstrating the dispersal that occurs upon dephosphorylation and proteolysis. In the middle region of the stratum corneum labeling is weaker, and finally absent in the cells near the surface, as filaggrin is modified and degraded to amino acids.

Regulation of Profilaggrin and Filaggrin Expression: Role of Calcium and Retinoids

The multiple events of morphologic and biochemical differentiation during the late stages of keratinocyte differentiation occur in a coordinate manner. Expression of profilaggrin occurs only in keratinizing epithelia in which the "differentiation-specific" keratins, K1 and K10, are expressed and occurs after the expression of K1/K10. In culture, profilaggrin is detected after confluence, when stratification, cornified envelope formation, and KHGs are first detectable by electron microscopy (Fleckman *et al.*, in preparation). Profilaggrin expression is enhanced under conditions that enhance morphologic and biochemical evidence of keratinization. Two proposed modulators of these events are intracellular calcium and retinoids. Calcium is required for expression of all currently known genes encoding differentiation-specific keratinocyte markers (Yuspa *et al.*, 1989; see also a review by Fuchs, 1990) including profilaggrin. Evidence suggests the presence of a calcium concentration gradient in the epidermis *in vivo* increasing from the basal to the granular cell layer (Menon *et al.*, 1985). This gradient may be in part responsible for the sequential expression of K1/K10 and profilaggrin. For example, human keratinocytes express profilaggrin only when grown in

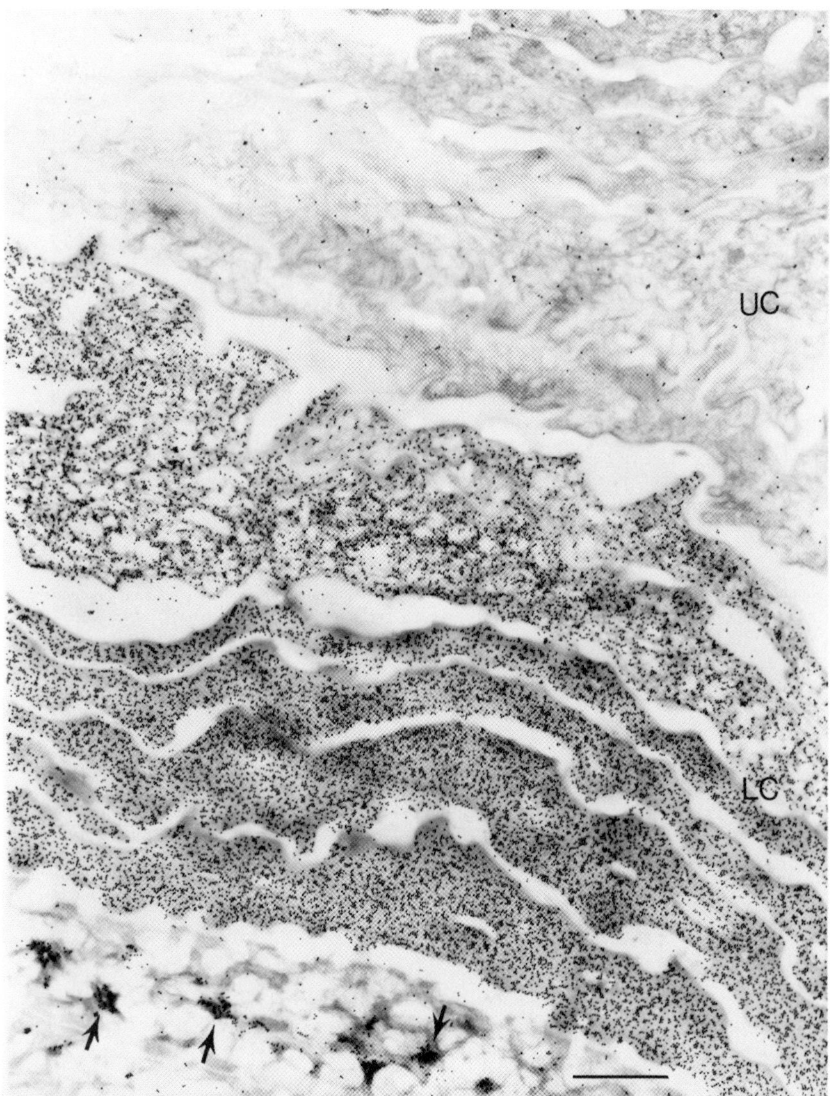

FIGURE 4 Ultrastructural immunolocalization of filaggrin-related proteins in human epidermis. Immunogold localization showing that filaggrin-related protein is present in keratohyalin (*arrows*) in the granular layer, distributed uniformly in lower cornified cells (LC), and that it disappears in upper cornified cells (UC). Bar = 1 μm. (Photo courtesy of M. Manabe.)

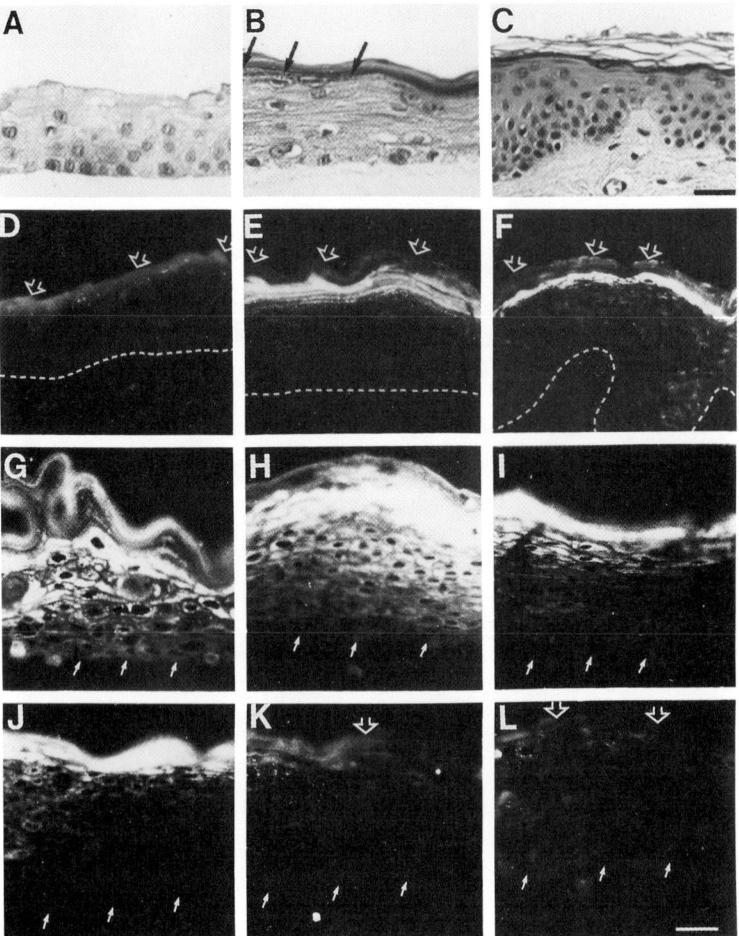

FIGURE 5 Morphology and expression of filaggrin in human keratinocyte cultures and tissue and variation with retinoic acid in the medium. (A–C) Histological paraffin section stained by haemalum–Phloxine–saffron. (A) Epidermal cells cultured on plastic (submerged). (B) Cultures grown on collagen lattices submerged for 1 week, then at the air–liquid interface for an additional week. (C) Skin tissue. (D–F) Immunofluorescence staining of similar sections to those in *A*, *B*, and *C* using a polyclonal antibody to human filaggrin. Note the improved filaggrin staining of cultures grown at the air–liquid interface, with staining similar to that of the granular layer in the tissue sample. Bar = 30 μm. (Adapted with permission from Asselineau *et al.*, 1990.) (G–L) Immunofluorescence staining as above on cultures grown on collagen lattices at the air–liquid interface in medium containing delipidized serum with retinoic acid added back at the following concentrations: *G*, 0 *M*; *H*, 10^{-10}*M*; *I*, 10^{-9}*M*; *J*, 10^{-8}*M*; *K*, 10^{-7}*M*; *L*, 10^{-6}*M*. (Adapted with permission from Asselineau *et al.*, 1989.)

medium containing an extracellular calcium concentration equal to or greater than that required for K1 and K10 expression (Yuspa *et al.*, 1989). Calcium is also required for posttranslational processing of profilaggrin in rat keratinocytes (as discussed later; see also Fig. 11) as well as transglutaminase-mediated crosslinking of cornified envelope precursors (loricrin, involucrin, and others) in the stratum corneum and of trichohyalin in the inner root sheath and medulla of hair follicles (Rogers *et al.*, 1977). The levels of calcium required in culture for these late stage events are higher than those required for expression of profilaggrin and K1/K10, supporting the proposal that a calcium gradient exists in this tissue and influences differentiation.

Reduced levels of retinoids are also critical for gene expression of differentiation-specific markers, K1, K10, loricrin, and profilaggrin (Fleckman *et al.*, 1984; Fuchs and Green, 1981; Hohl *et al.*, 1991a; reviewed in Lotan, 1980). Reduction of the level of retinoids in serum results in enhanced expression of these proteins, while excess retinoid (10^{-7} *M* and higher) abrogates their expression. Darmon (1991) emphasized the importance of a decreasing effective retinoic acid (RA) concentration gradient in the epidermis *in vivo* that facilitates the events of epidermal differentiation (Gilfix and Eckert, 1985; Vahlquist *et al.*, 1987).

The skin equivalent model system has been particularly useful in understanding the role of retinoids in profilaggrin expression and processing (Asselineau *et al.*, 1989; Asselineau *et al.*, 1990). In this model system, elevation of keratinocytes cultured on a collagen raft to the air–liquid interface results in enhanced morphologic differentiation and expression of profilaggrin (Fig. 5A–E). An optimal level of RA of $10^{-9}M$ was demonstrated for epidermal-type morphology; lower concentrations gave hyperkeratinization associated with excessive expression of profilaggrin/filaggrin, while higher concentrations resulted in a decrease or lack of profilaggrin (Fig. 5G–L). Profilaggrin is optimally expressed at RA concentrations below $10^{-8}M$, but processing to filaggrin occurs only if the RA level is even lower, on the order of $10^{-10}M$ (Asselineau *et al.*, 1990). Thus, one or more of the processing events may be even more sensitive to the level of RA than gene transcription itself.

Structure of the Profilaggrin Gene

Coding Sequences

The unusual repeating structure of the profilaggrin protein is reflected in the mRNA and gene structure. This was first suggested by *in vitro* translation of very large (>30S) rat epidermal mRNA isolated by sucrose density gradient centrifugation (Meek *et al.*, 1983) and confirmed by the sequencing of partial-length cDNA clones for mouse, rat, and human proteins (see Table 1; Haydock and Dale, 1986; Rothnagel *et al.*, 1987; McKinley-Grant *et al.*,

TABLE 1 **Interspecies Variation in Filaggrins**

	Human[a]	Rat[b]	Mouse[c]
mRNA size	13 kb	27 kb	17 kb
Repeat number	10–12	20 ± 2	>20
Repeat size	972 bp	1218 bp	750–65 bp
Profilaggrin repeat	324 aa	406 aa	250–55 aa
Filaggrin + linker	(307 + 17)[d]	390 + 26	224 + 26 224 + 31
Filaggrin	37 kDa	42 kDa	26 kDa

[a]From McKinley-Grant *et al.*, 1989; Gan *et al.*, 1990; and Presland *et al.*, 1992.
[b]From Haydock and Dale 1986, 1990.
[c]From Rothnagel *et al.*, 1987; Rothnagel and Steinert, 1990.
[d]Human linker size is based on homology with the mouse and rat sequences; this has not yet been confirmed biochemically.

1989; Haydock and Dale, 1990). In each species, the size of the repeat unit within the cDNA is equivalent to its size in the gene, suggesting that there are no introns in the coding sequence of the profilaggrin gene.

In order to investigate molecular aspects of the regulation of human profilaggrin gene expression in normal epidermis and in epidermal disorders, and to learn more about the structure of the gene and its encoded protein, genomic clones for human profilaggrin have been isolated and characterized in two laboratories. Figure 6 illustrates the structure of the human profilaggrin gene derived from restriction mapping and DNA sequence analysis, showing the presence of 11 complete filaggrin repeat units (Presland *et al.*, 1992). The number of profilaggrin repeat units in the human gene varies between 10 and 12 and is inherited in a Mendelian fashion (Gan *et al.*, 1990). Similarity between repeats is indicated by the presence of restriction sites at regular intervals of 0.97 kb, in agreement with the size of the human filaggrin repeat derived from the cDNA sequence (972 bp, McKinley-Grant *et al.*, 1989). However, filaggrin repeats show some heterogeneity as indicated by the digestion pattern with other restriction enzymes (e.g., *Eco*RI), which cut a limited number of repeats (Fig. 6). This variation was confirmed by a survey of a large number of human filaggrin repeats, which showed that the heterogeneity occurs at nearly 40% of the amino acid residues in the predicted sequence; this led to the enumeration of a "consensus" sequence for the human profilaggrin repeat unit (Gan *et al.*, 1990). Thus, the sequence of filaggrin may not be critical for its function in keratin aggregation. Indeed, this function may be more dependent on overall charge distribution than on a precise sequence. One region of 19 residues that shows little or no amino acid variability is the proposed linker region of human filaggrin (Gan *et al.*, 1990; Fleckman *et al.*, in preparation). The conformation of the linker region may be important for proteolytic processing of the profilaggrin precursor.

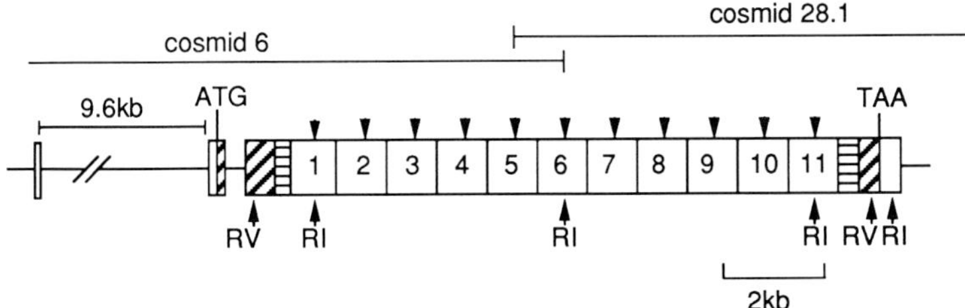

FIGURE 6 Genomic organization of the human profilaggrin gene. The gene is 23 kb in length and contains 11 complete filaggrin repeats of 972 bp (1–11) with truncated repeats (horizontal hatched boxes) and variant sequences (diagonal hatched boxes) at the 5′ and 3′ ends of the coding region. Open boxes represent the 5′ and 3′ noncoding sequences. The 5′ noncoding region is split into two exons separated by an intron of 9.6 kb. Scale bar represents 2 kb. Arrowheads above the filaggrin repeats indicate the location of conserved *Xba*I restriction sites. Restriction sites for *Xma*I, *Hgi*AI, *Sma*I, and *Hph*I are also conserved between repeats (*not shown*). Arrows below the diagram indicate *Eco*RI (RI) sites, which are not conserved between repeats and *Eco*RV (RV) sites, which delineate the boundaries of the coding region.

The coding region of the human profilaggrin gene contains four domains. The internal section with 11 repeating units is bounded at each end by truncated repeats and by unique "leader" and "tail" sequences that differ from filaggrin (Fig. 6; Presland *et al.*, 1992). The carboxy-terminal tail peptide is relatively hydrophobic compared to the hydrophilic character of the filaggrin unit and contains amino acids not seen in filaggrin. Interestingly, both rat and mouse profilaggrin also have a tail sequence (amino-terminal sequences of these species have not been reported). The tail sequence of approximately 20 amino acids shows striking conservation between species suggesting a functional significance for this peptide (e.g., a role in packing profilaggrin into keratohyalin granules or in initiating processing events within the transition cell layer of the epidermis).

Amino-Terminal Region of Human Profilaggrin and Identification of EF Hand Domains

The initiation codon is followed by a leader peptide that consists of two distinct domains summarized in diagrammatic form in Figure 7A. The first domain (residues 1–81) has significant homology to members of the S-100 family of calcium-binding proteins (Presland *et al.*, 1992).[1] These proteins include the S-100 α and β chains, p9Ka, calcyclin, intestinal calcium binding protein, and migration-related inhibitory factors 8 and 14 (MRP8 and 14,

[1]Initial sequence analysis by Gan and co-workers (Gan *et al.*, 1990) suggested that the amino-terminal unique region of profilaggrin was quite short (71 amino acids, beginning at the methionine at residue 223 in our sequence). However, further sequencing studies reveal that profilaggrin has a much longer amino-terminal region (293 amino acids; Presland *et al.*, 1992).

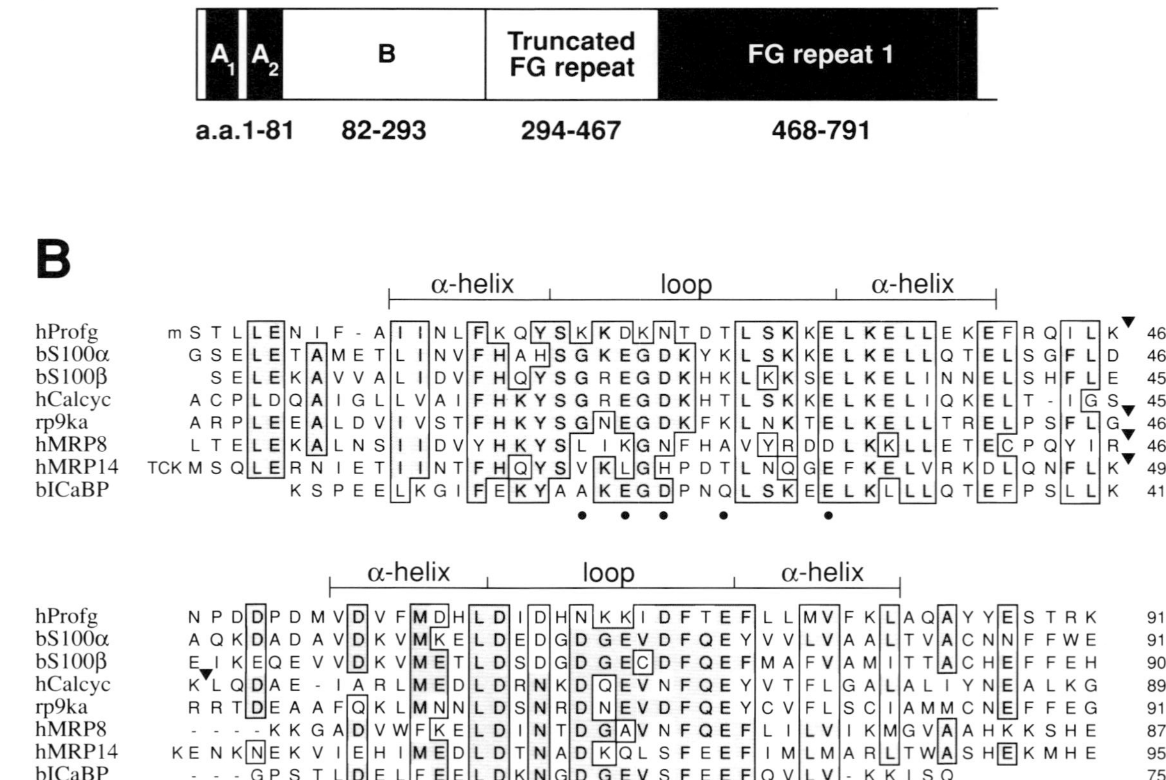
A
A1
A2
B
Truncated FG repeat
FG repeat 1
a.a.1-81
82-293
294-467
468-791
B
α-helix
loop
α-helix
hProfg m S T L L E N I F - A I I N L F K Q Y S K K D K N T D T L S K K E L K E L L E K E F R Q I L K 46
bS100α G S E L E T A M E T L I N V F H A H S G K E G D K Y K L S K K E L K E L L Q T E L S G F L D 46
bS100β S E L E K A V V A L I D V F H Q Y S G R E G D K H K L K K S E L K E L I N N E L S H F L E 45
hCalcyc A C P L D Q A I G L L V A I F H K Y S G R E G D K H T L S K K E L K E L I Q K E L T - I G S 45
rp9ka A R P L E E A L D V I V S T F H K Y S G N E G D K F K L N K T E L K E L L T R E L P S F L G 46
hMRP8 L T E L E K A L N S I I D V Y H K Y S L I K G N F H A V Y R D D L K K L L E T E C P Q Y I R 46
hMRP14 TCK M S Q L E R N I E T I I N T F H Q Y S V K L G H P D T L N Q G E F K E L V R K D L Q N F L K 49
bICaBP K S P E E L K G I F E K Y A A K E G D P N Q L S K E E L K L L L Q T E F P S L L K 41
α-helix
loop
α-helix
hProfg N P D D P D M V D V F M D H L D I D H N K K I D F T E F L L M V F K L A Q A Y Y E S T R K 91
bS100α A Q K D A D A V D K V M K E L D E D G D G E V D F Q E Y V V L V A A L T V A C N N F F W E 91
bS100β E I K E Q E V V D K V M E T L D S D G D G E C D F Q E F M A F V A M I T T A C H E F F E H 90
hCalcyc K L Q D A E - I A R L M E D L D R N K D Q E V N F Q E Y V T F L G A L A L I Y N E A L K G 89
rp9ka R R T D E A A F Q K L M N N L D S N R D N E V D F Q E Y C V F L S C I A M M C N E F F E G 91
hMRP8 - - - - K K G A D V W F K E L D I N T D G A V N F Q E F L I L V I K M G V A A H K K S H E 87
hMRP14 K E N K N E K V I E H I M E D L D T N A D K Q L S F E E F I M L M A R L T W A S H E K M H E 95
bICaBP - - - G P S T L D E L F E E L D K N G D G E V S F E E F Q V L V - K K I S Q 75

also known as calgranulin A and B, respectively). (Fig. 7B; for reviews, see Heizmann and Hunziker, 1990; Heizmann and Hunziker, 1991; Kligman and Hilt, 1988.) These proteins are characterized by the presence of two Ca^{2+}-binding domains known as "EF hands". An EF hand usually consists of a Ca^{2+}-binding loop of 12 amino acids that is flanked on each side by an α-helix; the loop region contains six oxygen-containing residues that together chelate a calcium ion (Kretsinger, 1987). EF hand-containing proteins are classified into three large families: S-100-like, calmodulin-like, and calbindin-D28-like, according to the number of EF hands and position of introns within the genes (Heizmann and Hunziker, 1990). The S-100 proteins have two EF hands, the amino-terminal one of which has a variant Ca^{2+}-binding loop with a low affinity for calcium (Mani *et al.*, 1983). Like other genes of the S-100 type, profilaggrin has two introns, one in the 5′ noncoding region and a second one between the EF hands in the same location as other S-100-like genes (Fig. 7B). Thus, profilaggrin is a new member of the S-100 family of calcium-binding proteins (Presland *et al.*, 1992).

Like profilaggrin, trichohyalin is an intermediate filament-associated protein; it is synthesized in the inner root sheath and medulla cells of mammalian hair follicles (Rothnagel and Rogers, 1986; Fietz *et al.*, 1990; O'Guin *et al.*, 1992). It also contains two EF hands at its amino terminus and has a similar exon/intron organization to profilaggrin and other genes of the S-100 family (Rogers *et al.*, 1991). Both epidermal and hair follicle cells contain calcium-dependent enzymes, which are activated late in differentiation and modify profilaggrin (see Section IV, B, 2) and trichohyalin (for a review, see Rogers *et al.*, 1991). Thus, the presence of Ca^{2+}-binding domains in these two intermediate filament-associated proteins suggests the

←

FIGURE 7 Structure of the human profilaggrin amino terminus and identification of an S-100-like calcium binding region. (A) Diagrammatic representation of the region. Amino acids 1–81 contain two EF hands (domains A1 and A2); amino acids 82–293 (domain B) contain the remainder of the leader peptide. Domain B has a high proportion of charged amino acids. The leader peptide is followed by a truncated filaggrin repeat of 174 amino acids, and then the first of eleven complete filaggrin repeats of 324 amino acids. (B) Comparison of the amino-terminal EF hands (domains A1 and A2) of human profilaggrin with S-100 protein family members (Heizmann and Hunziker, 1990; Lagasse and Clerc, 1988). Shown here are the sequences for bovine S-100α (bS100α) and -β (bS100β), human calcyclin (hCalcyc), rat p9ka (rp9ka), human MRP8 and -14 (hMRP8, hMRP14), and bovine intestinal calcium binding protein (bICaBP). Short gaps have been introduced where necessary to maximize alignment. The calcium binding loop and flanking α-helices are indicated above the sequence; residues considered essential for binding of calcium to bovine intestinal calcium binding protein (Van Eldik *et al.*, 1982) are shown by dots below the sequence. Identical residues are boxed and shaded; conservative changes are boxed (conservative changes included are oxygen-containing residues [EDQNST] and hydrophobic residues [LVIFM]). Residues are boxed only if four or more sequences have identical residues. The position of the intron between the EF hands, where known, is indicated by an arrowhead.

functional significance of calcium binding which may be required for granule formation or for activation, regulation, and/or coordination of post-translational modification events during differentiation.

5′ Noncoding Region and Identification of Possible cis-Acting Sequences

The human profilaggrin gene has a 5′ noncoding region of 75 bp that is divided into two exons separated by a large intron of 9.6 kb (Fig. 6; Presland *et al.*, 1992). The transcription initiation site was identified by primer extension analysis and the probable TATA box is located 27 bp 5′ of the mRNA start site (Fig. 8).

Sequence analysis of the region immediately upstream of the TATA box reveals several putative regulatory elements located within the 210 bp of the profilaggrin mRNA start site. These include an AP1 site (Angel *et al.*, 1987) at -76 and several possible GC and CAAT boxes. In addition, upstream of the translation initiation codon, within intron 1, there is an additional AP1 site and an overlapping cluster of seven sites resembling the response element recognized by the retinoic acid nuclear receptors (RARE) (Fig. 8; consensus AGGTCA; Näär *et al.*, 1991; Umesono *et al.*, 1988; Vasios *et al.*, 1989). These sites are speced 2-11 bp apart.

The functional significance of the AP1 and RARE-like sequences in the profilaggrin gene is under investigation. The phorbol ester, 12-O-tetradecanoyl-phorbol-13-acetate (TPA) activates genes containing an AP1 site by increasing the binding activity of *c-fos* and *c-jun* (Boyle *et al.*, 1991). TPA induces the expression of profilaggrin immunoreactive protein under certain culture conditions (Kim and Bernstein, 1987). We do not yet know if this response is directly mediated through the AP1 site; however, *c-fos* is expressed at high levels in transition cells between the granular and cornified layers (Fisher *et*

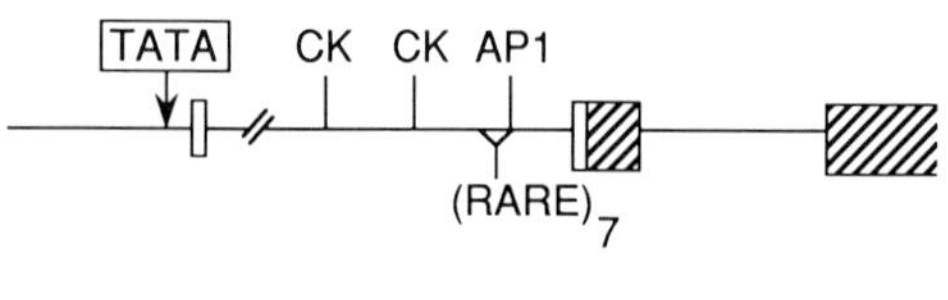

FIGURE 8 Human profilaggrin gene 5′ noncoding and upstream regulatory region. Exons are indicated by boxes and the two introns by a single line. The 5′ noncoding sequences (*open boxes*) are divided into two exons separated by an intron of 9.6kb. *Cis*-acting DNA sequence elements indicated are an AP1 site, a cluster of seven RARE-like elements that overlap at the 3′ end with the AP1 site, and two CK octamer sequences located within intron 1 (*see text*). The TATA box is located 27 bp 5′ of the transcription initiation site.

al., 1991). By contrast, several *in vitro* studies show that retinoids down-regulate profilaggrin mRNA and protein levels (Asselineau *et al.*, 1989, 1990; Hohl *et al.*, 1991a). Retinoids might act either *directly* through the RARE-like elements or *indirectly* by a mechanism involving antagonism of AP1-stimulated transcription. There are precedents for both types of mechanisms from studies of gene regulation by retinoic acid and glucocorticoids. [See Fuller (1991), for a review.] For example, in the case of the proliferin gene, repression of transcription by glucocorticoids requires the receptor to bind to the promoter and directly alters the AP1 function (Diamond *et al.*, 1990). In the collagenase and stromelysin genes, the retinoid receptor does not bind DNA, but represses gene expression indirectly by interfering with AP1 activity (Schüle *et al.*, 1991; Nicholson *et al.*, 1990). Either type of mechanism may explain regulation of profilaggrin gene transcription by retinoids.

Additional sequences of note in the profilaggrin sequence are two copies of the cytokeratin (CK) octamer sequence (Fig. 8; Blessing *et al.*, 1987). This sequence is present in the 5′ upstream region of many genes expressed in the epidermis, but its function is not yet understood (see Blessing *et al.*, 1989; Chin *et al.*, 1989).

Functional analysis of potential *cis*-acting elements will be necessary to understand appropriate tissue-specific and differentiation-specific expression of the human profilaggrin gene. Sequences in the 5′ and 3′ regions of the gene, as well as within the first intron, may be critical for regulation of expression. It will be of interest to see if there are any features in common with other differentiation markers, such as keratin K1 or loricrin. For instance, an enhancer sequence that shows the properties of calcium dependence and epidermal cell specificity in CAT constructs has been localized to the 3′-flanking DNA in the K1 gene (Rothnagel *et al.*, 1992). Such 3′ elements may be a feature of differentiation-specific markers in epidermis.

3′ Noncoding Region

Sequence analysis of the 3′ end of rat, mouse, and human profilaggrin genes demonstrates that they represent AT-rich sequences, quite distinct from that of the coding sequence (Gan *et al.*, 1990; Haydock and Dale, 1990; Rothnagel and Steinert, 1990; our unpublished data). In particular, the mouse and rat 3′ ends contain ATTTA sequences typical of short-lived mRNAs, such as those encoding cytokines or proto-oncogenes (Shaw and Kamen, 1986; Wreschner and Rechavi, 1988). This sequence is highly susceptible to RNase A-like nucleases (for a review, see Beutler, 1990), which may explain in part the apparent instability of profilaggrin mRNA *in vivo*. The observation that the bulk of profilaggrin mRNA lacks a discernible poly(A) tail is consistent with these findings (R. Presland, unpublished observation).

Evolutionary Considerations

From an evolutionary point of view, it is noteworthy that the human genes encoding loricrin, involucrin, trichohyalin, and profilaggrin have several structural features in common. First, they map to the same locus on chromosome 1q21 (McKinley-Grant *et al.*, 1989; Simon *et al.*, 1989; Yoneda *et al.*, 1992; Lee *et al.*, 1992), suggesting that they may have evolved from a common ancestral progenitor gene or have some common regulatory features. The proximity of these genes on chromosome 1 may facilitate their transcriptional activation as a single chromatin domain that would allow their coordinate or sequential expression in epidermal cells.

A second common feature is that they all contain an intron in the 5′ noncoding sequence (Eckert and Green, 1986; Yoneda *et al.*, 1992; Rogers *et al.*, 1991; Presland *et al.*, 1992). No function has yet been demonstrated for the introns in any of these genes, however, a role in transcriptional regulation is possible by analogy with studies of other systems. For instance, insertion of one or more introns into an "intronless" mouse growth hormone gene construct dramatically improved its expression in transgenic mice, leading to the proposal that the intronic sequences elevate *de novo* transcription by positively influencing transcription factor assembly and/or nucleosome phasing (Brinster *et al.*, 1988; Palmiter *et al.*, 1991). Third, their coding regions are largely composed of repeats or quasi-repeats with unique amino- and carboxyl-regions, suggesting they each arose by extensive internal duplication. However, these proteins have no significant sequence homology with each other. This is not surprising because they are each undergoing rapid divergent evolution, as evidenced by interspecies comparison of protein sequences for loricrin (Hohl *et al.*, 1991b; Mehrel *et al.*, 1990), mammalian involucrins (Djian and Green, 1991), and filaggrins (see Haydock and Dale, 1990).

Posttranslational Processing Events—Formation and Dissolution of Keratohyalin Granules

Although profilaggrin shows considerable heterogeneity between species (as evidenced by the size of filaggrin, the species-specificity of most antifilaggrin antibodies, and the amino acid sequence of the protein), filaggrins are similar in function, amino acid composition, and processing in all species investigated. Furthermore, some sequence features are common to all profilaggrins. Comparison of rat, mouse, and human sequences shows that processing sites are more conserved, including the region of the linker and probable casein kinase II (CKII) phosphorylation sites (Resing *et al.*, 1985; Resing *et al.*, 1989; Rothnagel and Steinert, 1990; Haydock and Dale, 1990). These common features suggest that constraints in the evolution of profilaggrin/filaggrin are more strongly related to the processing events than to protein function.

Phosphorylation of Nascent Profilaggrin and Formation of Keratohyalin Granules

The molecular events of profilaggrin processing are best understood for the mouse and rat proteins, but the general features are expected to apply to human profilaggrin processing. Profilaggrin is phosphorylated on multiple serine residues in each filaggrin domain and within each linker peptide between domains (Lonsdale-Eccles *et al.*, 1980; Resing *et al.*, 1985). There are also several potential phosphorylation sites in the amino- and carboxy-terminal portions of human profilaggrin. The phosphorylated form of the protein is insoluble and accumulates in the phosphorus-rich type of keratohyalin.

A number of kinases are responsible for profilaggrin phosphorylation. They can be conveniently studied using filaggrin as a substrate. Analysis of epidermal extracts on DE52 column chromatography yields three peaks of kinase activity (Fig. 9). The second and third peaks phosphorylate filaggrin,

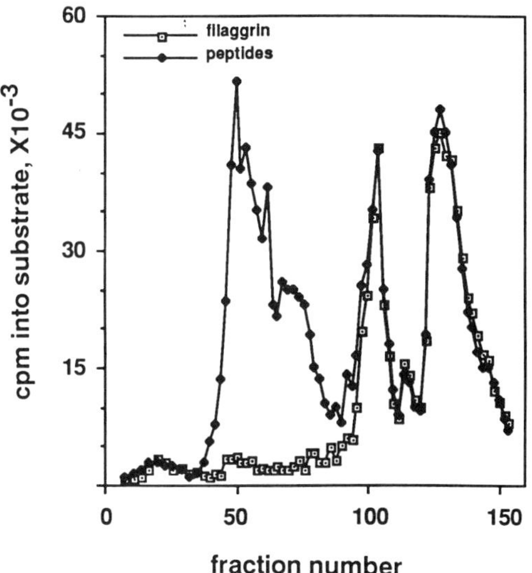

FIGURE 9 Rat epidermal profilaggrin kinases separated by DE52 ion-exchange chromatography. Fractions were assayed by incorporation of ^{32}P from γ-labeled ATP using filaggrin or filaggrin peptides as indicated for substrate. Note that three main peaks of kinase activityare present; the first (fractions 40–90) preferentially phosphorylates filaggrin peptides, while the others (fractions 90–115 and 116–150) phosphorylate both substrate preparations.

while the first peak preferentially phosphorylates large proteolytic fragments of filaggrin. Another kinase has also been identified that phosphorylates the linker region (*not shown*). The major component of the third peak has been identified as casein kinase II (CK II) by its ability to phosphorylate a synthetic peptide considered diagnostic for CK II as well as by inhibitors and chromatographic behavior (Kuenzel and Krebs, 1985). The major kinase of the second peak has not been characterized as a previously identified kinase. There are several kinase activities in the first peak: that these kinases cannot readily phosphorylate filaggrin but act on proteolytic fragments suggests that steric hindrance prevents their activity or that the order of phosphorylation is important for maximal phosphorylation. A similar situation is seen in other multiply-phosphorylated substrates, such as glycogen synthase, where phosphorylation at a CK II site facilitates phosphorylation at other sites (see review by Roach, 1990).

The large number of kinases identified may be relevant to the problem of packing the large profilaggrin molecule. One group of kinases may act cotranslationally, or immediately after translation, in forming an initial aggregate of profilaggrin at the keratohyalin surface. This first group of enzymes should readily phosphorylate filaggrin, as it would be most similar to the nascent profilaggrin, and would be similar to the second and third peaks (Fig. 9). This proposal is supported by the known association of CK II with ribosomes (Issinger, 1977). A second group of kinases would then complete the packing into the granule; these would be expected to be sensitive to steric formation as are enzymes in the first peak. This packing model is supported by the observation that the two types of profilaggrin immunoreactive granules present in rodent epidermis differ in their electron density (Manabe *et al.*, 1991). The large number of kinases and the intractable nature of the nascent substrate has made testing of this model difficult. However, mass spectrometry has proven useful in analyzing the phosphorylation sites on profilaggrin (K. Resing, unpublished data) and should aid in sorting out the details of this complicated system.

Keratohyalin Granule Dissolution

Unpacking profilaggrin from keratohyalin granules also presents a significant problem. Keratohyalin dissolution is associated with the processing of profilaggrin by both dephosphorylation and proteolysis. Correlation of the biochemical-processing events with the morphological events by *in vivo* labeling studies points to the transition cell as the site of the major events of profilaggrin processing. Transition cells occur at the interface between the granular and cornified layers; their keratohyalin is thinning and apparently dispersing through the cytoplasm, and they have a partially condensed keratin filament network.

Dephosphorylation

Studies of the processing events are complicated because profilaggrin is phosphorylated on multiple sites by several kinases, and it is possible that several phosphatases are involved in dephosphorylation. Although at least three acid phosphatases have been localized to this region of the epidermis (Freinkel and Traczyk, 1983; Hara *et al.*, 1985; Mäkinen and Mäkinen, 1981), only one can dephosphorylate protein substrates (Hara *et al.*, 1985) and thus appears to be a candidate for profilaggrin processing *in vivo*. Our experiments show that this acid phosphatase cannot dephosphorylate CK II-phosphorylated filaggrin (Kam *et al.*, 1992); however, the possibility remains that it may work on sites which have been phosphorylated by other kinases. A protein phosphatase with a neutral pH optimum, partially characterized from rat skin, has been shown to dephosphorylate profilaggrin and phosphofilaggrin (Haugen-Scofield *et al.*, 1988). This enzyme has been further characterized according to the recently developed scheme of classification of protein phosphatases (for review, see Cohen, 1989). Enzyme activity corresponds to a protein of approximately 40 kDa, which reacts with a monoclonal antibody directed to the catalytic subunit of Type 2A protein phosphatase (PP2A) (Fig. 10). Other properties of the enzyme are also consistent with its classification as a PP2A: enzyme activity is not dependent on divalent cations, it preferentially dephosphorylates the alpha subunit of phosphorylase kinase, and it is strongly inhibited by okadaic acid ($IC_{50} = 8 \times 10^{-11}$ *M*) (Kam *et al.*, 1992). This characterization opens new possibilities for further study of profilaggrin processing in cell culture systems. Although little is known about the regulation of the phosphatase *in vivo*, it is strongly inhibited by NaCl (IC_{50} = 1.5 m*M*) (Haugen-Scofield *et al.*, 1988; Kam *et al.*, 1992), suggesting that enzyme activation may be triggered by the decrease of sodium concentration when keratinocytes enter the granular layer (Grundin *et al.*, 1985; Wei *et al.*, 1982).

Proteolysis

Proteolytic cleavage of the large profilaggrin molecule is a two-stage process, first yielding intermediates composed of several filaggrin domains, then filaggrin (see Figs. 1 and 11; Harding and Scott, 1983; Resing *et al.*, 1984). For mouse filaggrin, the basis of this two-stage proteolysis lies in two distinct amino acid sequences of the linker peptide, shown both by protein sequence studies (Resing *et al.*, 1985) and by gene mapping (Rothnagel and Steinert, 1990). However, analysis of the linker regions of human and rat profilaggrin do not show such clear primary sequence differences, implicating the secondary structure in processing. The proteases are chymotrypsin-like enzymes that cleave at an aromatic residue, but their activities can be distinguished by inhibitors and activators *in vivo* and *in vitro*, suggesting

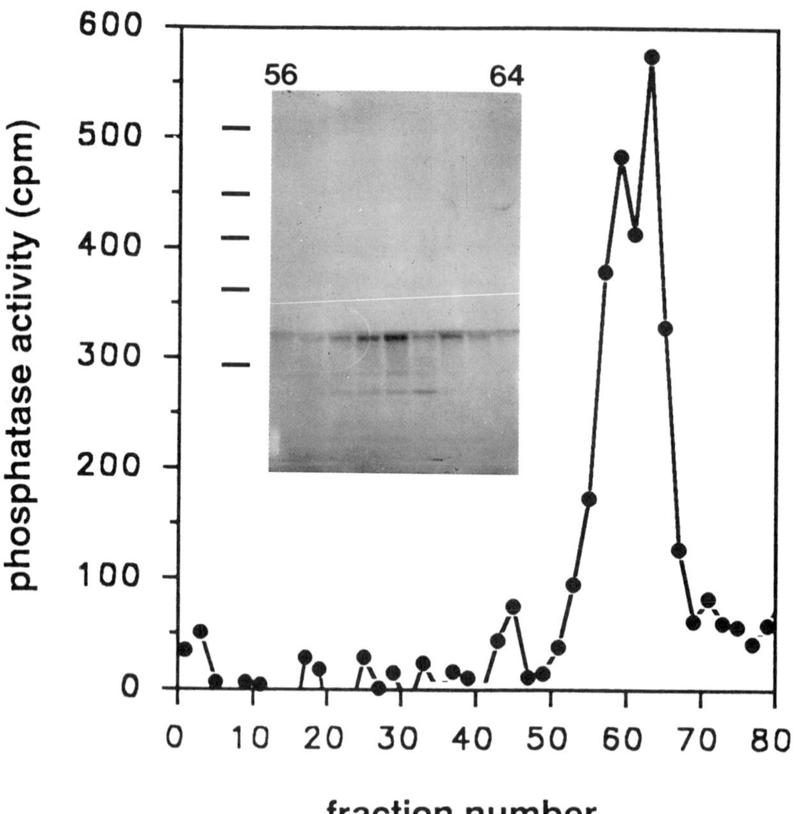

FIGURE 10 Rat epidermal profilaggrin phosphatase activity profile of fractions eluted from a Sephacryl S200 column. The phosphatase was partially purified by DE52 and hydroxyapatite columns then applied to a Sephacryl column. Fractions were assayed by release of ^{32}P from filaggrin labeled using [^{32}P]-ATP and CK II. The inset shows an immunoblot of fractions from the peak developed with antibody to the catalytic subunit of protein phosphatase 2A (PP2A; generously provided by M. Mumby). The active fractions contain the PP2A catalytic subunit of approximately 40 kDa. Molecular weight markers are 97, 68, 43, 29, 18, and 14 kDa.

that the two proteolytic events are independently regulated. Understanding the regulation of these processing events has been facilitated by the use of a rat epidermal keratinocyte cell line that converts profilaggrin to filaggrin (Kubilus and Baden, 1983). Profilaggrin is expressed when the cells reach confluence; processing to intermediates begins 24–36 hours later (stage one), and filaggrin appears at 48 hours (stage two). Processing is inhibited by chymostatin as expected from *in vitro* studies of the protease isolated from both rat and mouse epidermis (Resing *et al.*, 1989; Resing *et al.*, submitted). Stage-one protease isolated from mouse epidermis preferentially cleaves precipitated profilaggrin, which presumably is similar to the state of

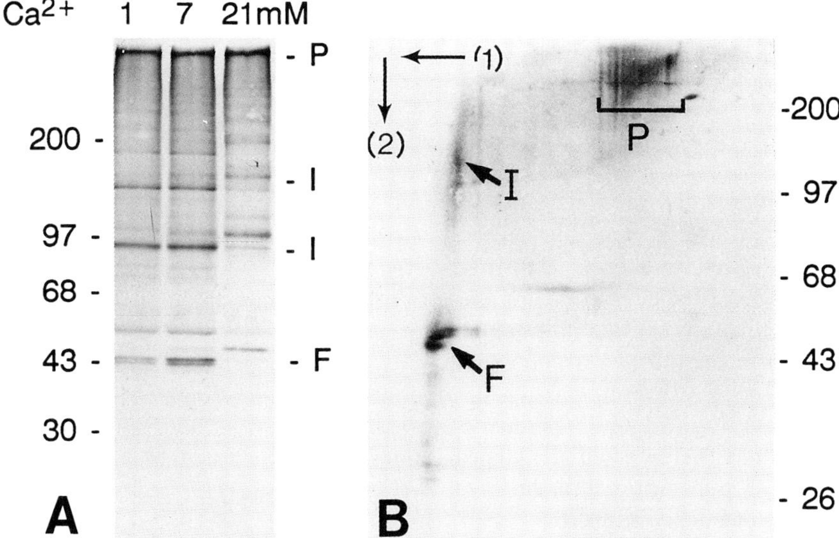

FIGURE 11 Processing of rat profilaggrin. (A) Rat keratinocytes grown in DMEM with fetal bovine serum were transferred at confluence to the same media with 1.6, 7, or 21 m*M* Ca^{2+} and harvested four days later. The immunoblot probe developed with antibody to rat filaggrin is shown; note that processing of profilaggrin (P) to intermediates (I) and filaggrin (F) is enhanced in 7 m*M* Ca^{2+} but inhibited at high levels. The optimum Ca^{2+} level is 5–10 m*M*. (B) Immunoblot of rat epidermal extract separated by two-dimensional gel electrophoresis and stained with polyclonal antibody to rat filaggrin. Proteins were separated by NPHGE in the first dimension (1) and SDS–PAGE in the second dimension (2). Note the relatively acidic, high-molecular-weight profilaggrin (P), and more cationic intermediates (I) and filaggrin (F).

profilaggrin in keratohyalin. Cleavage is at an aromatic residue within the linker region and peptide studies show strong preference (10-fold) for the unphosphorylated form of the linker. These results are consistent with the proposal that the protease recognizes secondary structure rather than primary structure. The products formed *in vitro* are 3–5 kDa larger than intermediates produced *in vivo,* suggesting that the linker segments are cleaved and subsequently "trimmed" by other enzyme(s). Analyses of rat and mouse profilaggrin cleavage sites suggest that both a carboxypeptidase and an aminopeptidase are required to produce filaggrin (Resing *et al.,* 1989; Resing *et al.,* submitted).

Stage-two processing requires calcium in the medium (Fig. 11A) and is inhibited by nifedipine (a calcium channel blocker), suggesting that calcium influx activates this event. Stage-two processing is also inhibited by the protease inhibitor leupeptin, with a concomitant increase in intermediates (Resing *et al.,* submitted). This implicates the neutral calcium-dependent protease, calpain (Miyachi *et al.,* 1986), which is present in an active state in differentiating keratinocytes (DeMartino *et al.,* 1986). However, calpain does not have the correct specificity for profilaggrin cleavage and therefore may play an indirect role in the processing (e.g., in zymogen activation of the profilaggrin protease).

Functional Correlates of Profilaggrin Processing

Phosphorylated profilaggrin is localized in keratohyalin granules. Dephosphorylation and stage-one proteolysis result in the formation of intermediates located in cells in transition between granular and cornified layers. These intermediates are composed of several filaggrin domains: they aggregate efficiently with keratin filaments (Harding and Scott, 1983) and probably form the initial partially condensed keratin filaments seen in transition cells. Stage-two proteolysis yields monomeric filaggrin domains that remain associated with keratin filaments in the densely packed lower layers of the stratum corneum in which filaggrin is homogeneously distributed (see Fig. 4).

The association of filaggrin with keratin filaments is limited to the lower portion of the stratum corneum, perhaps serving as a temporary scaffolding to orient the keratin filaments until they are covalently linked by disulfide bonds. In the upper layers, arginine residues in filaggrin are converted to citrulline (Harding and Scott, 1983) via an arginine deiminase found in epidermis and hair follicles (Fujisaki and Sugawara, 1981; Rogers *et al.*, 1977). This results in a change in overall charge of filaggrin and probable loosening of filaggrin from the keratin filaments (see Fig. 1). Finally, filaggrin is completely degraded to free amino acids in the upper layers. Enzymes implicated in this breakdown include a carboxypeptidase (Kikuchi *et al.*, 1989) and a protease whose activity is regulated by humidity (Scott and Harding, 1986). The resulting free amino acids, as well as some chemically modified amino acids, are hydrophilic and may serve as moisturizing components by binding water.

Summary and Future Directions

The emphasis in this chapter has been on recent work that improves understanding of the filaggrin pathway and regulation of profilaggrin and filaggrin expression at the level of the gene and its subsequent complex enzymatic processing. These advances will facilitate investigation of the molecular nature of disorders of keratinization in which filaggrin is poorly expressed, such as ichthyosis vulgaris (Sybert *et al.*, 1985; Fleckman *et al.*, 1987) and those in which its function or processing is altered, such as Harlequin ichthyosis and possibly epidermolytic hyperkeratosis (Dale *et al.*, 1990a). In addition, new information on the gene and its promoter structure, as well as identification of a calcium-binding domain within profilaggrin itself, opens new possibilities for the study of coordination and regulation of differentiation-specific events in keratinization.

References

Angel, P., Imagawa, M., Chiu, R., Stein, B., Imbra, R. J., Rahmsdorf, H. J., Jonat, C., Herrlich, P., and Karin, M. (1987). Phorbol ester-inducible genes contain a common *cis* element recognized by a TPA-modulated *trans* acting factor. *Cell* **49**, 729–739.

Asselineau, D., Bernard, B. A., Bailly, C., and Darmon, M. (1989). Retinoic acid improves epidermal morphogenesis. *Devel. Biol.* **133**, 322–35.

Asselineau, D., Dale, B. A., and Bernard, B. A. (1990). Filaggrin production by cultured human epidermal keratinocytes and its regulation by retinoic acid. *Differentiation* **45**, 221–9.

Ball, R. D., Walker, G. K., and Bernstein, I. A. (1978). Histidine-rich proteins as molecular markers of epidermal differentiation. *J. Biol. Chem.* **253**, 5861–5868.

Beutler, B. (1990). Regulation of cachectin biosynthesis occurs at multiple levels. *Prog. Clin. Biol. Res.* **349**, 229–40.

Blessing, M., Zentgraf, H., and Jorcano, J. L. (1987). Differentially expressed bovine cytokeratin genes. Analysis of gene linkage and evolutionary conservations of 5′-upstream sequences. *EMBO. J.* **6**, 567–575.

Blessing, M., Jorcano, J. L., and Franke, W. W. (1989). Enhancer elements directing cell-type-specific expression of cytokeratin genes and changes of the epithelial cytoskeleton by transfections of hybrid cytokeratin genes. *EMBO. J.* **8**, 117–126.

Boyle, W. J., Smeal, T., Defize, L. H. K., Angel, P., Woodgett, J. R., Karin, M., and Hunter, T. (1991). Activation of protein kinase C decreases phosphorylation of *c-jun* at sites that negatively regulate its DNA-binding activity. *Cell* **64**, 573–584.

Brinster, R. L., Allen, J. M., Behringer, R. R., Gelinas, R. E., and Palmiter, R. D. (1988). Introns increase transcriptional efficiency in transgenic mice. *Proc. Natl. Acad. Sci. U.S.A.* **85**, 836–840.

Chin, M. T., Broker, T. R., and Chow, L. T. (1989). Identification of a novel constitutive enhancer element and an associated binding protein: implications for human papillomavirus type II enhancer regulation. *J. Virol.* **63**, 2967–2976.

Cohen, P. (1989). The structure and regulation of protein phosphatase. *Annu. Rev. Biochem.* **58**, 453–508.

Dale, B. A., and Ling, S. Y. (1979). Immunologic cross-reaction of stratum corneum basic protein and keratohyalin granule protein. *J. Invest. Dermatol.* **72**, 257–261.

Dale, B. A., Holbrook, K. A., and Steinert, P. M. (1978). Assembly of stratum corneum basic protein and keratin filaments in macrofibrils. *Nature* **276**, 729–731.

Dale, B. A., Resing, K. A., and Lonsdale-Eccles, J. D. (1985a). Filaggrin: a keratin filament-associated protein. *Ann. N.Y. Acad. Sci.* **455**, 330–42.

Dale, B. A., Holbrook, K. A., Kimball, J. R., Hoff, M., and Sun, T. T. (1985b). Expression of epidermal keratins and filaggrin during human fetal skin development. *J. Cell Biol.* **101**, 1257–69.

Dale, B. A., Resing, K. A., and Haydock, P. V. (1990a). Filaggrins. *In* "Cellular and Molecular Biology of Intermediate Filaments" (R. D. Goldman and P. M. Steinert, eds.) pp. 393–412. Plenum, New York.

Dale, B. A., Holbrook, K. A., Kimball, J. R., Fleckman, P., Brumbaugh, S., and Sybert, V. P. (1990b). Genetic heterogeneity in harlequin ichthyosis, an inborn error of epidermal keratinization: variable morphology and expression of structural proteins. *J. Invest. Dermatol.* **94**, 6–18.

Darmon, M. (1991). Retinoic acid in skin and epithelia. *Semin. Develop. Biol.* **2,** 219–228.

DeMartino, G. N., Huff, C. A., and Croall, D. E. (1986). Autoproteolysis of the small subunit of calcium-dependent protease II activates and regulates protease activity. *J. Biol. Chem.* **261,** 12047–12052.

Diamond, M. I., Minor, J. N., Yoshinaga, S. K., and Yamamoto, K. R. (1990). Transcription factor interactions: selectors of positive or negative regulation from a single DNA element. *Science* **249,** 1266–1272.

Djian, P., and Green, H. (1991). Involucrin gene of tarsioids and other primates: Alternatives in evolution of the segment of repeats. *Proc. Natl. Acad. Sci. U.S.A.* **88,** 5231–5321.

Eckert, R., and Green, H. (1986). Structure and evolution of the human involucrin gene. *Cell* **46,** 583–589.

Fietz, M. J., Presland, R. B., and Rogers, G. E. (1990). The cDNA-deduced amino acid sequence of trichohyalin, a differentiation marker in the hair follicle contains a 23 amino acid repeat. *J. Cell Biol.* **110,** 427–436.

Fisher, C., Haydock, P. V., and Dale, B. A. (1987). Localization of profilaggrin mRNA in newborn rat skin by *in situ* hybridization. *J. Invest. Dermatol.* **88,** 661–4.

Fisher, C., Byers, M. R., Iadarola, M. J., and Powers, E. A. (1991). Patterns of epithelial expression of *fos* protein suggest important role in the transition from viable to cornified cell during keratinization. *Development* **111,** 253–8.

Fleckman, P., Haydock, P., Blomquist, C., and Dale, B. A. (1984). Profilaggrin and the 67-kDa keratin are coordinately expressed in cultured human epidermal keratinocytes. *J. Cell Biol.* **99,** 315A.

Fleckman, P., Dale, B. A., and Holbrook, K. A. (1985). Profilaggrin, a high-molecular-weight precursor of filaggrin in human epidermis and cultured keratinocytes. *J. Invest. Dermatol.* **85,** 507–12.

Fleckman, P., Holbrook, K. A., Dale, B. A., and Sybert, V. P. (1987). Keratinocytes cultured from subjects with ichthyosis vulgaris are phenotypically abnormal. *J. Invest. Dermatol.* **88,** 640–5.

Fleckman, P., Haydock, P. V., Dale, B. A., Grant, F., Kindsvogel, W., Blomquist, C., and Brumbaugh, S. (in preparation). Expression of profilaggrin and keratin #1 are linked at the time cultured human epidermal keratinocytes reach confluence.

Freinkel, R. K., and Traczyk, T. N. (1983). Acid hydrolases of the epidermis: Subcellular localization and relationship to cornification. *J. Invest. Dermatol.* **80,** 441–446.

Fuchs, E. (1990). Epidermal differentiation: the bare essentials. *J. Cell Biol.* **111,** 2807–2814.

Fuchs, E., and Green, H. (1981). Regulation of terminal differentiation of cultured human keratinocytes by vitamin A. *Cell* **25,** 617–25.

Fujisaki, M., and Sugawara, K. (1981). Properties of peptidylarginine deiminase from the epidermis of newborn rats. *J. Biochem.* **89,** 257–263.

Fukuyama, K., and Epstein, W. L. (1975). A comparative autoradiographic study of keratohyalin granules containing cystine and histidine. *J. Ultrastruct. Res.* **51,** 314–325.

Fuller, P. J. (1991). The steroid receptor superfamily: mechanisms of diversity. *FASEB J.* **5,** 3092–3099.

Gan, S. Q., McBride, O. W., Idler, W. W., Markova, N., and Steinert, P. M. (1990). Organization, structure, and polymorphisms of the human profilaggrin gene. *Biochemistry* **29,** 9432–40.

Gan, S. Q., Markova, N., Korge, B., Kim, I. G., and Steinert, P. (1991). The correct expression of a human mini profilaggrin gene in transgenic mice. *J. Invest. Dermatol.* **96,** 534A.

Gilfix, B. M., and Eckert, R. L. (1985). Coordinate control by vitamin A of keratin gene expression in human keratinocytes. *J. Biol. Chem.* **260,** 14026–14029.

Grundin, T. C., Roomans, G. M., Forslind, B., Londberg, M., and Werner, Y. (1985). X-ray microanalysis of psoriatic skin. *J. Invest. Dermatol.* **85,** 378–380.

Hara, A., Kato, T., Sawada, H., Fukuyama, K., and Epstein, W. L. (1985). Characterization of

Fe^{2+}-activated acid phosphatase in rat epidermis. *Comp. Biochem. Physiol.* **82B**, 269–274.

Harding, C. R., and Scott, I. R. (1983). Histidine-rich proteins (filaggrins): Structural and functional heterogeneity during epidermal differentiation. *J. Mol. Biol.* **170,** 651–673.

Haugen-Scofield, J., Resing, K. A., and Dale, B. A. (1988). Characterization of an epidermal phosphatase specific for filaggrin phosphorylated by casein kinase II. *J. Invest. Dermatol.* **91,** 553–9.

Haydock, P. V., and Dale, B. A. (1986). The repetitive structure of the profilaggrin gene as demonstrated using epidermal profilaggrin cDNA. *J. Biol. Chem.* **261,** 12520–5.

Haydock, P. V., and Dale, B. A. (1990). Filaggrin, an intermediate filament-associated protein: structural and functional implications from the sequence of a cDNA from rat. *DNA Cell Biol.* **9,** 251–61.

Heizmann, C. W., and Hunziker, W. (1990). Intracellular calcium-binding molecules. *In* "Intracellular Calcium Regulation" (F. Bonner, ed.), pp. 211–248. Wiley-Liss, New York.

Heizmann, C. W., and Hunziker, W. (1991). Intracellular calcium-binding proteins: more sites than insights. *TIBS* **16,** 98–103.

Hohl, D., Lichti, U., Breitkreutz, D., Steinert, P. M., and Roop, D. R. (1991a). Transcription of the human loricrin gene *in vitro* is induced by calcium and cell density and suppressed by retinoic acid. *J. Invest. Dermatol.* **96,** 414–8.

Hohl, D., Mehrel, T., Lichti, U., Turner, M. L., Roop, D. R., and Steinert, P. M. (1991b). Characterization of human loricrin. Structure and function of a new class of epidermal cell envelope proteins. *J. Biol. Chem.* **266,** 6626–36.

Holbrook, K. A. (1989). Biologic structure and function: perspectives on morphologic approaches to the study of the granular layer keratinocyte. *J. Invest. Dermatol.* **92,** 84s–104s.

Huff, C. A., Rosenthal, D. S., Yuspa, S. H., and Roop, D. R. (1991). Identification and characterization of a complex regulatory array 3′ the gene encoding human keratin 1. *J. Invest. Dermatol.* **96,** 550A.

Issinger, O. G. (1977). Purification and properties of a ribosomal casein kinase from rabbit reticulocytes. *Biochem. J.* **165,** 511–518.

Jessen, H., Peters, P. D., and Hall, T. A. (1976). Sulphur in epidermal keratohyalin granules: a quantitative assay by X-ray microanalysis. *J. Cell Sci.* **22,** 161–171.

Kam, E., Resing, K. A., Lim, S. K., and Dale, B. A. (1992). Characterization of rat epidermal profilaggrin phosphatase as a type 2A protein phosphatase. *submitted.*

Kikuchi, M., Fukuyama, K., Hirayama, K., and Epstein, W. L. (1989). Purification and characterization of carboxypeptidase from terminally differentiated rat epidermal cells. *Biochim. Biophys. Acta* **991,** 19–24.

Kim, H. J., and Bernstein, I. A. (1987). Exposure to 12-O-tetradecanoylphorbol-13-acetate (TPA) induces the synthesis of histidine-rich protein (filaggrin) in monolayer cultures of rat keratinocytes. *J. Invest. Dermatol.* **88,** 624–9.

Kligman, D., and Hilt, D. C. (1988). The S-100 protein family. *TIBS* **13,** 437–443.

Kretsinger, R. H. (1987). Calcium coordination and the calmodulin fold: divergent versus convergent evolution. *Cold Spring Harbor Symp. Quant. Biol.* **52,** 499–510.

Kubilus, J., and Baden, H. P. (1983). Growth and differentiation of cultured newborn rat keratinocytes. *J. Invest. Dermatol.* **80,** 124–130.

Kuenzel, E. A., and Krebs, E. G. (1985). A synthetic peptide substrate specific for casein kinase II. *Proc. Natl. Acad. Sci. U.S.A.* **82,** 737–741.

Lagasse, E., and Clerc, R. G. (1988). Cloning and expression of two human genes encoding calcium-binding proteins that are regulated during myeloid differentiation. *Mol. Cell. Biol.* **8,** 2402–2410.

Lee, S.-C., Kim, L. G., McBride, O. W., Compton, J. G., O'Keefe, E., and Steinert, P. M. (1992). The human trichohyalin gene. *J. Invest. Dermatol.* **98,** 626.

Lonsdale-Eccles, J. D., Haugen, J. A., and Dale, B. A. (1980). A phosphorylated keratohyalin-derived precursor of epidermal stratum corneum basic protein. *J. Biol. Chem.* **255**, 2235–2238.

Lotan, R. (1980). Effects of vitamin A and its analogs (retinoids) on normal and neoplastic cells. *Biochim. Biophys. Acta* **605**, 33–91.

Mäkinen, P. L., and Mäkinen, K. K. (1981). Purification and properties of rat skin acid phosphatases. *Int. J. Pept.* **18**, 352–369.

Manabe, M., Sanchez, M., Sun, T. T., and Dale, B. A. (1991). Interaction of filaggrin with keratin filaments during advanced stages of normal human epidermal differentiation and in Ichthyosis vulgaris. *Differentiation* **48**, 43–50.

Mani, R., Shelling, F., Sykes, B., and Kay, C. (1983). Spectral studies on the calcium binding properties of bovine brain S-100b protein. *Biochemistry* **22**, 1734–1740.

McKinley-Grant, L. J., Idler, W. W., Bernstein, I. A., Parry, D. A., Cannizzaro, L., Croce, C. M., Huebner, K., Lessin, S. R., and Steinert, P. M. (1989). Characterization of a cDNA clone encoding human filaggrin and localization of the gene to chromosome region 1q21. *Proc. Natl. Acad. Sci. U.S.A.* **86**, 4848–52.

Meek, R. L., Lonsdale-Eccles, J. D., and Dale, B. A. (1983). Epidermal filaggrin is synthesized on a large messenger ribonucleic acid as a high-molecular-weight precursor. *Biochemistry* **22**, 4867–71.

Mehrel, T., Hohl, D., Rothnagel, J. A., Longley, M. A., Bundman, D., Cheng, C., Lichti, U., Bisher, M. E., Steven, A. C., Steinert, P. M., Yuspa, S. H., and Roop, D. R. (1990). Identification of a major keratinocyte cell envelope protein, loricrin. *Cell* **61**, 1103–12.

Menon, G. P., Grayson, S., and Elias, P. M. (1985). Ionic calcium reservoirs in mammalian epidermis: ultrastructural localization by ion-capture cytochemistry. *J. Invest. Dermatol.* **84**, 508–512.

Miyachi, Y., Yoshimura, N., Suzuki, S., Hamakubo, T., Kannagi, R., Imamura, S., and Murachi, T. (1986). Biochemical demonstration and immunohistochemical localization of calpain in human skin. *J. Invest. Dermatol.* **86**, 346–349.

Murozuka, T., Fukuyama, K., and Epstein, W. L. (1979). Immunochemical comparison of histidine-rich protein in keratohyalin granules and cornified cells. *Biochim. Biophys. Acta* **579**, 334–345.

Näär, A., Boutin, J. M., Lipkin, S. M., Yu, V. C., Holloway, J. M., Glass, C. K., and Rosenfeld, M. G. (1991). The orientation and spacing of core DNA-binding motifs dictate selective transcriptional responses to three nuclear receptors. *Cell* **65**, 1267–1279.

Nicholson, R. C., Mader, S., Nagpal, S., Leid, M., Rochette-Egly, C., and Chambon, P. (1990). Negative regulation of the rat stromelysin gene promoter by retinoic acid is mediated by an AP1 binding site. *EMBO J.* **9**, 4443–4454.

O'Guin, M., Sun, T.-T., and Manabe, M. (1992). Interaction of trichohyalin with intermediate filaments: three immunologically defined stages of trichohyalin maturation. *J. Invest. Dermatol.* **98**, 24–32.

Palmiter, R. D., Sandgren, E. P., Avarbock, M. R., Allen, D. D., and Brinster, R. L. (1991). Heterologous introns can enhance expression of transgenes in mice. *Proc. Natl. Acad. Sci. U.S.A.* **88**, 478–482.

Presland, R. B., Haydock, P. V., Fleckman, P., Nirinsuksiri, W., and Dale, B. A. (1992). Characterization of the human epidermal profilaggrin gene: genomic organization and identification of an S-100-like calcium-binding domain at the amino-terminus. *J. Biol. Chem.* **267**, 23772–23781.

Resing, K. A., Walsh, K. A., and Dale, B. A. (1984). Identification of two intermediates during processing of profilaggrin to filaggrin in neonatal mouse epidermis. *J. Cell Biol.* **99**, 1372–1378.

Resing, K. A., Dale, B. A., and Walsh, K. A. (1985). Multiple copies of phosphorylated filaggrin in epidermal profilaggrin demonstrated by analysis of tryptic peptides. *Biochemistry* **24**, 4167–75.

Resing, K. A., Walsh, K. A., Haugen, S. J., and Dale, B. A. (1989). Identification of proteolytic cleavage sites in the conversion of profilaggrin to filaggrin in mammalian epidermis. *J. Biol. Chem.* **264**, 1837–45.

Resing, K. A., Al-Alawi, N., Fleckman, P., Blomquist, C., and Dale, B. A. (submitted). Calcium induces the second stage of profilaggrin proteolytic processing: hypothesis for regulation by plasma membrane Ca^{2+}-channels and calpain I.

Rheinwald, J. G., and Green, H. (1975). Serial cultivation of strains of human epidermal keratinocytes: the formation of keratinizing colonies from single cells. *Cell* **6**, 331–344.

Roach, P. J. (1990). Control of glycogen synthase by hierarchal protein phosphorylation. *FASEB J.* **4**, 2961–2968.

Rogers, G. E., Harding, H. W., and Llewellyn-Smith, I. J. (1977). The origin of citrulline-containing proteins in the hair follicle and the chemical nature of trichohyalin, an intracellular precursor. *Biochim. Biophys. Acta* **495**, 159–175.

Rogers, G. E., Fietz, M. J., and Fratini, A. (1991). Trichohyalin and matrix proteins. *Ann. N.Y. Acad. Sci.* **642**, 64–81.

Rothnagel, J. A., and Rogers, G. E. (1986). Trichohyalin, an intermediate filament-associated protein of the hair follicle. *J. Cell Biol.* **102**, 1419–1429.

Rothnagel, J. A., and Steinert, P. M. (1990). The structure of the gene for mouse filaggrin and a comparison of the repeating units. *J. Biol. Chem.* **265**, 1862–5.

Rothnagel, J. A., Mehrel, T., Idler, W. W., Roop, D. R., and Steinert, P. M. (1987). The gene for mouse epidermal filaggrin precursor. Its partial characterization, expression, and sequence of a repeating filaggrin unit. *J. Biol. Chem.* **262**, 15643–8.

Rothnagel, J. A., Greenhalgh, D. A., Gagne, T., Longley, M. A., Horak, D. L., Lu, B., and Roop, D. R. (1992). Identification of the calcium-sensitive regulatory unit of the gene encoding human keratin 1. *J. Invest. Dermatol.* **98**, 568A.

Schüle, R., Rangarajan, P., Yang, N., Kliewer, S., Ransone, L. J., Bolado, J., Verma, I. M., Evans, R. M. (1991). Retinoic acid is a negative regulator of AP1-responsive genes. *Proc. Natl. Acad. Sci. U.S.A.* **88**, 6092–6096.

Scott, I. R., and Harding, C. R. (1986). Filaggrin breakdown to water binding compounds during development of the rat stratum corneum is controlled by the water activity of the environment. *Devel. Biol.* **115**, 84–92.

Shaw, G., and Kamen, R. (1986). A conserved AU sequence from the 3′-untranslated region of GM-CSF mRNA mediates selective mRNA degradation. *Cell* **46**, 659–667.

Sibrack, L. A., Gray, R. H., and Bernstein, I. A. (1974). Localization of the histidine-rich protein in keratohyalin: a morphological and macromolecular marker of epidermal differentiation. *J. Invest. Dermatol.* **62**, 394–405.

Simon, M., Phillips, M., Green, H., Stroh, H., Glat, K., Bruns, G., and Latt, S. A. (1989). Absence of a single repeat from the coding region of the human involucrin gene leading RFLP. *Am. J. Hum. Genet.* **45**, 910–916.

Steinert, P. M., Cantieri, J. S., Teller, D. C., Lonsdale-Eccles, J. D., and Dale, B. A. (1981). Characterization of a class of cationic proteins that specifically interact with intermediate filaments. *Proc. Natl. Acad. Sci. U.S.A.* **78**, 4097–101.

Steven, A. C., Bisher, M. E., Roop, D. R., and Steinert, P. M. (1990). Biosynthetic pathways of filaggrin and loricrin—two major proteins expressed by terminally differentiated epidermal keratinocytes. *J. Struct. Biol.* **104**, 150–62.

Sybert, V. P., Dale, B. A., and Holbrook, K. A. (1985). Ichthyosis vulgaris: identification of a defect in synthesis of filaggrin correlated with an absence of keratohyalin granules. *J. Invest. Dermatol.* **84**, 191–194.

Umesono, K., Giguére, V., Glass, C. K., Rosenfeld, M. G., and Evans, R. M. (1988). Retinoic acid and thyroid hormone induce gene expression through a common responsive element. *Nature* **336**, 262–265.

Vahlquist, A., Stenstrom, E., and Torma, H. (1987). Vitamin A and β-carotene concentrations at different depths of the epidermis. A preliminary study of the cow snout. *Ups. J. Med. Sci.* **92**, 253–258.

Van Eldik, L. J., Zendegui, J. G., Marshak, D. R., and Watterson, D. M. (1982). Calcium-binding proteins and the molecular basis of calcium action. *Int. Rev. Cytol.* 77, 1–61.

Vasios, G. W., Gold, J. D., Petkovich, M., Chambon, P., and Gudas, L. J. (1989). A retinoic acid-responsive element is present in the 5′-flanking region of the laminin B1 gene. *Proc. Natl. Acad. Sci. U.S.A.* **86**, 9099–9103.

Wei, X., Roomans, G. M., and Forslind, M. D. (1982). Elemental distribution in guinea pig as revealed by x-ray microanalysis in the scanning transmission microscope. *J. Invest. Dermatol.* **79**, 167–169.

Wreschner, D. H., and Rechavi, G. (1988). Differential mRNA stability to reticulocyte ribonuclease correlates with 3′ noncoding (U)nA sequences. *Eur. J. Biochem.* **172**, 333–340.

Yoneda, K., Hohl, D., McBride, O. W., Wang, M., Cehrs, K. U., Idler, W., and Steinert, P. M. (1992). The human loricrin gene. *J. Biol. Chem.* **267**, 18060–18066.

Yuspa, S. H., Kilkenny, A. E., Steinert, P. M., and Roop, D. R. (1989). Expression of murine epidermal differentiation markers is tightly regulated by restricted extracellular calcium concentrations *in vitro*. *J. Cell Biol.* **109**, 1207–17.

Note Added in Proof

Studies on the structure of the profilaggrin gene and S-100-like domain have been confirmed by Markova, N. G., Marekov, L. N., Chipev, D. C., Gan, S. -Q., Idler, W. W., and Steinert, P. M. (1993). Profilaggrin is a major epidermal calcium-binding protein. *Mol. and Cell. Biol.* **13**, 613–625.

4

The Cornified Envelope: A Key Structure of Terminally Differentiating Keratinocytes

Uwe Reichert, Serge Michel, and Rainer Schmidt

Introduction

The skin is the interface of the body with the environment, and its uppermost layer, the stratum corneum, is subject to continuous abrasion by chemical and physical injury. To protect the body against invasion of microorganisms and toxic agents as well as against loss of essential body fluids, the horny layer is perpetually renewed by the cornification of living epidermal keratinocytes to form dead corneocytes.

The cornification process requires about one month under non-pathological conditions and is characterized by a series of morphological and biochemical changes. These changes are highly coordinated in space and time and give rise to typical arrangement of epidermal cells in several strata with an increasing degree of differentiation (for review see Fuchs, 1990). The first layer, the stratum basale, is attached to the dermo–epidermal junction and supplies (by mitosis) successors for the cells lost at the skin surface. Once cells leave the basal layer to enter the stratum spinosum, they lose the capacity to divide, they increase in size, flatten, and their water content diminishes. The synthesis of the proliferation-specific keratins K5 and K14 is interrupted, and keratins K1 and K10 are expressed instead and aggregate to form filaments. Later, in the stratum granulosum, these filaments are further packed together with filaggrin, a protein newly synthesized in this layer. This process results in the formation of macrofibrils. Further changes take place in the biosynthesis of other proteins, lipids, and cell surface carbohydrates. In parallel, proteins that are no longer used, nucleic acids, and even entire cell organelles such as nuclei, mitochondria, and plasma membranes are successively destroyed. New organelle-like structures are formed such as lamellar bodies, which share some properties with lysosomes, and a stable protein envelope is also formed. The synthesis of this envelope just beneath the disintegrating plasma membrane is catalyzed by Ca^{2+}-dependent transglutaminase(s). The ultimate product of keratinocyte differentiation, the corneocyte, consists essentially of the cornified envelope filled with keratin bundles. Ceramides and fatty acids originating from the lamellar bodies are attached to the surface of the envelope, a substantial part of them covalently.

The content of the corneocytes has for a long time been considered to be the main protective agent of the horny layer and has thus attracted biochemists, biophysicists, dermatologists, and molecular biologists to study the composition and structure of the different keratins and the regulatory mechanisms involved in their synthesis. The cornified envelope, on the other hand, as been neglected as marginal, not only morphologically ("marginal band," Hashimoto, 1969) but also functionally, although recognition of its existence may be traced back to the first third of this century (Szodoray, 1930; for a full account of early morphological and biochemical studies see Matoltsy, 1977). This situation has recently changed, and a series of recent reviews documents the increasing interest in this corneocyte structure and the mechanisms involved in its formation (Hohl, 1990; Greenberg *et al.*, 1991; Polakowska and Goldsmith, 1991; Rice *et al.*, 1992).

We summarize here current knowledge regarding the morphology, chemical composition, and molecular building blocks of the cornified envelope as well as the enzymes and modulating factors participating in its synthesis. When available, data dealing with human skin are given preferential consideration.

Morphology and Composition of the Cornified Envelope

Morphological Aspects

In ultrathin skin sections, the cornified envelope appears as an electron-dense band, about 15 nm thick,[1] at the periphery of the upper epidermal layers (Matoltsy and Balsamo, 1955; Farbman, 1966; Hashimoto, 1969). An electron-lucent zone separates this band from the intercellular space and was previously regarded as being the corneocyte plasma membrane (Martinez and Peters, 1971). However, later results strongly suggested that the 4-nm thick lucent band (Lavker, 1976) consists of a monomolecular ω-hydroxyceramide layer covalently attached to the protein envelope (Swartzendruber *et al.*, 1987). In transmission electron micrographs, isolated envelopes look fluffy at their inner surface and smooth at the extracellular side (Haftek *et al.*, 1991; Serre *et al.*, 1991).

Under the Nomarski contrast microscope, cornified envelopes purified from human epidermis appear in two different forms (Fig. 1): a fragile, irregularly shaped type of envelope and a rigid, polygonal type (Michel *et al.*, 1988b). Both types can easily be distinguished by their reactivity with tetramethylrhodamine isothiocyanate: the rigid type gives rise to a bright yellow-orange fluorescence whereas the fragile type appears fainter (Michel and Reichert, 1992). The monoclonal antibodies G36-19 and B17-21 raised against human plantar stratum corneum react specifically with envelopes of the fragile type prepared from plantar corneocytes (Serre *et al.*, 1991). Human keratinocytes in submerged culture are only able to produce the fragile type. Specimens from healthy human epidermis contain largely rigid envelopes. Samples obtained from different patients with different disorders of keratinization show both types in varying proportions (see p. 113). Tape-stripping experiments with normal epidermis obtained from plastic surgery show that envelopes of the fragile type are found almost exclusively in the lower horny layer at the interface with the stratum granulosum, whereas the rigid type is mainly recovered from the more distal stratum corneum. This distinct spatial distribution probably indicates different stages of maturation (Michel *et al.*, 1988b).

The immunohistological results of Hohl *et al.* (1991b) point in the same direction. By using the polyclonal antibody SAF-102 elicited against the carboxy-terminal sequence of the envelope precursor loricrin, these authors show a brick wall-like peripheral staining of cells in the stratum granulosum and lower stratum corneum, which diminishes in the upper horny layer, perhaps due to masking of the epitope in the course of further crosslinking.

[1]Early electron microscopic studies suggest an increasing thickening of the cornified envelope during keratinocyte differentiation (Farbman, 1966).

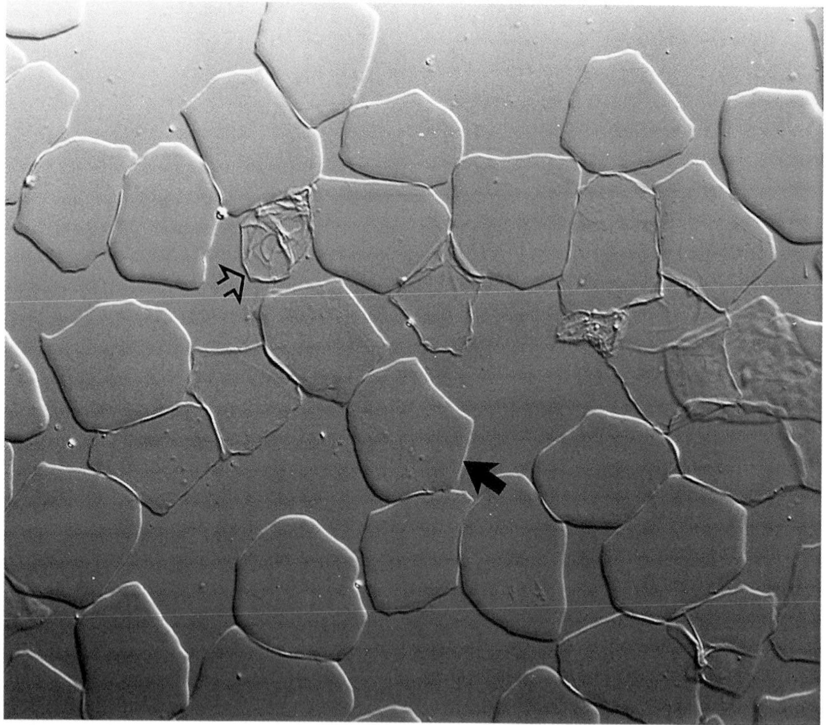

FIGURE 1 Nomarski contrast micrographs of the cornified envelopes from normal human skin. The filled and open arrows point to a rigid and fragile envelope, respectively.

Biochemical Composition

The cornified envelope constitutes about 5–7% of the total dry mass of the horny layer (Matoltsy and Matoltsy, 1966; Swartzendruber *et al.*, 1988) and consists of about 90% protein and 10% lipids (Swartzendruber *et al.*, 1988). It is the most insoluble structure of the corneocyte and can be recovered after extensive extraction of horny layer samples or cultured keratinocytes with alkali (Matoltsy and Balsamo, 1955) or with the combination of a denaturant [urea or sodium dodecylsulfate (SDS)] and a reducing agent such as β-mercaptoethanol (Manabe *et al.*, 1981; Sun and Green, 1976). Table 1 compares the amino acid composition of human cornified envelopes from different origins. The data for the three callus samples suggest that different extraction techniques may lead to divergent results. However, these divergences are relatively small compared to those obtained with different tissues or cells. For example, the glycine content varies from 7.2 mol% (psoriatic scales) to 9.2 mol% (cultured foreskin keratinocytes after induction of the crosslinking process with a calcium ionophore) to 33.94 mol% (foreskin

TABLE 1 **Amino Acid Composition (in mol%) of Human Cornified Envelopes from Different Origins**

Amino acid	Callus[a]	Callus[b]	Callus[c]	Body epidermis[d]	Foreskin[e]	Culture[f]	Psoriatic scales[c]
Ala [A]	5.2	4.7	3.8	2.8	3.44	7.0	6.4
Arg [R]	4.6	5.0	3.2	4.1	1.79	4.3	3.4
Asx [B = D + N]	5.8	6.0	5.0	5.4	1.73	9.1	8.3
Cys [C]	4.9	4.3	---[g]	---	4.56	2.5	---
Glx [Z = E + Q]	14.0	13.3	17.0	16.6	8.85	15.9	24.0
Gly [G]	14.1	18.0	21.0	23.1	33.94	9.2	7.3
His [H]	2.1	2.2	2.0	2.0	0.64	1.9	2.4
Ile [I]	3.3	3.0	2.0	2.1	1.97	3.5	2.2
Leu [L]	5.8	5.9	3.3	3.1	1.54	7.9	4.0
Lys [K]	6.6	5.5	6.0	5.6	4.77	7.6	18.5
Met [M]	n.d.[h]	1.2	0.7	0.2	0.27	1.5	n.d.
Phe [F]	2.3	2.8	1.0	0.7	2.42	2.9	1.5
Pro [P]	13.7	6.8	8.2	9.6	---	7.8	12.2
Ser [S]	7.4	11.5	16.0	13.7	20.35	7.2	1.8
Thr [T]	3.7	3.5	3.7	3.2	2.08	4.8	2.1
Trp [W]	---	0.3	---	---	---	---	---
Tyr [Y]	1.1	3.0	1.6	1.9	2.25	1.8	---
Val [V]	5.2	4.3	5.6	5.9	3.63	5.1	6.0

[a]NaOH extraction (Matoltsy and Matoltsy, 1966).
[b]Urea–mercaptoethanol extraction (Manabe *et al.*, 1981).
[c]Urea–mercaptoethanol–SDS extraction (Martinet *et al.*, 1990).
[d]Urea–mercaptoethanol–SDS extraction of epidermis from legs, feet, or breasts (Martinet *et al.*, 1990).
[e]SDS–mercaptoethanol extraction (Hohl *et al.*, 1991b).
[f]SDS–mercaptoethanol extraction of foreskin keratinocytes treated with a calcium ionophore to induce envelope formation (Rice and Green, 1979).
[g]---, Not determined.
[h]n.d., Not detected.

epidermis). This variability points in the direction of a certain flexibility in the molecular composition of the cornified envelope. The extraction technique most frequently used is the SDS/mercaptoethanol procedure of Sun and Green (1976) or variations thereof.

Characteristic differences in the molecular composition of human cornified envelopes from different sources are also observed after cleavage of electrodialytically purified envelopes with cyanogen bromide, and separation of the resulting peptides by SDS–polyacrylamide gel electrophoresis (Michel *et al.*, 1988b; Michel and Juhlin, 1990). There are no striking differences in the peptide maps of fragile and rigid envelopes isolated from the tape-stripped horny layer of normal human skin (Michel *et al.*, 1988b).

However, slight inter- and intraindividual variations have been described (Legrain *et al.*, 1991).

The capability of cultured normal human keratinocytes to spontaneously synthesize envelopes is poor. However, these cells may be rendered "envelope competent" by incubation with a calcium ionophore such as A23187 (Rice and Green, 1979; Cline and Rice, 1983). The envelopes obtained under these conditions resemble the fragile type of the normal human epidermis but differ in their peptide map essentially by the appearance of a smear of bands between about 4 and 15 kDa (Michel *et al.*, 1987).

Transformation of human keratinocytes by Simian Virus 40 almost completely abolishes the capacity to differentiate (Taylor-Papadimitriou *et al.*, 1982). They cannot even be made envelope-competent with a calcium ionophore under standard culture conditions (1.8 m*M* Ca^{2+}). However, increasing the Ca^{2+} concentration to 5 m*M* forces them to make envelope-like structures ("pseudo-envelopes"; Schmidt *et al.*, 1988a), which exhibit, under Nomarski contrast microscopy, a morphology different from the irregularly shaped, fragile envelope type normally found in cultured keratinocytes. Pseudo-envelopes are round and appear to be much thicker and more rigid. While the peptide patterns of envelopes produced by normal human keratinocytes *in vitro* contain at least some well-defined bands, the peptide maps of pseudo-envelopes remain a continuous unresolved smear, even at higher sample dilutions.

The shift from a limited number of well-defined bands in the peptide maps of normal envelopes to an unresolved smear in the peptide maps of pseudo-envelopes suggests that when envelope formation is induced under nonphysiological conditions, there is a loss of ordered structure. This loss of structure may be due to the incorporation of nonspecific proteins into the envelope; it has been observed that this happens following ionophore treatment in fibroblasts, which normally do not synthesize envelopes (Simon and Green, 1984). As discussed later, a similar phenomenon can be seen in certain pathological situations.

The existence of ϵ-(γ-glutamyl)lysine crosslinks (Fig. 2) in mammalian stratum corneum was first described by Goldsmith *et al.* in 1974, very soon after the discovery of these isopeptide bonds in fibrin clots polymerized by factor XIII (Lorand *et al.*, 1968; Matacic and Loewy, 1968; Pisano *et al.*, 1969) and in digests of wool keratin (Asquith *et al.*, 1970; Harding and Rogers, 1971). In 1977, R. Rice and H. Green found a massive concentration of ϵ-(γ-glutamyl)lysine bonds in the envelopes isolated from human callus samples. The crosslinked lysine accounted for about 18% of total lysine, in contrast to the soluble proteins of cultured keratinocytes and fibroblasts, which were only found to have 0.3% of their lysine residues crosslinked. In a similar but more recent study, Martinet *et al.* (1990) recovered (3.85 $\pm$ 0.45) ϵ-(γ-glutamyl)lysine crosslinks per (60 $\pm$ 6) lysine residues, which corresponds to about 6.5%, a value that is considerably lower

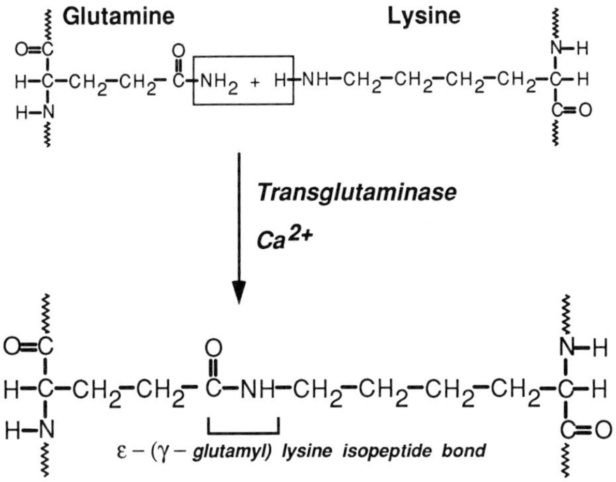

FIGURE 2 Schematic presentation of ϵ-(γ-glutamyl)lysine crosslinking between two proteins.

than that reported by Rice and Green. On the other hand, by extensive proteolytic digestion of envelope fragments from freshly excised human foreskin that had been metabolically labeled with $[^3H]$lysine, Hohl *et al.* (1991b) recovered 38% of the total $[^3H]$lysine label in the form of ϵ-(γ-glutamyl)lysine and only 22% in the form of free lysine, indicating that the total percentage of crosslinked lysine may even approach 60%.

Based on earlier observations (Piacentini *et al.*, 1988b), Martinet *et al.* (1990) provide evidence that, in addition to ϵ-(γ-glutamyl)lysine isopeptide bonding, bis(γ-glutamyl)polyamine crosslinking of glutamine residues (Fig. 3) contributes significantly to the stabilization of the cornified envelope (Table 2). The polyamine recruited for this purpose is almost exclusively spermidine. Taking into account both types of crosslinks, one can calculate an average peptide size of about 20 kDa per crosslink in the envelopes of normal human body epidermis (see Table 2).

Pathological Aspects

Michel and Juhlin (1990) analyzed the cornified envelopes from patients with various congenital hyperkeratotic disorders. Based on envelope morphology, they could divide these diseases in two major groups. The first group [including keratoderma palmoplantare, KID (keratitis, ichthyosis, deafness) syndrome, congenital pachyonichia, erythrokeratoderma variabilis and parapsoriasis], exhibits a more-or-less normal envelope morphology. Among these diseases, parapsoriasis differs slightly in that nucleus-like inclusions within some envelopes are a characteristic feature; this charac-

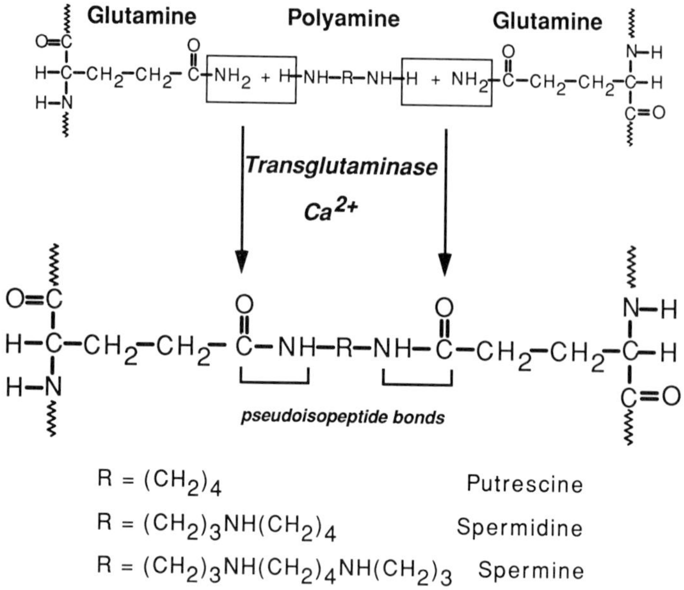

FIGURE 3 Schematic presentation of bis(γ-glutamyl)polyamine crosslinking between two proteins.

TABLE 2 **Crosslink Types and Levels in Cornified Envelopes Purified from Normal Human Body Epidermis and Psoriatic Scales**[a]

Crosslink type	Epidermis	Crosslinks (per 1000 amino acid residues)	Crosslinks (nmol per mg envelope protein)	kDa Protein per crosslink
ϵ-(γ-Glutamyl)lysine	normal	4.16 ± 0.38	37 ± 4	27
	psoriatic	5.35 ± 1.10	47 ± 10	21
Mono-(γ-glutamyl)putrescine	normal	n.d.[b]		
	psoriatic	traces		
Mono-(γ-glutamyl)spermidine	normal	traces		
	psoriatic	<0.2		
Bis-(γ-glutamyl)putrescine	normal	traces		
	psoriatic	<0.6		
Bis-(γ-glutamyl)spermidine	normal	1.66 ± 0.22	15 ± 2	67
	psoriatic	9.84 ± 1.40	87 ± 12	11.5
Total	normal	≃6	≃52	≃20
	psoriatic	≃15	≃134	≃7.5

[a] According to Martinet *et al.* (1990).
[b] n.d., Not detected.

teristic readily distinguishes this hyperkeratosis from psoriasis vulgaris. The second group, with a significantly altered envelope morphology, comprises psoriasis vulgaris, Darier's disease, ichthyosis vulgaris and ichthyosis brittle hair syndrome. Significant differences are also seen in the peptide patterns obtained after cyanogen bromide cleavage of electrodialytically purified envelopes from the different lesions (Michel and Juhlin, 1990).

Qutaishat and Kumar (1988) detected deviations in antigenic properties and amino acid composition of envelopes obtained from psoriatic scales and normal callus. Similar differences in the amino acid composition were also reported by Martinet *et al.* (1990; see Table 1). Moreover, these authors found, compared to normal human body epidermis, a large increase in bis(γ-glutamyl)spermidine linkage in the envelopes isolated from psoriatic scales, where the average peptide size per crosslink diminished from 20 to 7.5 kDa (Table 2).

Precursor Proteins

Overview

Since there are still no chemical or enzymatic means available to specifically cleave ϵ-(γ-glutamyl)lysine isopeptide bonds, the molecular building blocks of the cornified envelope cannot be recovered in original form after crosslinking has taken place. Thus, the disclosure of potential envelope precursors has to rely on indirect techniques such as:

1. the identification of natural transglutaminase substrates in the epidermis (Buxman *et al.*, 1976, 1980; Hanigan and Goldsmith, 1978; Rice and Green, 1979; Kubilus and Baden, 1982; Lobitz and Buxman, 1982; Zettergren *et al.*, 1984; Simon and Green, 1984, 1985, 1988; Richards *et al.*, 1988; Takahashi and Tezuka, 1989; Mehrel *et al.*, 1990; Phillips, S. B. *et al.*, 1990);
2. the *in vivo* incorporation of a selectively labeled protein into the insoluble envelope (Richards *et al.*, 1988);
3. the disappearance of proteins from the extractable protein fraction after induction of the crosslinking process in cell culture (Simon and Green, 1984, 1985; Baden *et al.*, 1987b; Michel *et al.*, 1987);
4. the reaction of a protein with antibodies raised against cornified envelopes (Baden *et al.*, 1987a,b; Kubilus *et al.*, 1987; Michel *et al.*, 1987; Phillips, S. B. *et al.*, 1990), or, as a corollary,
5. the reaction of antibodies raised against the putative envelope precursor with the corneocyte periphery *in situ* (Kubilus and Baden, 1982; Lobitz and Buxman, 1982; Warhol *et al.*, 1985; Tezuka and Takahashi, 1987; Richards *et al.*, 1988; Steven *et al.*, 1990; Hohl *et al.*, 1991b) or, preferably, with purified envelopes (Rice and Green, 1979; Mehrel *et al.*, 1990; Haftek *et al.*, 1991).

TABLE 3 **Amino Acid Composition (in mol%) of Potential Envelope Precursors**

Amino acid	Involucrin[a]	Involucrin[b]	CREP[c]	Loricrin[d]	Keratolinin[e]
Ala [A]	1.5	1.20	5.3	0.95	8.44
Arg [R]	0.7	0.51	6.3	0.0	2.46
Asp [D]		1.37		0.32	
Asn [N]		0.68		0.0	
Asx [B = D + N]	2.8	(2.05)	8.3	(0.32)	8.71
Cys [C]	0.3	0.34	4.3	6.01	0.97
Glu [E]		19.83		0.0	
Gln [Q]		25.64		4.44	
Glx [Z = E + Q]	45.8	(45.47)	12.7	(4.44)	13.46
Gly [G]	6.7	6.50	18.3	46.84	8.80
His [H]	4.7	4.96	2.3	0.32	1.13
Ile [I]	0.4	0.17	3.7	1.58	3.14
Leu [L]	14.6	15.21	6.8	0.0	8.19
Lys [K]	7.4	7.69	9.1	2.22	5.65
Met [M]	0.9	1.03	n.d.[f]	0.0	0.63
Phe [F]	0.6	0.17	2.9	2.85	2.67
Pro [P]	5.7	7.35	n.d.	2.85	4.50
Ser [S]	1.6	1.20	9.8	22.78	10.43
Thr [T]	1.6	1.37	5.4	2.22	9.22
Trp [W]	0.2	0.34	n.d.	0.32	0.70
Tyr [Y]	0.8	0.34	n.d.	2.53	1.63
Val [V]	3.7	4.10	5.2	3.48	6.32

[a]Human; from amino acid hydrolysis (Rice and Green, 1979).
[b]Human; from cDNA sequence (Eckert and Green, 1986).
[c]Human; from amino acid hydrolysis (Tezuka and Takahashi, 1987).
[d]Human; from cDNA sequence (Hohl *et al.*, 1991b).
[e]Bovine; contains in addition 2.38 mol% citrulline; from amino acid hydrolysis (Lobitz and Buxman, 1982).
[f]n.d., traces or not detected.

A substantial number of keratinocyte proteins have been suspected to serve as envelope precursors. However, only a few of them such as involucrin, keratolinin, loricrin, and the cysteine-rich envelope protein CREP meet the majority of the above-described criteria and have been further characterized by amino acid analysis. These proteins will be discussed in some detail in the following sections. Each of them has some peculiarities with regards to its amino acid composition (Table 3). CREP and loricrin provide the relatively high level of cysteine found in epidermal envelopes (see Table 1). All of them, with the exception of loricrin, meet the requirement for an elevated amount of glutamine/glutamate and lysine residues. Furthermore, CREP and loricrin are very rich in glycine residues, whereas involucrin matches the proline content of purified envelopes. The amino acid

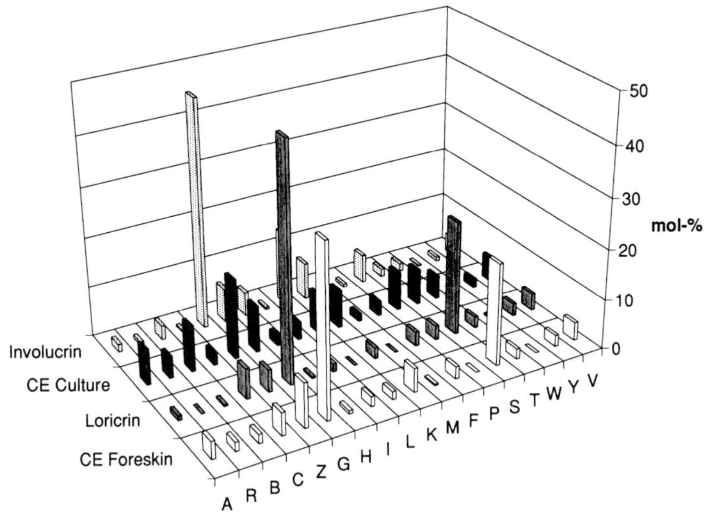

FIGURE 4 Amino acid composition of human cornified envelopes (CE) and precursor proteins.

profile (Fig. 4) of loricrin corresponds qualitatively to the composition of envelopes from foreskin epidermis, whereas the profile of involucrin is closer to the composition of envelopes from cultured keratinocytes.

Involucrin

In 1979, R. Rice and H. Green described a soluble precursor of the cornified envelope that had been identified in crude extracts of cultured keratinocytes by its ability to incorporate [^{14}C]-2-aminoethanol or dansylcadaverine, known amine donors for transglutaminase-catalyzed reactions. Antisera were elicited against the purified protein and the specific antibodies were completely absorbed by washed envelopes. By indirect immunofluorescence, it was shown that the antibodies decorate isolated envelopes and in histological sections label the stratum corneum as well as the living layers from the outer half of the stratum spinosum upward. In the lower layers, the precursor was localized in the cytoplasm but became concentrated at the cell periphery before crosslinking took place at the site of stratum corneum formation. Similar results were reported in a more recent ultrastructural study using immunogold-labeling techniques (Warhol *et al.*, 1985).

The precursor, later termed involucrin (from the Latin *involucrum*, envelope; Watt and Green, 1981), comprised between 5 and 10% of the soluble protein in cultured keratinocytes and had an isoelectric point of pI 4.5 ± 0.3. Different physical techniques resulted in diverging molecular mass estimates (over 500 kDa by size exclusion chromatography, 92–140 kDa by

SDS–polyacrylamide gel electrophoresis, and 83 kDa by sedimentation) suggesting that involucrin assumes a rod-like shape (Rice and Green, 1979; Simon and Green, 1984). Simon and Green (1985) found that this precursor with its unusual amino acid content of about 45 mol% glutamate and glutamine residues (see Table 3) promotes the crosslinking of proteins in the particulate fraction of cultured keratinocytes.

Involucrin can easily be enriched from cultured human keratinocytes by heating crude extracts to 95°C. Most of the proteins precipitate, but involucrin remains in solution accounting for more than 90% of the protein content (Etoh *et al.*, 1986).

The gene of human involucrin has been cloned and sequenced (Eckert and Green, 1986). It consists of a 5′ exon of 43 bp and a 3′ exon of 2107 bp separated by an intron of 1188 bp in the 5′-untranslated sequence. The coding region is found in the second exon. It consists of a central segment containing 39 repeats of 30 nucleotides, which is flanked by two segments with low homology to the consensus sequence of the central repeats. The ten amino acids encoded by the consensus sequence contain three glutamines and two glutamic acids at conserved positions. The complete amino acid sequence of involucrin (585 residues) deduced from the nucleotide sequence is given in Figure 5. The molecular mass of 68 kDa as calculated from the deduced amino acid sequence is significantly lower than the values obtained by physical techniques as discussed earlier.

In order to identify the reactive glutamines involved in crosslinking, Simon and Green (1988) incubated keratinocyte cytosol and particulates with [^{14}C]glycine ethyl ester, a commonly used amine donor in transglutaminase-catalyzed reactions. They purified and then tryptically digested involucrin from the reaction mixture. After separation of the labeled fragments, they found that 90% of the label was located in a single glutamine residue (496) and 10% in the adjacent glutamine residue (495). Additional glutamine residues became reactive when purified involucrin was fragmented by cyanogen bromide cleavage before incubation with [^{14}C]glycine ethyl ester and keratinocyte particulates. The authors suggested that "in the cell the initial crosslinking of involucrin takes place at residue 496. The other glutamines could become reactive by a number of mechanisms. As the transglutaminase is membrane associated, the first crosslink of involucrin to a membrane protein should increase the concentration of other potentially reactive glutamines in the neighborhood of the enzyme. The conformation of involucrin might also be altered by reaction of the first glutamine residue, thereby changing the accessibility of other glutamine residues to the active site of the transglutaminase." Thus, it is possible that involucrin could serve as a scaffold for an ordered crosslinking between different proteins by sequentially exposing reactive glutamine residues to the transglutaminases, thus setting in motion a 'zipper mechanism.'

The central segment of tandem repeats in the human gene is of recent

Flanking segment:

1	M S Q Q H T L P V T L S P A L S Q E L L K T V P P P V N T H Q E Q M K Q P T P L P P P
44	C Q K V P V E L P V E V P S K Q E E K H M T A V K G L P E Q E C E Q Q Q K E P Q E Q E
87	L Q Q Q H W E Q H E E Y Q K A E N P E Q Q L K Q E K T Q R D Q Q L N K Q L E E E K K L
130	L D Q Q L D Q E L V K R D E Q L G M K K E Q L

Central segment:

153	**L** E L P **E Q**	**Q E G** H **L K H L E Q**	**Q E G Q L K H** P **E Q**	**Q E G Q L** E L P **E Q**
189	**Q E G Q L** E L P **E Q**	**Q E G Q L** E L P **E Q**	**Q E G Q L** E L P **E Q**	**Q E G Q L** E L P Q **Q**
229	**Q E G Q L** E L S **E Q**	**Q E G Q L** E L S **E Q**	**Q E G Q L** E L S **E Q**	**Q E G Q L K H L E** H
269	**Q E G Q L** E V P **E** E	**Q** M **G Q L K** Y **L E Q**	**Q E G Q L K H L** D **Q**	**Q E** K **Q** P E L P **E Q**
309	**Q** M **G Q L K H L E Q**	**Q E G Q** P **K H L E Q**	**Q E G Q L** E Q **L E** E	**Q E G Q L K H L E Q**
349	**Q E G Q L** E **H L E** H	**Q E G Q L** G L P **E Q**	**Q** V L **Q L K** Q **L E** K	**Q** Q **G Q** P **K H L E** E
389	E **E G Q L K H L** V **Q**	**Q E G Q L K H L** V **Q**	**Q E G Q** **L E Q**	**Q E** R **Q** V E **H L E Q**
426	**Q** V **G Q L K H L E** E	**Q E G Q L K H L E Q**	**Q** Q **G Q L** E V P **E Q**	**Q** V **G Q** P **K N L E Q**
466	E **E** K **Q L** E L P **E Q**	**Q E G Q** V **K H L E** K	**Q E** A **Q L** E L P **E Q**	**Q** V **G Q** P **K H L E Q**
506	**Q E** K H **L** E **H** P **E Q**	**Q** D **G Q L K H L E Q**	**Q E G Q L K** D **L E Q**	**Q** K **G Q L** E

Flanking segment:

542	Q P V F A P A P G Q V Q D I Q P A L P T K G E V L L P V E H Q Q Q K Q E V Q W P P K H
585	K

FIGURE 5 Deduced amino acid sequence of human involucrin (Eckert and Green, 1986). Amino acids of the consensus sequence in the central repeat segment are given in bold face. The unique glutamine residue preferred by keratinocyte transglutaminase as initial amide donor is boxed.

origin and the result of gene duplication and multiple mutations from the ancestral flanking region (Eckert and Green, 1986; Tseng and Green, 1988). Antisera elicited against human involucrin react strongly with similar protein fractions of higher primates but only weakly with those found in lemurs (Parenteau *et al.*, 1987). Prosimians such as the lemurs and tarsiers possess involucrins with a structure similar to human protein. However, the central repeat segment of these involucrins is derived from another sequence at a different location in the ancestral segment (Tseng and Green, 1988; Djian and Green, 1991). More recently, involucrin-like molecules have also been detected in nonprimates such as cow, dog, rabbit, sheep, guinea pig, rat, and finback whale by using antibodies directed against either fragments of the ancestral segment (Simon and Green, 1989) or full-length human involucrin (Kubilus *et al.*, 1990). Subsequently, the genes for dog and pig involucrin have been cloned and sequenced (Tseng and Green, 1990). Like the corresponding genes of the prosimians, each contains a homologous segment of short tandem repeats at the same position in the coding region. However, the codon sequence of repeats in prosimians differs significantly from that of the nonprimate mammals. These results suggest a more ubiquitous species distribution of involucrin-like molecules than originally thought.

Keratolinin

In 1976, Buxman *et al.* isolated from bovine epidermal extracts a soluble transglutaminase substrate with a molecular mass of 150 kDa. Later, Buxman *et al.* (1980) found that antisera directed against this protein react with a 36-kDa protein, which is readily converted by epidermal extracts in the presence of Ca^{2+} ions to polymers ranging from 75 kDa to over 200 kDa. Subsequently, these intermediates aggregate to an insoluble amorphous material. ϵ-(γ-glutamyl)lysine isopeptide bonds have been identified in the polymers, but not in the 36-kDa precursor. Polyclonal antibodies have been raised against the purified precursor. Using Ouchterlony double-diffusion techniques, the initial substrate and the soluble crosslinked product have been shown to be immunologically identical (Lobitz and Buxman, 1982). Indirect immunofluorescence with freeze-dried sections of bovine snout epidermis demonstrate the presence of the antigen throughout the cytoplasm and inner cell membrane of granular cells. In the stratum corneum, however, fluorescence is limited to the inner cell periphery.

Zettergren *et al.* (1984) proposed "keratolinin" (from the Greek *keratos,* horny tissue; *lininos,* to cover the inner surface of) as a suitable name for the 36 kDa protein, which can be further dissociated into noncovalent subunits by chaotropic agents or detergent. The molecular mass of the subunits is 6.0–6.2 kDa. They can be separated by isoelectric focusing into two moieties with isoelectric points of pI 6.0 and 6.3.

Keratolinin can also be purified from heat-separated human epidermis (Zettergren *et al.*, 1984). The human protein exhibits similar biochemical, but different immunological properties. The pI values for the two subunits of the human protein are 5.0 and 5.4. Antiserum to human keratolinin fails to crossreact with the bovine protein counterpart and *vice versa*. There is also no crossreaction with antibodies raised against involucrin. Since the amino acid composition of human keratolinin has not yet been published, the composition of the bovine protein is given in Table 3.

Loricrin

A separate chapter (Chapter 5) is dedicated to loricrin (from the Latin *lorica,* a protective shell or cover), hence only some of the major properties of this newly detected envelope presursor (Mehrel *et al.*, 1990; Hohl *et al.*, 1991b) are summarized here.

The human loricrin is a cationic protein particularly rich in glycine, serine, and cysteine residues and completely lacking in arginine, asparagine, glutamic acid, leucine, and methionine (see Table 3). This protein and envelopes purified from foreskin epidermis possess a very similar qualitative amino acid profile, which is quite distinct from the profile found in involucrin and envelopes produced in culture after ionophore treatment (Fig. 4). Loricrin has a molecular mass of 25.8 kDa as calculated from its amino acid sequence (315 residues) deduced from a full-length cDNA clone. Its mRNA comprises 1.25 kb and is as abundant as the mRNAs of other major keratinocyte differentiation products such as keratins 1 and 10. In contrast to the other potential envelope precursors discussed so far, loricrin cannot be extracted from epidermis without SDS and a sulfhydryl-reducing agent, suggesting that it becomes precrosslinked by disulfide bonds.

The loricrin molecule can be subdivided into glutamine–lysine-rich terminal regions embedding three expanded glycine–serine-rich domains,

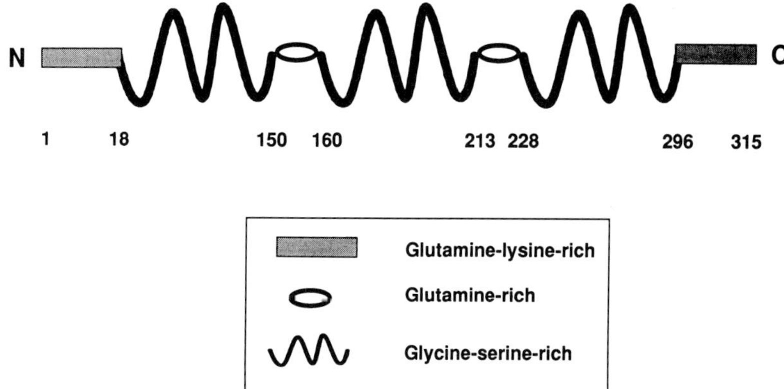

FIGURE 6 Domain structure of human loricrin (simplified from Hohl *et al.*, 1991b).

which are separated by two small glutamine-rich sequences (Fig. 6). At least two lysine residues (88, 315), one of which (88) is positioned in a glycine–serine-rich domain, and four glutamine residues (153, 157, 215, 219) of loricrin have been shown to be involved in protein crosslinking. The cysteine residues are spread evenly throughout the molecule. Glycine–serine-rich sequences are also found on keratins 1 and 10, where they are reported to form an Ω-loop-like structure (Zhou *et al.*, 1988). It is thus conceivable that loricrin crosslinked via ε-(γ-glutamyl)lysine isopeptide bonds into the cornified envelope forms similar Ω loops, which serve as "docking stations" for the keratins.

Using monospecific antisera to loricrin, Steven *et al.* (1990) demonstrated by immunoelectron microscopy the passage of loricrin from keratohyalin L-granules in the stratum granulosum of mouse epidermis to the periphery of cells throughout the stratum corneum, where the epitope is exposed only at the inner envelope surface.

Cysteine-rich Envelope Protein

Tezuka and Takahashi (1987) extracted a cysteine-rich envelope protein (CREP) from the "membrane-region" of human epidermal corneocytes. This protein, whose amino acid composition is given in Table 3, has a molecular mass of 16 kDa and an isoelectric point of pI 4.8. A polyclonal antibody raised against the purified protein decorated the periphery of cells in the stratum spinosum, stratum granulosum, and the lower stratum corneum of both plantar and forearm epidermis. Immunoblotting experiments revealed the 16 kDa protein in extracts from the stratum granulosum as well as the inner, middle, and outer part of the horny layer. Besides the 16 kDa protein, four other proteins (64, 57, 48, and 42 kDa) from the inner and middle part of the stratum corneum reacted with the antibody.

The same laboratory (Takahashi and Tezuka, 1989) subsequently reported that a similar (or the same ?) 15 kDa protein may be crosslinked by transglutaminases. However, the fact that all these proteins are extractable from horny layer samples by simple buffer casts a doubt upon their role as constituents of the cornified envelope in the strict sense.

Other Precursor Proteins

None of the keratinocyte proteins discussed so far is membrane-bound in its free, not yet crosslinked form. However, the participation of membrane-associated precursors appears to be essential for fixing the growing envelope at the cell periphery. Thus special attention should be given to a study by Simon and Green (1984) showing that when transglutaminase is activated in cultured keratinocytes, at least six membrane-associated proteins become nonextractable with SDS–mercaptoethanol. Two of these proteins, with molecular masses of 195 and 210 kDa, are specific for differentiating keratinocytes, whereas the others can also be found in fibroblasts.

Antisera raised against both proteins, when isolated after two-dimensional electrophoresis, react only with suprabasal cells in confluent keratinocyte cultures and are specifically absorbed by exhaustively washed keratinocyte envelopes. The 195 kDa protein is located primarily at the cell periphery, whereas the 210 kDa protein is uniformly dispersed in the cytoplasm. By using a monoclonal antibody (AE 11) raised against trypsinized epidermal cells, Ma and Sun (1986) have shown that the 195 kDa protein precursor exists in a soluble form in basal cells and becomes insolubilized at the cell periphery during keratinocyte differentiation. Furthermore, Baden *et al.* (1987a) have shown that 4 out of 31 clones of monoclonal antibodies generated to the envelopes of cultured human keratinocytes label, in immunoblots, the 195 kDa membrane-associated protein. It is thus possible that at least the 195 kDa keratinocyte-specific membrane protein is destined to anchor the growing envelope at the cell periphery.

Of the 31 hybridoma culture supernatants (Baden *et al.*, 1987a), 5 react with involucrin, 12 with keratins, 6 with an 82 kDa protein not further characterized, and only one (HEC-2) with basic polypeptides of molecular masses 14.9, 16.8, and 24.8 kDa, which had previously been shown by two-dimensional electrophoresis to have an isoelectric point close to pI 9.0 (Baden *et al.*, 1987b). These basic polypeptides are transglutaminase substrates, and possess antigenic determinants that are released by V8-protease treatment of cornified envelopes. They are widespread in the epidermis of many mammals. Because of this ubiquitous distribution, they have been named "pancornulins" (Phillips, S. B. *et al.*, 1990).

The relatively high percentage of monoclonal antibodies obtained using cornified envelopes as antigen, but which react with keratins (Baden *et al.*, 1987b) may be not accidental, but rather may reflect the property of transglutaminase(s) to tightly fasten cytoskeletal and desmosomal components to this highly insoluble structure. In a recent study using indirect immunofluorescence microscopy and immunogold electron microscopy, Haftek *et al.* (1991) demonstrated the persistence of keratin 10, desmoplakins I and II, the intracellular epitope of desmoglein, intercellular corneodesmosomal proteins, and filaggrin on electrodialytically purified envelopes. The existence of ϵ-(γ-glutamyl)lysine crosslinks between keratins has been known for a long time (Abernethy *et al.*, 1977), and Richards *et al.* (1988) reported evidence for filaggrin as a component of the cell envelope of the newborn rat.

In experiments similar to those of Simon and Green (1984) but using the transformed keratinocyte line SV-K14, which can only be rendered envelope-competent after serum starvation, Michel *et al.* (1987) detected five potential precursor proteins, which disappeared from the extractable cellular pool after induction of the crosslinking process with a calcium ionophore. With the exception of involucrin, which was present in only trace amounts, the molecular masses of these proteins were distinct from those found by Simon and Green (1984) in normal cultured keratinocytes. Only one of them was in part associated with the cell membrane, and all of them could also be detected in envelope-noncompetent cells.

The heterogeneity of potential envelope precursors found *in vivo* and *in vitro* as well as the presence of some of these proteins in cells that do not normally form an envelope or envelope-like structure has led to the idea that the molecular composition of this structure may not be strictly determined. It may rather be subject to variations depending on the availability of potential substrate proteins at the precise moment when the crosslinking transglutaminase becomes activated (Michel *et al.*, 1987, 1988b). It is possible that besides specific proteins such as involucrin or loricrin, which may have a scaffolding function, nonspecific waste proteins, which are generated in the final stages of keratinocyte differentiation by cell organelle destruction, are also incorporated into the cornified envelope ('dust-bin hypothesis'). The participation of envelope-nonspecific transglutaminase substrates in the crosslinking process could become particularly important under nonphysiological or pathological conditions, especially when the availability of the specific precursors is limited.

Associated Lipids

As previously discussed on p. 109, an electron lucent zone, about 4-nm thick, separates the electron-dense protein envelope from the intercellular space in transmission electron micrographs. A series of papers (Swartzendruber *et al.*, 1987; Wertz *et al.*, 1989a,b) strongly suggests that this translucent band reflects a monolayer of lipids, which is covalently linked via ester bonds (Fig. 7) to acyl groups of the cornified envelope. In human stratum corneum, the major components are two hydroxyceramides having different mobilities on thin-layer chromatograms and comprising 53.3 and 24.8% (w/w) of the lipid content of the envelope. Further constituents are fatty acids (12.7%) and ω-hydroxy acids (9.4%). The above-mentioned ceramides consist of ω-hydroxy acids linked by amide bonds to sphingosine. The predominant ω-hydroxy acids in both ceramides and in the hydroxy-acid fraction are the 30-carbon saturated as well as 32- and 34-carbon monounsaturated species. Most of the bound fatty acids are saturated and contain 14–22 carbons, but significant portions of monoenoic acids and linoleic acid are also detected. Interdigitation of the sphinogosine chains between adjacent envelopes has been suggested to play a role in stratum corneum cohesion (Wertz *et al.*, 1989b).

Figure 7 represents two possibilities for the attachment of hydroxyceramides to protein envelopes. It should be mentioned, however, that the first possibility in the scheme does not allow for the attachment of O-acylceramides, which contain an ω-hydroxy-linked fatty acid (predominantly linoleic acid) and are thought to play an essential role in epidermal

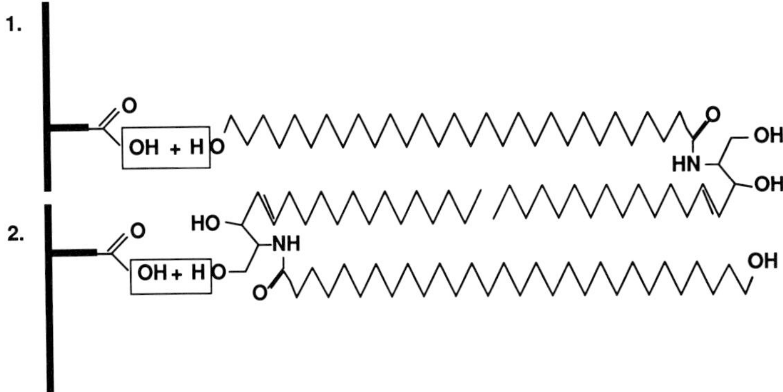

FIGURE 7 Two possibilities for covalent attachment by ester formation of a representative ω-hydroxyacyl sphingosine to free carboxyl groups (e.g., γ-glutamyl residues) of the cornified envelope. If sufficient carboxyl groups are available, alternation of these two arrangements will result in a densely packed hydroxyceramide envelope (according to Swartzendruber *et al.*, 1987).

integrity (Wertz *et al.*, 1987). In view of the observation that the covalent lipid fraction amounts to about 10% of dry weight of envelopes purified from pig skin (Swartzendruber *et al.*, 1988), and assuming that the same holds true for human epidermis, a substantial number of free carboxyl groups are necessary at the external surface of the protein envelope to covalently bind the ceramides. Of the two cloned envelope-precursor molecules, involucrin contains 124 aspartate and glutamate residues, while loricrin contains only 1 aspartate residue. Thus, it is evident that involucrin (and not loricrin) is the more likely candidate for serving as a matrix for the covalent attachment of the lipid envelope.

Transglutaminases

Overview

Transglutaminases, a family of enzymes classified under the category EC 2.3.2.13, catalyze in a Ca^{2+}-dependent manner the crosslinking between proteins and the incorporation of polyamines into suitable proteins (Folk, 1980, 1983; Folk *et al.*, 1980). First, the cysteine in the active center of the enzyme forms a thioester with the γ-amide of the peptide bound glutamine. After the release of ammonia, the transient acyl enzyme intermediate reacts with any nucleophilic primary amine group forming either an isopeptide bond or a γ-glutamyl polyamine bond (Fig. 8). The specificity of the transglutaminase-catalyzed reaction is primarily based on the amino acid

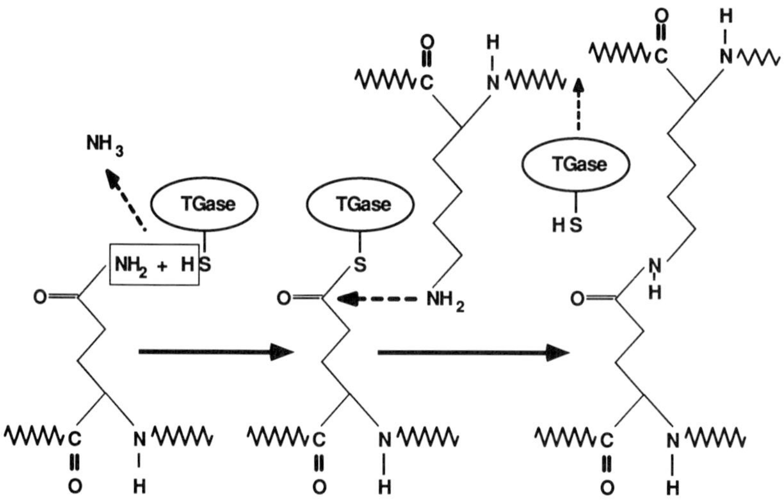

FIGURE 8 Crosslinking of peptide-bound glutamine and lysine residues via an intermediate thioesteracyl–transglutaminase complex.

composition and charge distribution in the neighborhood of the protein glutamine residue (Folk, 1980), whereas a broad affinity exists for peptide bound lysines. This means that almost any protein and polyamine can participate in the crosslinking process (Folk and Finlayson, 1977; Beniati and Folk, 1988). Furthermore, it has been reported that transglutaminases covalently incorporate themselves into high-molecular-weight complexes with other proteins (Barsigian *et al.*, 1991).

The resulting covalent isopeptide crosslinks are very stable and resistant to proteolytic attack, an essential prerequisite for the physiological products of transglutaminase-catalyzed reactions.

Factor XIII, one of the best-characterized transglutaminases, is a plasma protein that circulates in the blood as a tetramer composed of two catalytic subunits *a* and two noncatalytic subunits *b* (Schwartz *et al.*, 1973; Chung *et al.*, 1974). Factor XIII is a proenzyme and is activated by thrombin to factor XIII*a*, which is involved in the polymerization of fibrin to form the insoluble fibrin clot (Lorand, 1986). Deficiency of factor XIII can, among other symptoms, result in a lifelong tendency toward bleeding.

The dimer of subunit *a* also exists in other tissues and cells (e.g., placenta, uterus, monocytes, and hepatocytes) (Weisberg *et al.*, 1987; Wolpl *et al.*, 1987). Human factor XIII*a* consists of 731 amino acid residues, has a molecular mass of 83.15 kDa, and has been cloned by Grundmann *et al.* (1986), Ichinose *et al.* (1986), and Takahashi *et al.* (1986).

Transglutaminase K, the plasma membrane-associated keratinocyte transglutaminase (Thacher and Rice, 1985; Schmidt *et al.*, 1985) is involved in the formation of the cornified envelope (Rice and Green, 1977, 1978, 1979).

Tissue transglutaminase (transglutaminase C) seems to be a ubiquitous cytosoluble enzyme in animal tissues. Recently, its involvement in apoptosis, the physiological process of programmed cell death, has been suggested (Fesus *et al.*, 1987; Knight *et al.*, 1991; Piacentini *et al.*, 1991a,b). Upon activation, transglutaminase C crosslinks the majority of cellular proteins and thus inactivates enzymes that could otherwise cause damage to neighboring cells upon lysis of the plasma membrane. The resulting protein clot is called an “apoptotic body.”

Molecular cloning of factor XIII, transglutaminase K, and transglutaminase C allows a direct comparison of their primary structures. The deduced amino acid sequences (Fig. 9) demonstrate that these transglutaminases are related but distinct proteins with a high degree of homology in the active site region. The amino acid sequence of the recently cloned erythrocyte band 4.2 (Korsgren *et al.*, 1990; Korsgren and Cohen, 1991; Sung *et al.*, 1990) has been included in Figure 9 because of its homology (approximately 30%) with transglutaminase C and subunit *a* of factor XIII. The absence of transglutaminase activity in this protein is explained by the replacement of one active site cysteine by alanine (Korsgren *et al.*, 1990).

Table 4 lists synonyms used in the literature for transglutaminases, and Table 5 summarizes some characteristic properties of the different transglutaminases and erythrocyte band 4.2.

Transglutaminase C

The cDNA coding for guinea pig liver (Ikura *et al.*, 1988), mouse macrophage, and human endothelial cell (Gentile *et al.*, 1991) tissue transglutaminase (transglutaminase C) has provided detailed information about its amino acid composition. An mRNA of 3.6 kb encodes an enzyme of 77 kDa containing 690 amino acid residues. Even though transglutaminase C contains six potential glycosylation sites, the enzyme is not glycosylated (Ikura *et al.*, 1988). Another peculiarity of the enzyme is that no regions homologous to GTP binding proteins (Gentile *et al.*, 1991) can be identified in the amino acid sequence, even though it binds GTP as a regulator (Achyuthan and Greenberg, 1987; Schmidt *et al.*, 1988b).

An immunohistochemical study has revealed the constitutive expression of transglutaminase C in a large variety of human cell types (Thomazy and Fesus, 1989). For some time, no exact role could be attributed to this cytosolic enzyme. However, its affinity for constituents of the extracellular matrix like fibronectin (Upchurch *et al.*, 1987), fibrin (Achyuthan *et al.*, 1988), and type I collagen (Juprelle-Soret *et al.*, 1988) indicates that it could play a role in wound healing after its release from injured cells. In erythrocytes, activation of transglutaminase C leads to changes in membrane proteins

```
K      MMDGPRSDVGRWGGNPLQPPTTPSPEPEPEPEPDGRSRRGGGRSFWARCCGCCSCRNAADDDWGPEPSDSRGRGSSSGTRRPGSRGSDSRRPVSRGSGVNAAGD------   102
XIII                                                                                 MSETSRTAF---GGRRAVPPNNS-NAAEDDLPTVE    31

K      -------GTIREGMLVVNGVDLLSSRSDQNRREHHTDEYEYDELIVRRGQPFHMLLLL---SRTYESS-DRITLELLIGNNPEVGKGTHVIIPVGK-GGSGGWKAQVV   198
C                MAEELVLERCDLE---LETNGRDHHTADLCREKLVVRRGQPFWLTLHFEG--RNYEASVDSLTFSVVTGPAPSQEAGTKARFPLRDAVEEGDWTATVV    93
XIII   LQGVVPRGVNLQEFLNVTSVHLFKERWDTNKVDHHTDKYENNKLIVRRGQSFYVQIDF---SRPYDPRRDLFRVEYVIGRYPQENKGTYIPVPIVSELQSGKWGAKIV   136
4.2              MGQALGIKSCDFQAAR---NNEEHHTKALSSRRLFVRRGQPFTIILYFRAPVRAFLPALKKVALTAQTGEQPSKINRTQATFPISSLGDRKWWSAVVE    95

K      KASGQNLNLRVHTSPNAIIGKFQFTVRTQSDAGEFQLPFDPRNEIYILFNPWCPEDIVYVDHEDWRQEYVLNESGRIYYGTEAQIGERTWNYGQFDHGVLDACLYILD   306
C      DQQDCTLSLQLTTPANAPIGLYRLSLEASTGYQGSSFVLG---HFILLFNAWCPADAVYLDSEEERQEYVLTQQGFIYQGSAKFIKNIPWNFGQFQDGILDICLILLD   198
XIII   MREDRSVRLSIQSSPKCIVGKFRMYVAVWTPYGVLRTSRNPETDTYILFNPWCEDDAVYLDNEKEREEYVLNDIGVIFYGEVNDIKTRSWSYGQFEDGILDTCLYVMD   244
4.2    ERDAQSWTISVTTPADAVIGHYSLLLQVS---GRKQLLLG---QFTLLFNPWNREDAVFLKNEAQRMEYLLNQNGLIYLGTADCIQAESWDFGQFEGDVIDLSLRLLS   197

K      R-------RGMPYGGRGDPVNVSRVISAMVNSLDDNGVLIGNWSGDYSRGTNPSAWVGSVEILLSYLRTGY-SVPY-GQCWVFAGVTTTVLRCLGLATRTVTNFNSAH   405
C      VNPKFLKNAGRDCSRRSSPVYVGRVGSGMVNCNDDQGVLLGRWDNNYGDGVSPMSWIGSVDILRRWKNHGCQRVKY-GQCWVFAAVACTVLRCLGIPTRVVTNYNSAH   305
XIII   R-------AQMDLSGRGNPIKVSRVGSAMVNAKDDEGVLVGSWDNIYAYGVPPSAWTGSVDILLEYRS-SENPVRY-GQCWVFAGVFNTFLRCLGIPARIVTNYFSAH   343
4.2    K--------DKQVEKWSQPVHVARVLGALLHFLKEQRVLPTPQTQATQEGALLNKRRGSVPILRQWLTGRGRPV-YDGQAWVLAAVACTVLRCLGIPARVVTTFASAQ   296
                                                                                                  ****!****
```

```
K      DTDTSLTMDIYFDENMKPLEHLNH-DSVWNFHVWNDCWMKRPDLPSGFDGWQVVDATPQETSSGIFCCGPCSVESIKNGLVYMKYDTPFIFAEVNSDKVYWQRQDDGS   512
C      DQNSNLLIE-YFRNEFGEIQGDKS-EMIWNFHCWVESWMTRPDLQPGYEGWQALDPTPQEKSEGTYCCGPVPVRAIKEGDLSTKYDAPFVFAEVNADVVDWIQQDDGS   411
XIII   DNDANLQMD-IFLEEDGNVNSKLTKDSVWNYHCWNEAWMTRPDLPVGFGGWQAVDSTPQENSDGMYRCGPASVQAIKHGHVCFQFDAPFVFAEVNSDLIYITAKKDGT   450
4.2    GTGGRLLIDEYYNEE-GLQNGEGQRGRIWIFQTSTECWMTRPALPQGYDGWQILDPSAPNGGGVLGSCDLVPVRAVKEGTVGLTPAVSDLFAAINASCVVWKCCEDGT   403

K      -FKIVYVEEKAIGTLIVTKAISSNMREDITYLYKHPEGSDAERKAVETAAAHGSKP---NVYANRGSAED---VAMQVE-AQDAVMGQDLMVSVMLINHSSSRRTVKL   612
C      -VHESINRSLIVGLKISTKSVGRDEREDITHTYKYPEGSSEEREAFTRAN-------HLNKLAEK----EETGMAMRIRVGQSMNMGSDFDVFAHITNNTAEEYVCRL   507
XIII   HVVENVDA-THIGKLIVTKQIGGDGMMDITDTYKFQEGQEEERLALETALMYGAKKP-LNTEGVMKSR---SNVDMDFE-VENAVLGKDFKLSITFRNNSHNRYTITA   552
4.2    -LELTDSNTKYVGNNISTKGVGSDRCEDITQNYKYPEGSLQEKEVLERVEKEKMEREKDNGIRPPSLETASPLYLLLKAPSSLPLRG-DAQISVTLVNHSEQEKAVQL   509
                                  @@@@@@@@@@

K      HLYLSVTFYTGVSGTIFKETK-KEVELAPGASDRVTMPVAYKEYRPHLVDQGAMLLNVSGHVKESGQ---VLAKQHTFRLRTPDLSLTLLGAAVVGQECEVQIVFKNP   716
C      LLCARTVSYNGILGPECGTKYLLNLTLEPFSEKSVPLCILYEKYRDCLTESN--LIKVRALLVEPVINSYLLAERDLY-LENPEIKIRILGEPKQKRKLVAEVSLQNP   612
XIII   YLSANITFYTGVPKAEFKKETF-DVTLEPLSFKKEAVLIQAGEYMGQLLEQASLHFFVTARINETRD---VLAKQKSTVLTIPEIIIKVRGTQVVGSDMTVTVQFTNP   656
4.2    AIGVQAVHYNGVLAAKLWRKKL-HLTLSANLEKIITIGLFFSNFERNPPENT--FLRLTAMATHSESNLSCFA-QEDIAICRPHLAIKMPEKAEQYQPLTASVSLQNS   613

K      LPWTLTNVVFRLEGSGLQRP-KILNVGD-IGGNETVTLRQSFVPVRPGPRQLIASLDSPQLSQVHGVIQVDVAPAPGDGGFFSDAGGDSHLGETIPMASRGGA   817
C      LPVALEGCTFTVEGAGLTEEQKTVEIPDPVEAGEEVKVRMDLVPLHMGLHKLVVNFESDKLKAVKGFRNVII--GPA   687
XIII   LKETLRNVWVHLDGPGVTRPMKKM-FRE-IRPNSTVQWEEVCRPWVSGHRKLIASMSSDSLRHVYGELDVQIQRRPSM   732
4.2    LDAPMEDCVISILGRGLIHRERSYRFRSVWPEN-TMCAKFQFTPTHVGLQRLTVEVDCNMFQNLTNYKSVTVVAPELSA   691
```

FIGURE 9 Deduced amino acid sequences of human transglutaminases and erythrocyte band 4.2. K, transglutaminase K; C, transglutaminase C; XIII, factor XIII*a*. *, active site region containing a cysteine (!) involved in the formation of the thioesteracyl–transglutaminase intermediate complex; @, putative Ca^{2+} binding site Amino acids conserved at the same position in the four proteins are given in bold face.

TABLE 4 **Synonyms for the Different Transglutaminases**

Type	Synonym
Transglutaminase C	Tissue transglutaminase; cellular transglutaminase; cytoplasmic transglutaminase; endothelial transglutaminase; type II transglutaminase; liver transglutaminase; liver surface transglutaminase; erythrocyte transglutaminase
Transglutaminase K	Keratinocyte transglutaminase; particulate transglutaminase; type I transglutaminase; membrane-associated transglutaminase
Transglutaminase E	Epidermal transglutaminase; callus transglutaminase; bovine snout transglutaminase
Factor XIIIa	Plasma transglutaminase; Laki-Lorand factor; fibrinoligase; fibrin-stabilizing factor

with extensive crosslinking of spectrin, ankyrin, band 3, and actin (Lorand *et al.*, 1983).

A later theory postulates that the principal role of transglutaminase C may be its participation in apoptosis, which occurs, for example, during embryogenesis, normal tissue turnover, and tumor regression (Fesus *et al.*, 1989, 1991; Knight *et al.*, 1991) and shares some common features with terminal keratinocyte differentiation (Reichert and Fesus, 1991). In hepatocytes and two human cancer cell lines, induction and activation of transglutaminase C has been directly correlated with the formation of apoptotic bodies (Fesus *et al.*, 1987; Piacentini *et al.*, 1991a,b).

The elevated levels of transglutaminase C crosslinked β-crystallin subunits in cataract tissue of the human lens (Lorand *et al.*, 1981; Lorand, 1988) can be regarded as a physiological "false step" indicating that a precise regulation of transglutaminase C activation is required to avoid acci-

TABLE 5 **Some Characteristic Properties of the Different Transglutaminases and Erythrocyte Band 4.2**

Type	Molecular mass (kDa)	Number of amino acids	mRNA (kb)	Chromosomal location
Transglutaminase C	84[a], 77[b]	690	3.6	?
Transglutaminase K	90[a]	788	2.9	14
Transglutaminase E	72[a]	?	3.3	?
Factor XIII*a*	83[b]	731	3.8	6p24–25
Band 4.2	72[a], 77[b]	691	2.35	?

[a]From SDS–polyacrylamide electrophoresis.
[b]From deduced amino acid sequence.

dents. Interestingly, this enzyme is the only transglutaminase whose activity is (in addition to Ca^{2+}) also regulated by GTP (Achyuthan *et al.*, 1987; Schmidt *et al.*, 1988b) and other nucleotides (Bergamini *et al.*, 1987; Kawashima, 1991). GTP binding to transglutaminase C causes a conformational change, which inactivates the enzyme and prevents its proteolysis by trypsin. That Ca^{2+} ions interfere negatively with GTP binding suggests that both participate in the control of transglutaminase C activity (Achyuthan and Greenberg, 1987; Bergamini *et al.*, 1987).

In vitro activation of transglutaminase C in a transformed human keratinocyte cell line (SV-K14 cells) by artificially increasing intracellular Ca^{2+} levels to nonphysiological levels with a Ca^{2+}-ionophore, results in the formation of intracellular crosslinked protein conglomerates, called "pseudo-envelopes," and is accompanied by cell death (Schmidt *et al.*, 1988a).

Transglutaminase K

Transglutaminase K (keratinocyte transglutaminase) has been cloned by several groups (Phillips, M. A. *et al.*, 1990; Kim *et al.*, 1991; Polakowska *et al.*, 1991a; Yamanishi *et al.*, 1991) and the corresponding gene is localized in chromosome 14 (Polakowska *et al.*, 1991b). Its deduced amino acid sequence (788 residues) shows a homology with related transglutaminases (see Fig. 9). Transglutaminase K was first identified in a particulate fraction of epidermal keratinocytes (Sun and Green, 1976) before experiments to determine its precise location confirmed its association with the plasma membrane (Schmidt *et al.*, 1985; Thacher and Rice, 1985). Pulse–chase experiments have shown that transglutaminase K is first synthesized as a cytosoluble enzyme (Rice *et al.*, 1992) before fatty acid acylation (with palmitate and myristate) provides a hydrophobic anchor for its association with the plasma membrane (Chakravarty and Rice, 1989). Acylation occurs at a cluster of five cysteine residues near the amino terminus (Phillips, M. A. *et al.*, 1990); this has been confirmed by site-directed mutagenesis. Deletion or replacement of these cysteines by alanine or serine leave the enzyme in its soluble state (Rice *et al.*, 1992). Mild trypsinization of purified plasma membranes releases a catalytically active proteolytic fragment with a molecular mass of 10 kDa smaller than the Triton X-100 solubilized transglutaminase K (Schmidt *et al.*, 1987; 1988b). The remaining 10-kDa membrane fragment was later identified as the transglutaminase K membrane anchorage region (Chakravarty and Rice, 1989; Rice *et al.*, 1990). The membrane anchorage region of transglutaminase K is subject to phorbol ester stimulated phosphorylation, which occurs at serine residues near the site of fatty acid acylation and could affect its interaction with substrates (Chakravarty *et al.*, 1990; Phillips, M. A. *et al.*, 1990).

The primary physiological role of transglutaminase K, which is expressed in many stratifying epithelia (Parenteau *et al.*, 1986), is the synthesis of cornified envelopes during the process of terminal keratinocyte differentiation (Schmidt *et al.*, 1985; Thacher and Rice, 1985). In cultured keratinocytes, transglutaminase K expression (and thus envelope formation) is affected by retinoids (Lichti *et al.*, 1985; Schmidt *et al.*, 1985; Rubin and Rice, 1986), which control transglutaminase K synthesis at the pretranslational level (Floyd and Jetten, 1989; Michel *et al.*, 1989).

In normal human epidermis, transglutaminase K is expressed and can be detected by immunofluorescence staining in the plasma membrane of keratinocytes ranging from the middle spinous layer to the beginning of the stratum corneum. Enzymatic activity, however, is restricted to one or at the most two cell layers at the interface of the stratum granulosum and stratum corneum (Michel and Démarchez, 1988). In psoriatic lesions (Bernard *et al.*, 1986) as well as in epidermis reconstructed *in vitro* (Asselineau *et al.*, 1989) one observes the precocious appearance of transglutaminase K in the first suprabasal layers. Esmann *et al.* (1989) described increased transglutaminase activity in psoriatic skin.

Table 6 compares some characteristic properties of transglutaminases C and K.

Transglutaminase E

Transglutaminase E ("epidermal transglutaminase") was first isolated and purified from bovine snout epidermis (Buxman and Wuepper, 1975) and from human hair follicle-free epidermis (Ogawa and Goldsmith, 1976). The enzyme has an apparent molecular mass of about 50 kDa and no crossreac-

TABLE 6 **A Comparison between Transglutaminases C and K**

	Transglutaminase C	Transglutaminase K
Localization	Cytosol	Plasma membrane
Function	Apoptosis	Envelope formation
Immunoreactivity	Anti-GPL[a]	B.C1[b]
Molecular mass	84 kDa[c]	92 kDa[c]
Activation by Ca^{2+}	Sigmoidal	Hyperbolic
AC_{50}[d]	235 m*M*	75 m*M*
Regulation by GTP	Yes	No

[a]Rabbit polyclonal antibody raised against guinea pig liver transglutaminase C (C. Miller, London).

[b]Monoclonal antibody raised against human transglutaminase K (S. Thacher, Texas A & M College).

[c]Molecular mass estimation by SDS–polyacrylamide electrophoresis using phosphorylase *b* (97.5 kDa) and BRL prestained bovine serum albumin (67 kDa) as the standards.

[d]Ca^{2+} concentration eliciting 50% of maximum transglutaminase activity.

tion of anti-human transglutaminase E antibodies with human hair follicle transglutaminase is observed (Ogawa and Goldsmith, 1977).[2] Different mechanisms of regulation (Negi *et al.*, 1981) and activation (Negi and Ogawa, 1981) were proposed until it became evident that the 50-kDa protein is the proteolytic cleavage product of transglutaminase E, which possesses an apparent molecular mass of 72 kDa in humans (Negi *et al.*, 1985), 77.8 $\pm$ 0.7 kDa in guinea pigs (Kim *et al.*, 1990), and 90 kDa in mice (Martinet *et al.*, 1988). The zymogen form of guinea pig transglutaminase E (protransglutaminase E) has been further characterized by Kim *et al.* (1990). These authors report that treatment of the proenzyme with dispase, proteinase K, trypsin, or thrombin gives rise to the active form of the enzyme. Transglutaminase E formed by the action of dispase is indistinguishable in size from the zymogen. Under denaturing conditions, however, transglutaminase E dissociates into 50 and 27 kDa fragments. Since reducing agents are not needed for its dissociation, a noncovalent association of the two peptide chains is proposed. Negi *et al.* (1990) suggest that the 72-kDa transglutaminase E and its 50-kDa proteolytic cleavage product may mutually regulate enzymatic activity.

Since in cultured human keratinocytes no expression of transglutaminase E is detectable even though these cells are able to synthesize cornified envelopes (Park *et al.*, 1988), the exact role of this enzyme, which shows high affinity for loricrin (P. Steinert, personal communication), in envelope formation has yet to be identified (for one hypothesis see earlier discussion).

Inhibitors of Transglutaminase-Catalyzed Protein Crosslinking

Primary amines such as methylamine and dansylcadaverine, which act as competitive inhibitors, have been used to inhibit transglutaminase activity (Fig. 10) (Chuang, 1981; Slaughter *et al.*, 1982). However, conditions are not ideal—they interfere with other enzymes; they are active only at elevated concentrations (in the millimolar range); and they inhibit cell attachment (Cornwell *et al.*, 1983).

The active-site-directed inhibitor cystamine has been used to block envelope formation in keratinocytes (Rice and Green, 1979). Birckbichler *et al.* (1981) observed an increase in proliferation markers after transglutaminase inhibition with cystamine in WI-38 human lung cells.

Other observations of inhibition include the following.

1. Alkyl isocyanates are covalent inhibitors of transglutaminases (Gross *et al.*, 1975) but have the disadvantage of lacking specificity.

[2]Later studies of Martinet *et al.* (1988), however, suggest that epidermal and hair follicle transglutaminase in newborn mice are two different processing products of the same protein and that production of the proper forms of the enzymes may be essential to the formation of mature cornified envelopes and hair shafts, respectively.

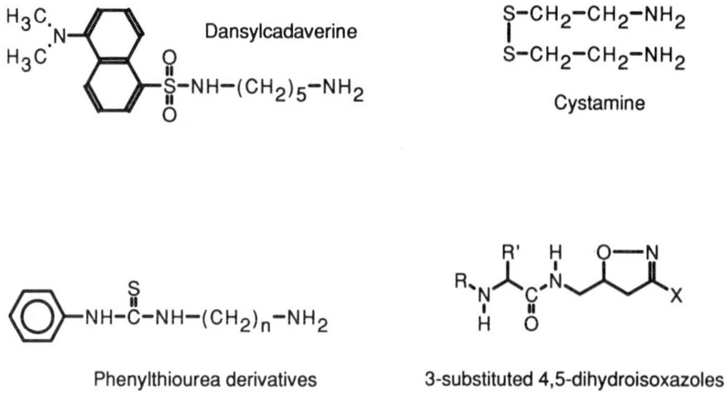

FIGURE 10 Some representative inhibitors of transglutaminase-mediated protein crosslinking.

2. Lee *et al.* (1985) developed a series of α,ω-diaminoalkane derivatives of phenylthiourea that are competitive inhibitors of transglutaminase C and factor XIII; Lorand *et al.* (1987) reported that 2-[3-(diallylamino) propionyl]-benzothiophene inhibits protein crosslinking in erythrocytes and platelets.
3. A new class of mechanism-based inhibitors has been described (Goldsmith *et al.*, 1988, 1991; Killackey *et al.*, 1989). These tyrosinamidomethyl dihydrohaloisoxazoles inhibit transglutaminases irreversibly and thus reduce envelope formation in human keratinocytes *in vivo* and *in vitro*.

Figure 10 presents the representative structures of compounds that inhibit transglutaminase-catalyzed protein crosslinking.

Factors Modulating Envelope Formation

Calcium Ions

Very little is known about the regulation of envelope formation *in vivo*. Experiments performed by Michel and Démarchez (1988) provide some useful information about distribution and activity of transglutaminase K in human epidermis. Indirect immunofluorescence staining of human skin sections using a monoclonal antibody raised against purified transglutaminase K reveal its presence from the middle spinous layer to the beginning of the stratum corneum. Incubation of the sections with dansylcadaverine (a fluorescent transglutaminase K substrate) (Buxman and Wuepper, 1978) in the presence of Ca^{2+} ions shows that under these conditions the pattern of transglutaminase K distribution and activity are identical. Subcutaneous injection of dansylcadaverin into human skin transplanted on nude mice, however, demonstrates that despite the presence of transglutaminase K in the lower layers of the epidermis, its activity *in vivo* is restricted to one or at most two cell layers at the interface of the stratum granulosum and stratum corneum. The absence of activity is not due to a lack of substrate molecules but very likely caused by intracellular Ca^{2+} levels that are insufficient to activate the enzyme. These findings are in good agreement with the report of Steinert and Idler (1979) who demonstrated that only traces of ϵ-(γ-glutamyl)lysine bonds are detectable in the lower epidermis.

In vitro experiments reveal a distinct response of transglutaminases K and C to Ca^{2+} activation (Schmidt *et al.*, 1988a). Transglutaminase C exhibits sigmoidal activation kinetics with an AC_{50} of about 0.2 m*M*, whereas the kinetics for transglutaminase K is hyperbolic with an AC_{50} of 0.07 m*M*.

In cultured keratinocytes, changes in Ca^{2+} concentration do not only affect cell proliferation and differentiation but also modulate transglutaminase K expression (Hennings *et al.*, 1981) and thus envelope formation. In low Ca^{2+} medium, transglutaminase expression is either greatly reduced or the enzyme is not expressed at all (Lichti *et al.*, 1985; Rubin and Rice, 1986). A shift from high (1–2 m*M*) to low (0.025 m*M*) Ca^{2+} in the medium causes a gradual depletion of transglutaminase K, whereas a change from low to high Ca^{2+} stimulates expression (Rubin and Rice, 1988). Pillai and Bikle (1991) have shown that increasing the Ca^{2+} content of the medium, which induces keratinocyte differentiation and envelope formation, is associated with an increase in intracellular Ca^{2+} levels.

Loricrin, a major component of the cornified envelope, is expressed late in epidermal differentiation. Its expression in cultured keratinocytes is Ca^{2+} dependent (Magnaldo *et al.*, 1990; Hohl *et al.*, 1991a). Ca^{2+} concentrations above 0.1 m*M* are required for the expression of loricrin mRNA; maximum mRNA levels are observed at 0.35 m*M* (Hohl *et al.*, 1991a).

Vitamin A and Its Derivatives

Vitamin A and its derivatives, the retinoids, are important modulators of cell proliferation and differentiation (Scott *et al.*, 1982; Roberts and Sporn, 1984; Jetten, 1985). Skin is a major target organ for retinoids, which are actually used for the treatment of diseased skin (Peck, 1984; Schaefer and Reichert, 1990). In cultured human epidermal keratinocytes, retinoids modulate (among other things) keratin expression and envelope formation (Fuchs and Green, 1981; Green and Watt, 1982; Yuspa *et al.*, 1982,1983; Nagae *et al.*, 1987).

In 1982, S. H. Yuspa *et al.* reported the paradoxical finding that in cultured keratinocytes, retinoic acid increases transglutaminase activity while at the same time inhibits envelope formation (Yuspa *et al.*, 1982). The identification of distinct transglutaminases in keratinocytes and their differential regulation by retinoids explained this observation (Lichti *et al.*, 1985; Jetten and Shirley, 1986). In the presence of retinoic acid, transglutaminase C synthesis in cultured keratinocytes and many other cell types is upregulated (Davies *et al.*, 1985; Piacentini *et al.*, 1988a; Rubin and Rice, 1986; Gil *et al.*, 1990; Cai *et al.*, 1991), whereas transglutaminase K and thus envelope formation is suppressed in keratinocytes *in vitro* (Lichti *et al.*, 1985; Schmidt *et al.*, 1985; Thacher *et al.*, 1985; Rubin and Rice, 1986). Retinoic acid, whose action is mediated by the corresponding nuclear receptors (Denning and Verma, 1991), augments transcription of the transglutaminase C gene (Chiocca *et al.*, 1988; 1989; Suedhoff *et al.*, 1990) and represses transcription of the transglutaminase K gene (Floyd and Jetten, 1989; Michel *et al.*, 1989). The fact that retinoids control transglutaminase K expression and thus envelope formation in cultured keratinocytes has led to the development of bioassays to determine biological activity using either envelope formation (Michel *et al.*, 1988a; Hough-Monroe and Milstone, 1991) or transglutaminase K expression (Michel *et al.*, 1991) as indicator.

At concentrations of 1–100 n*M*, retinoic acid completely blocks the Ca^{2+}-induced synthesis of loricrin in cultured keratinocytes when the two are administered simultaneously. Its addition to cultures already exposed to high Ca^{2+} results in complete depletion of loricrin mRNA within 48 to 72 hours (Hohl *et al.*, 1991a). Similar results are obtained when human epidermis (reconstructed *in vitro* by growing keratinocytes on dermal equivalents) is exposed to 1 μ*M* retinoic acid (Magnaldo *et al.*, 1991). Interestingly, involucrin, another major envelope precursor protein, is not at all affected by retinoids (Green and Watt, 1982).

Other Modulators

Treatment of epidermal keratinocytes with tumor promotors of the phorbol ester family stimulates differentiation in low Ca^{2+} medium and induces transglutaminase K (Jeng *et al.*, 1985; Jetten *et al.*, 1989; Lichti *et al.*, 1985). The addition of promotor to keratinocytes cultured in high Ca^{2+} medium (1–2 m*M*) increases phosphorylation of transglutaminase K (Chakravarty *et al.*, 1990).

Sodium butyrate, a potent inducer of morphological and biochemical cell differentiation (Wright, 1973; Leder and Leder, 1975; Prasad, 1980) increases the rate of spontaneous envelope formation in cultured normal human keratinocytes (Schmidt *et al.*, 1989; Staiano-Coico *et al.*, 1989) via induction of transglutaminase K. At the same time, sodium butyrate completely abolishes the inhibitory effect of retinoic acid on transglutaminase K expression and envelope formation without affecting other markers of keratinocyte differentiation such as keratins and the enzyme cholesterol sulfotransferase, which are also under the control of retinoids (Schmidt *et al.*, 1989; Jetten *et al.*, 1990). In PC 12 pheochromocytoma cells, sodium butyrate induces transglutaminase C (Byrd and Lichti, 1987).

In mouse keratinocytes, α-1,25-dihydroxyvitamin D_3, the active form of vitamin D_3, induces transglutaminase via alternate mechanisms to those of retinoic acid and in this way modifies epidermal differentiation (Lee *et al.*, 1989). Similar observations have been made in chick epidermal cells with the glucocorticoid hydrocortisone (Obinata and Endo, 1977, 1979).

In cultured human keratinocytes, the ability to spontaneously synthesize cornified envelopes is related to cellular cholesterol content, which increases shortly after culture confluence parallel to the observed increase in spontaneous envelope formation. Supplementation of culture medium with inhibitors of cholesterologenesis suppresses envelope formation without affecting transglutaminase K expression or activity (Schmidt *et al.*, 1991).

Summary

Although our knowledge of the molecular mechanisms involved in envelope formation is still rudimentary and to a large extent based on indirect evidence, some major actors in the scenario come to light:

1. a few principal precursor proteins such as involucrin and loricrin, which may have an organizing function due to their unique physicochemical properties;
2. two distinct transglutaminases (K and E), which may act sequentially;
3. Ca^{2+}, whose intracellular level may tune the activity and substrate specificity of these transglutaminases and thus their concerted action; and
4. specific lipids, which are covalently attached to the protein envelope and may play a role in the organization of the intercellular lipid layer and in corneocyte cohesion.

The sketch in Figure 11 outlines a possible scheme. In an initiation step activated by Ca^{2+} ions, transglutaminase K crosslinks a membrane-bound anchorage protein (e.g., the 195 kDa protein) with a cytosolic precursor protein (e.g., involucrin) or with a polyamine (e.g., spermidine). In the prolongation phase, further membrane-bound and cytosolic presursor molecules are crosslinked in a two-dimensional fashion until the inner surface of the keratinocyte membrane is covered by protein plaques, which in turn are crosslinked with each other to result in a primary envelope. Depending on the conditions (i.e., physiological or pathological state), nonspecific proteins ("dust-bin proteins") may be incorporated into this network, particularly when the availability of specific precursors is limited. Furthermore, transglutaminase K becomes more and more immobilized or even becomes integrated into the envelope, while the original plasma membrane is progressively destroyed. It is very possible that envelope formation finishes at this stage in conventionally cultured keratinocytes and gives rise to the fragile type of envelope.

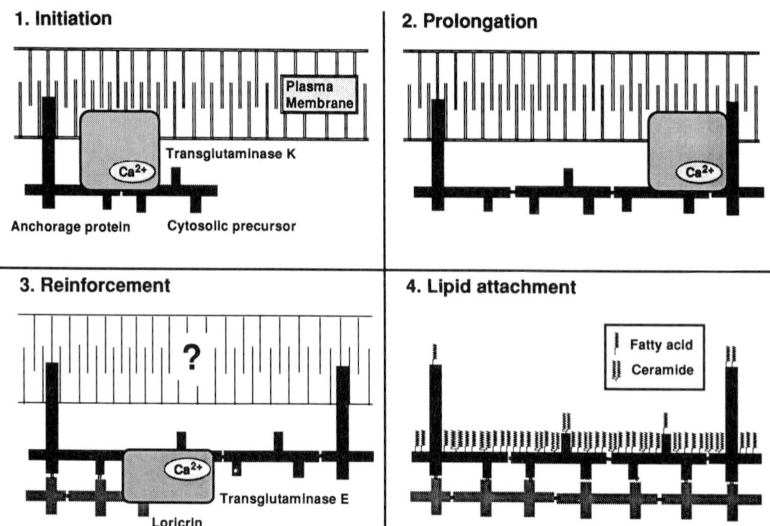

FIGURE 11 A working model for cornified envelope formation.

In vivo, however, two additional processes take place that lead to the mature rigid envelope type: (1) the reinforcement of the inner surface of the envelope by the attachment of loricrin and perhaps other proteins such as keratolinin; as well as (2) the addition of a covalent lipid envelope to the outer surface.

The cytosoluble transglutaminase E catalyzes the first process after the conversion of the zymogen to the active enzyme by limited proteolysis. As a result of the high cysteine content of loricrin, the inner belt of the envelope is stabilized by a large number of disulfide bonds. Keratin fibers are connected to this belt either covalently by isopeptide bonding or noncovalently by interaction between the Ω loops of keratin and loricrin.

The enzymes involved in the covalent linkage of hydroxyceramides and fatty acids to the outer surface of the protein envelope have not yet been identified. However, it is likely that they originate from lamellar bodies and are secreted together with lipids. Because of its high content of glutamate and aspartate residues, involucrin is an ideal candidate to accept the huge number of lipid molecules esterified to the envelope. Furthermore, esterification may mask reactive epitopes at the envelope surface and thus lead to the weak response with anti-involucrin antibodies observed in histological skin sections.

This working model, though preliminary, should provide a useful basis for further experiments in the elucidation of the molecular mechanisms and regulation of envelope formation.

Addendum

Since this chapter has gone to press, *cornifin,* a new protein precursor of the cornified envelope has been detected by the isolation of and analysis of several cDNA clones that encode mRNAs highly expressed in differentiating rabbit tracheal epithelial (RbTE) cells (Marvin *et al.*, 1992). Among them was clone SQ37 which, in the predicted coding region, exhibited a high degree of homology with the sequence of the human small proline-rich (*spr-1*) protein 1 (Kartasova and van de Putte, 1988; Kartasova *et al.*, 1988).

Based on the deduced amino acid sequence, cornifin has a molecular mass of 14 kDa. Polyclonal antibodies raised against the repeated sequence of the C terminus (SQ37A) and against a unique N-terminal peptide (SQ37B) reacted with a single protein after the electrophoretic separation and immunoblot analysis of total cellular proteins from squamous-differentiated RbTE cells. The apparent molecular mass was 15 kDa under nonreducing conditions and about 23 kDa under reducing conditions, indicating conformational changes after cleavage of intramolecular disulfide bonds. The corresponding molecular masses of the SQ37 homologue from normal human keratinocytes were 10 and 15 kDa, respectively.

Cornifin is particularly rich in proline (31%), glutamine (20%), lysine (13%) and cysteine (11%). At its C terminus, cornifin contains 13 repeats of the consensus octapeptide *glu-pro-cys-gln-pro-lys-val-pro*. The high content in lysine and glutamine makes cornifin a potential candidate for transglutaminase-catalysed protein crosslinking. Furthermore, its high proline content could very well compensate for a certain deficit in proline that is found when the amino acid composition of the known envelope precursors (see Table 3, p. 116), particularly loricrin, is compared with that of assembled cornified envelopes (see Table 1, p. 111).

Several lines of experimental evidence support the view that cornifin is a transglutaminase substrate and cornified envelope precursor:

1. The polyclonal antibodies SQ37A and SQ37B decorated the cytoplasm of the spinous and granular layers in human epidermis and rabbit esophagus and tongue, but not the basal and cornified layers. SQ37A reacted with fragmented crosslinked envelopes.
2. Induction of protein crosslinking in confluent RbTE cells with a Ca^{2+} ionophore shifted the immunoreactivity of electrophoretically separated proteins to high molecular mass species. At the same time a loss of total immunoreactivity was observed, which may indicate masking of the epitope after protein crosslinking. Addition of EDTA or putrescine inhibited these processes.
3. When total extracts of human keratinocytes or RbTE cells were incubated with dansyl cadaverine, this amine and cornifin were covalently linked.

Backendorf and Hohl (1992) discuss the possibility that the SPR proteins involucrin and loricrin may be evolutionally related. All of them have "a similar genomic organization: short first exon, one single intron and a second exon, which contains the entire open reading frame. The internal domains, specific for each of these genes, are characterized by the multiple reiteration of specific peptide motifs." All of them map to human chromosome 1q21. "This clustered organization might indicate that these genes were created by gene duplication of a common ancestor and have diverged by evolving internal domains specific for each protein."

References

Abernethy, J. L., Hill, R. L., and Goldsmith, L. A. (1977). ϵ-(γ-Glutamyl)lysine crosslinks in human stratum corneum. *J. Biol. Chem.* **252,** 1837–1839.

Achyuthan, K. E., and Greenberg, C. S. (1987). Identification of a guanosine triphosphate-binding site on guinea pig liver transglutaminase. *J. Biol. Chem.* **262,** 1901–1906.

Achyuthan, K. E., Mary, A., and Greenberg, C. S. (1988). The binding sites on fibrin(ogen) for guinea pig transglutaminase are similar to those of blood coagulation factor XIII. Characterization of the binding of liver transglutaminase to fibrin. *J. Biol. Chem.* **263,** 14296–14301.

Asquith, R. S., Otterburn, M. S., Buchanan, J. H., Cole, M., Fletcher, J. C., and Gardner, K. L. (1970). The identification of *N*-(γ-glutamyl)-L-lysine crosslinks in native wool keratins. *Biochim. Biophys. Acta* **207,** 342–348.

Asselineau, D., Bernard, B. A. Bailly, C., and Darmon, M. (1989). Retinoic acid improves epidermal morphogenesis. *Dev. Biol.* **133,** 322–335.

Backendorf, C., and Hohl, D. (1992). A common origin of cornified envelope proteins. *Nature Genetics* **2,** 91.

Baden, H. P., Kubilus, J., and Phillips, S. B. (1987a). Characterization of monoclonal antibodies generated to the cornified envelope of human cultured keratinocytes. *J. Invest. Dermatol.* **89,** 454–459.

Baden, H. P., Kubilus, J., Phillips, S. B., Kvedar, J. C., and Tahan, S. R. (1987b). A new class of soluble basic protein precursors of the cornified envelope of mammalian epidermis. *Biochim. Biophys. Acta* **925,** 63–73.

Barsigian, C., Stern, A. M., and Martinez, J. (1991). Tissue (type II) transglutaminase covalently incorporates itself, fibrinogen, or fibronectin into high-molecular-weight complexes on the extracellular surface of isolated hepatocytes. Use of 2-[(2-oxopropyl)thio] imidazolium derivatives as cellular transglutaminase inactivators. *J. Biol. Chem.* **266,** 22501–22509.

Beninati, S., and Folk, J. E. (1988). Covalent polyamine–protein conjugates: analysis and distribution. *Adv. Exp. Med. Biol.* **231,** 79–94.

Bergamini, C. M., Signorini, M., and Poltronieri, L. (1987). Inhibition of erythrocyte transglutaminase by GTP. *Biochim. Biophys. Acta* **916,** 149–151.

Bernard, B. A., Reano, A., Darmon, Y. M., and Thivolet, J. (1986). Precocious appearance of involucrin and epidermal transglutaminase during differentiation of psoriatic skin. *Br. J. Dermatol.* **114,** 279–283.

Birckbichler, P. J., Orr, G. R., Conway, E., and Patterson, M. K. (1977). Transglutaminase activity in normal and transformed cells. *Cancer Res.* **37,** 1340–1344.

Birckbichler, P. J., Orr, G. R., Patterson, M. K., Conway, E., and Carter, H. A. (1981). Increase in proliferative markers after inhibition of transglutaminase. *Proc. Natl. Acad. Sci. U.S.A.* **78,** 5005–5008.

Buxman, M. M., and Wuepper, K. D. (1975). Keratin crosslinking and epidermal transglutaminase. A review with observations on the histochemical and immunochemical localization of the enzyme. *J. Invest. Dermatol.* **65,** 107–112.

Buxman, M. M., and Wuepper, K. D. (1976). Isolation, purification and characterization of bovine epidermal transglutaminase. *Biochim. Biophys. Acta* **452,** 356–369.

Buxman, M. M., and Wuepper, K. D. (1978). Cellular localization of epidermal transglutaminase: a histochemical and immunological study. *J. Histochem. Cytochem.* **26,** 340–348.

Buxman, M. M., Buehner, G. E., and Wuepper, K. D. (1976). Isolation of substrates of epidermal transglutaminase from bovine epidermis. *Biochem. Biophys. Res. Commun.* **73,** 470–478.

Buxman, M. M., Lobitz, C. J., and Wuepper, K. D. (1980). Epidermal transglutaminase. Identification and purification of a soluble substrate with studies of *in vitro* cross-linking. *J. Biol. Chem.* **255,** 1200–1203.

Byrd, J. C., and Lichti, U. (1987). Two types of transglutaminase in the PC12 pheochromocytoma cell line. Stimulation by sodium butyrate. *J. Biol. Chem.* **262,** 11699–11705.

Cai, D., Ben, T., and De Luca, L. M. (1991). Retinoids induce tissue transglutaminase in NIH-3T3 cells. *Biochem. Biophys. Res. Commun.* **175,** 1119–1124.

Chakravarty, R., and Rice, R. H. (1989). Acylation of keratinocyte transglutaminase by palmitic and myristic acids in the membrane anchorage region. *J. Biol. Chem.* **264,** 625–629.

Chakravarty, R., Rong, X., and Rice, R. H. (1990). Phorbolester-stimulated phosphorylation of keratinocyte transglutaminase in the membrane anchorage region. *Biochem. J.* **271,** 25–30.

Chiocca, E. A., Davies, P. J. A., and Stein, J. P. (1988). The molecular basis of retinoic acid action. Transcriptional regulation of tissue transglutaminase gene expression in macrophages. *J. Biol. Chem.* **263,** 11584–11589.

Chiocca, E. A., Davies, P. J. A., and Stein, J. P. (1989). Regulation of tissue transglutaminase gene expression as a molecular model for retinoid effects on proliferation and differentiation. *J. Cell. Biochem.* **39,** 293–304.

Chuang, D. M. (1981). Inhibitors of transglutaminase prevent agonist-mediated internalization of beta-adrenergic receptors. *J. Biol. Chem.* **256,** 8291–8293.

Chung, S. I., Lewis, M. S., and Folk, J. E. (1974). Relationships of the catalytic properties of human plasma and platelet transglutaminases (activated blood coagulation factor XIII) to their subunit structures. *J. Biol. Chem.* **249,** 940–950.

Cline, P. R., and Rice, R. H. (1983). Modulation of involucrin and envelope competence in human keratinocytes by hydrocortisone, retinyl acetate, and growth arrest. *Cancer Res.* **43,** 3203–3207.

Cornwell, M. M., Juliano, R. L., and Davies, P. J. (1983). Inhibition of the adhesion of Chinese hamster ovary cells by the naphthylsulfonamides dansylcadaverine and *N*-(6-aminohexyl)-5-chloro-1-naphthylenesulfonamide (W7). *Biochim. Biophys. Acta* **762,** 414–419.

Davies, P. J. A., Murtaugh, M. P., Moore, W. T., Johnson, G. S., and Lucas, D. (1985). Retinoic acid-induced expression of tissue transglutaminase in human promyelocytic leukemia (HL-60) cells. *J. Biol. Chem.* **260,** 5166–5174.

Denning, M. F., and Verma, A. K. (1991). Involvement of retinoic acid nuclear receptors in retinoic acid-induced tissue transglutaminase gene expression in rat tracheal 2C5 cells. *Biochem. Biophys. Res. Commun.* **175,** 344–350.

Djian, P., and Green, H. (1991). Involucrin gene of tarsioids and other primates: Alternatives in evolution of the segment of repeats. *Proc. Natl. Acad. Sci. U.S.A.* **88,** 5321–5325.

Eckert, R. L., and Green, H. (1986). Structure and evolution of the human involucrin gene. *Cell* **46,** 583–589.

Esmann, J., Voorhees, J. J., and Fisher, G. J. (1989). Increased membrane-associated transglutaminase activity in psoriasis. *Biochem. Biophys. Res. Commun.* **164,** 219–224.

Etoh, Y., Simon, M., and Green, H. (1986). Involucrin acts as a transglutaminase substrate at multiple sites. *Biochem. Biophys. Res. Commun.* **136**, 51–56.

Farbman, A. I. (1966). Plasma membrane changes during keratinization. *Anat. Rev.* **156**, 269–282.

Fesus, L., Thomazy, V., and Falus, A. (1987). Induction and activation of tissue transglutaminase during programmed cell death. *FEBS (Fed. Eur. Biochem. Soc.) Lett.* **224**, 104–108.

Fesus, L., Thomazy, V., Autuori, F., Ceru, M. P., Tarcsa, E., and Piacentini, M. (1989). Apoptotic hepatocytes become insoluble in detergents and chaotropic agents as a result of transglutaminase action. *FEBS (Fed. Eur. Biochem. Soc.) Lett.* **245**, 150–154.

Fesus, L., Tarcsa, E., Kedei, N., Autuori, F., and Piacentini, M. (1991). Degradation of cells dying by apoptosis leads to accumulation of ε(γ-glutamyl)lysine isodipeptide in culture fluid and blood. *FEBS (Fed. Eur. Biochem. Soc.) Lett.* **284**, 109–112.

Floyd, E. E., and Jetten, A. M. (1989). Regulation of type I (epidermal) transglutaminase mRNA levels during squamous differentiation: down regulation by retinoids. *Mol. Cell. Biol.* **9**, 4846–4851.

Folk, J. E. (1980). Transglutaminases. *Annu. Rev. Biochem.* **49**, 517–531.

Folk, J. E. (1983). Mechanism and basis for specificity of transglutaminase-catalyzed ε-(γ-glutamyl)lysine bond formation. *Adv. Enzymol.* **54**, 1–56.

Folk, J. E., and Finlayson, J. S. (1977). The ε-(γ-glutamyl)lysine crosslink and the catalytic role of transglutaminases. *Adv. Protein Chem.* **31**, 1–133.

Folk, J. E., Park, M. H., Chung, S. I., Schrode, J., Lester, E. P., and Cooper, H. L. (1980). Polyamines as physiological substrates for transglutaminases. *J. Biol. Chem.* **255**, 3695–3700.

Fuchs, E. (1990). Epidermal differentiation. *Curr. Opinion Cell. Biol.* **2**, 1028–1035.

Fuchs, E., and Green, H. (1981). Regulation of terminal differentiation of cultured human keratinocytes by vitamin A. *Cell* **25**, 617–625.

Gentile, V., Saydak, M., Chiocca, E. A., Akande, O., Birckbichler, P. J., Lee, K. N., Stein, J. P., and Davies, P. J. A. (1991). Isolation and characterization of cDNA clones to mouse macrophage and human endothelial cell tissue transglutaminases. *J. Biol. Chem.* **266**, 478–483.

Gil, D., Chandraratna, R., Breen, T., Arefieg, T., Marler, D., Henry, E., Basilion, J., and Davies, P. (1990). Selective induction of transglutaminase by retinoid analogs. *J. Cell Biol.* **344a**, 1921–1939.

Goldsmith, L. A., Baden, H. P., Roth, S. I., Colman, R., Lee, L., and Fleming, B. (1974). Vertebral epidermal transamidases. *Biochim. Biophys. Acta* **351**, 113–125.

Goldsmith, L. A., Falciano, V., and DeYoung, L. M. (1988). Novel transglutaminase inhibitors decrease cell envelopes in human keratinocyte cultures. *J. Invest. Dermatol.* **90**, 564.

Goldsmith, L. A., DeYoung, L. M., Falciano, V., Ballaron, S. J., and Akers, W. (1991). Inhibition of human epidermal transglutaminases *in vitro* and *in vivo* by tyrosinamidomethyl dihydrohaloisoxazoles. *J. Invest. Dermatol.* **97**, 156–158.

Green, H., and Watt, F. M. (1982). Regulation by vitamin A of envelope crosslinking in cultured keratinocytes derived from different human epithelia. *Mol. Cell. Biol.* **2**, 1115–1117.

Greenberg, C. S., Birckbichler, P. J., and Rice, R. H. (1991). Transglutaminases: Multifunctional crosslinking enzymes that stabilize tissues. *FASEB J.* **5**, 3071–3077.

Gross, M., Whetzel, N. K., and Folk, J. E. (1975). Alkyl isocyanates as active site-directed inactivators of guinea pig liver transglutaminase. *J. Biol. Chem.* **250**, 7693–7699.

Grundmann, U., Amann, E., Zettlmeissl, G., and Kupper, H. A. (1986). Characterization of cDNA coding for human factor XIII*a*. *Proc. Natl. Acad. Sci. U.S.A.* **83**, 8024–8028.

Haftek, M., Serre, G., Mils, V., and Thivolet, J. (1991). Immunocytochemical evidence for a possible role of crosslinked envelopes in stratum corneum cohesion. *J. Histochem. Cytochem.* **39**, 1531–1538.

Hanigan, H., and Goldsmith, L. A. (1978). Endogenous substrates for epidermal transglutaminase. *Biochim. Biophys. Acta* **522,** 589–601.

Harding, H. W. J., and Rogers, G. E. (1971). ε-(γ-Glutamyl)lysine crosslinkage in citrulline-containing protein fractions from hair. *Biochemistry* **10,** 624–630.

Hashimoto, K. (1969). Cellular envelopes of keratinized cells of the human epidermis. *Arch. Klin. Exp. Derm.* **235,** 374–385.

Hennings, H., Steinert, P., and Buxman, M. M. (1981). Calcium induction of transglutaminase and the formation of ε-(γ-glutamyl)lysine crosslinks in cultured mouse epidermal cells. *Biochem. Biophys. Res. Commun.* **102,** 739–745.

Hohl, D. (1990). Cornified cell envelope. *Dermatologica* **180,** 201–211.

Hohl, D., Lichti, U., Breitkreutz, D., Steinert, P. M., and Roop, D. R. (1991a). Transcription of the human loricrin gene *in vitro* is induced by calcium and cell density and suppressed by retinoic acid. *J. Invest. Dermatol.* **96,** 414–418.

Hohl, D., Mehrel, T., Lichti, U., Turner, M. L., Roop, D. R., and Steinert, P. M. (1991b). Characterization of human loricrin. Structure and function of a new class of epidermal cell envelope proteins. *J. Biol. Chem.* **266,** 6626–6636.

Hough-Monroe, L., and Milstone, L. M. (1991). Quantitation of crosslinked protein: An alternative to counting cornified envelopes as an index of keratinocyte differentiation. *Anal. Biochem.* **199,** 25–28.

Ichinose, A., Hendrickson, L. E., Fujikawa, K., and Davie, E. W. (1986). Amino acid sequence of the *a* subunit of human factor XIII. *Biochemistry* **25,** 6900–6906.

Ikura, K., Nasu, T., Yokota, H., Tsuchiya, Y., Sasaki, R., and Chiba, H. (1988). Sequence of guinea pig transglutaminase from cDNA sequence. *Biochemistry* **27,** 2898–2905.

Jeng, A. Y., Lichti, U., Strickland, J. E., and Blumberg, P. M. (1985). Similar effects of phospholipase C and phorbol ester tumor promoters on primary mouse epidermal cells. *Cancer Res.* **45,** 5714–5721.

Jetten, A. M. (1985). Retinoids and their modulation of cell growth. *In* "Growth and Maturation Factors" (G. Guroff, ed.), Vol. 3, pp. 221–293. Wiley Interscience, New York.

Jetten, A. M., and Shirley, J. E. (1986). Characterization of transglutaminase activity in rabbit tracheal epithelial cells. Regulation by retinoids. *J. Biol. Chem.* **261,** 15097–15101.

Jetten, A. M., George, M. A., Pettit, G. R., and Rearick, J. I. (1989). Effects of bryostatins and retinoic acid on phorbol ester- and diacylglycerol-induced squamous differentiation in human tracheobronchial epithelial cells. *Cancer Res.* **49,** 3990–3995.

Jetten, A. M., George, M. A., and Rearick, J. I. (1990). Down-regulation of squamous cell-specific markers by retinoids: Transglutaminase type I and cholesterol sulfotransferase. *Methods Enzymol.* **190,** 42–48.

Juprelle-Soret, M., Wattiaux-De Coninck, S., and Wattiaux, R. (1988). Subcellular localization of transglutaminase. Effect of collagen. *Biochem. J.* **250,** 421–427.

Kartasova, T. and van de Putte, P. (1988). Isolation, characterization, and UV-stimulated expression of two families of genes encoding polypeptides of related structure in human epidermal keratinocytes. *Mol. Cell. Biol.* **8,** 2195–2203.

Kartasova, T., van Muijen, G. N. P., van Pelt-Heerschap, H., and van de Putte, P. (1988). Novel protein in human epidermal keratinocytes: regulation of expression during differentiation. *Mol. Cell. Biol.* **8,** 2204–2210.

Kawashima, S. (1991). Inhibition of rat liver transglutaminase by nucleotides. *Experientia* **47,** 709–712.

Killackey, J. J. F., Bonaventura, B. J., Castelhano, A. L., Billedeau, R. J., Farmer, W., Deyoung, L., Krantz, A., and Pliura, D. H. (1989). A new class of mechanism-based inhibitors of transglutaminase enzymes inhibits the formation of crosslinked envelopes by human malignant keratinocytes. *Mol. Pharmacol.* **35,** 701–706.

Kim, H. C., Lewis, M. S., Gorman, J. J., Park, S. C., Girard, J. E., Folk, J. E., and Chung, S. I. (1990). Protransglutaminase E from guinea pig skin—Isolation and partial characterization. *J. Biol. Chem.* **265,** 21971–21978.

Kim, H. C., Idler, W. W., Kim, I. G., Han, J. H., Chung, S., and Steinert, P. M. (1991). The complete amino acid sequence of the human transglutaminase K enzyme deduced from the nucleic acid sequences of cDNA clones. *J. Biol. Chem.* **266,** 536–539.

Knight, C. R. L., Rees, R. C., and Griffin, M. (1991). Apoptosis: a potential role for cytosolic transglutaminase and its importance in tumour progression. *Biochim. Biophys. Acta.* **1096,** 312–318.

Korsgren, C., and Cohen, C. M. (1991). Organization of the gene for human erythrocyte membrane protein 4.2: structural similarities with the gene for the *a* subunit of factor XIII. *Proc. Natl. Acad. Sci. U.S.A.* **88,** 4840–4844.

Korsgren, C., Lawler, J., Lambert, S., Speicher, S., and Cohen, C. M. (1990). Complete amino acid sequence and homologies of human erythrocyte band 4.2. *Proc. Natl. Acad. Sci. U.S.A.* **87,** 613–617.

Kubilus, J., and Baden, H. P. (1982). Isolation of two immunologically related transglutaminase substrates from cultured human keratinocytes. *In Vitro* **18,** 447–455.

Kubilus, J., Kvedar, J., and Baden, H. P. (1987). Identification of new components of the cornified envelope of human and bovine epidermis. *J. Invest. Dermatol.* **89,** 44–50.

Kubilus, J., Phillips, S. B., Goldaber, M. A., Kvedar, J. C., and Baden, H. P. (1990). Involucrin-like proteins in nonprimates. *J. Invest. Dermatol.* **94,** 210–215.

Lavker, R. M. (1976). Membrane coating granules: the fate of the discharged lamellae. *J. Ultrastruct. Res.* **55,** 79–86.

Leder, A., and Leder, P. (1975). Butyric acid, a potent inducer of erythroid differentiation in cultured erythroleukemic cells. *Cell* **5,** 319–322.

Lee, K. N., Fesus, L., Yancey, S. T., Girard, J. E., and Chung, S. I. (1985). Development of selective inhibitors of transglutaminase. *J. Biol. Chem.* **260,** 14689–14694.

Lee, S. C., Ikai, K., Ando, Y., and Imamura, S. (1989). Effects of 1α,25-dihydroxyvitamin D_3 on the transglutaminase activity of transformed mouse epidermal cells in culture. *J. Dermatol.* **16,** 7–11.

Legrain, V., Michel, S., Ortonne, J. P., and Reichert, U. (1991). Intra- and inter-individual variations of cornified envelope peptide composition in normal and psoriatic skin. *Arch. Dermatol. Res.* **283,** 512–515.

Lichti, U., Ben, T., and Yuspa, S. H. (1985). Retinoic acid-induced transglutaminase in mouse epidermal cells is distinct from epidermal transglutaminase. *J. Biol. Chem.* **260,** 1422–1426.

Lobitz, C. J., and Buxman, M. M. (1982). Characterization and localization of bovine epidermal transglutaminase substrate. *J. Invest. Dermatol.* **78,** 150–154.

Lorand, L. (1986). Activation of blood coagulation factor XIII. *Ann. N.Y. Acad. Sci.* **485,** 144–158.

Lorand, L. (1988). Transglutaminase-mediated crosslinking of proteins and cell aging: The erythrocyte and lens models. *In* "Advances in Post-Translational Modifications of Proteins and Aging" (V. Zappia, P. Galletti, R. Porta, and F. Wold, eds.), Vol. 231, pp. 79–94. Plenum Press, New York.

Lorand, L., Downey, J., Gotoh, T., Jacobson, A., and Tokura, S. (1968). The transpeptidase system which crosslinks fibrin by γ-glutamyl-ϵ-lysine bonds. *Biochem. Biophys. Res. Commun.* **31,** 222–229.

Lorand, L., Hsu, L. K. H., Siefring, G. E., and Rafferty, N. S. (1981). Lens transglutaminase and cataract formation. *Proc. Natl. Acad. Sci. U.S.A.* **78,** 1356–1360.

Lorand, L., Bjerrum, O. J., Hawkins, M., Lowe-Krentz, L., and Siefring, G. E. Jr. (1983). Degradation of transmembrane proteins in ca^{2+}-enriched human erythrocytes. An immunochemical study. *J. Biol. Chem.* **258,** 5300–5305.

Lorand, L., Barnes, N., Brunar-Lorand, J. A., Hawkins, M., and Michaslka, M. (1987). Inhibition of protein crosslinking in Ca^{2+}-enriched human erythrocytes and activated platelets. *Biochemistry* **26,** 308–313.

Ma, A. S. P., and Sun, T. T. (1986). Differentiation-dependent changes in the solubility of a 195-kDa protein in human epidermal keratinocytes. *J. Cell. Biol.* **103,** 41–48.

Magnaldo, T., Pommes, L., Asselineau, D., and Darmon, M. (1990). Isolation of a CG-rich cDNA identifying mRNA present in human epidermis and modulated by calcium and retinoic acid in cultured keratinocytes. Homology with murine loricrin mRNA. *Mol. Biol. Rep.* **14,** 237–249.

Magnaldo, T., Bernerd, F., Asselineau, D., and Darmon, M. (1991). Expression of loricrin is negatively controlled by retinoic acid in human epidermis reconstructed *in vitro. Differentiation,* **49,** 39–46.

Manabe, M., Hirotani, T., Negi, M., Hattori, M., and Ogawa, H. (1981). Isolation and characterization of the membraneous fraction in human stratum corneum. *J. Dermatol.* **8,** 329–333.

Martinet, N., Kim, H. C., Girard, J. E., Nigra, T. P., Strong, D. H., Chung, S. I., and Folk, J. E. (1988). Epidermal and hair follicle transglutaminases. *J. Biol. Chem.* **263,** 4236–4241.

Martinet, N., Beninati, S., Nigra, T. P., and Folk, J. E. (1990). *N*-(1,8)-Bis(γ-glutamyl)spermidine crosslinking in epidermal cell envelopes. Comparison of crosslink levels in normal and psoriatic cell envelopes. *Biochem. J.* **271,** 305–308.

Martinez, I. R., and Peters, A. (1971). Membrane-coating granules and membrane modifications in keratinizing epithelia. *Am. J. Anat.* **130,** 93–120.

Marvin, K. W., George, M. D., Fujimoto, W., Saunders, N. A., Bernacki, S. H., and Jetten, A. M. (1992). Cornifin, a cross-linked envelope precursor in keratinocytes that is down-regulated by retinoids. *Proc. Natl. Acad. Sci. U.S.A.* **89,** 11026–11030.

Matacic, S., and Loewy, A. G. (1968). The identification of isopeptide crosslinks in insoluble fibrin. *Biochem. Biophys. Res. Commun.* **30,** 356–362.

Matoltsy, A. G. (1977). The membrane of horny cells. *In* "Biochemistry of Cutaneous Epidermal Differentiation" (M. Seiji and I. A. Bernstein, eds.), pp. 93–109, University of Tokyo Press, Tokyo.

Matoltsy, A. G., and Balsamo, C. A. (1955). A study of the components of the cornified epithelium in the skin. *J. Biophys. Biochem. Cytol.* **1,** 339–360.

Matoltsy, A. G., and Matoltsy, M. N. (1966). The membrane proteins of horny cells. *J. Invest. Dermatol.* **46,** 127–129.

Mehrel, T., Hohl, D., Rothnagel, J. A., Longley, M. A., Bundman, D., Cheng, C., Lichti, U., Bisher, M. E., Steven, A. C., Steinert, P. M., Yuspa, S. H., and Roop, D. R. (1990). Identification of a major keratinocyte cell envelope protein, loricrin. *Cell* **61,** 1103–1112.

Michel, S., and Démarchez, M. (1988). Localization and *in vivo* activity of epidermal transglutaminase. *J. Invest. Dermatol.* **90,** 472–474.

Michel, S., and Juhlin, L. (1990). Cornified envelopes in congenital disorders of keratinization. *Br. J. Dermatol.* **122,** 15–21.

Michel, S., and Reichert, U. (1992). L'enveloppe cornée: une structure caractéristique des cornéocytes. *Rev. Eur. Dermatol. MST,* **4,** 9–17.

Michel, S., Schmidt, R., Robinson, S. M., Shroot, B., and Reichert, U. (1987). Identification and subcellular distribution of cornified envelope precursor proteins in the transformed human keratinocyte line SV-K14. *J. Invest. Dermatol.* **88,** 301–305.

Michel, S., Regnier, M., and Shroot, B. (1988a). Cornified envelope formation, an *in vitro* assay of retinoid activity. *Models Dermatol.* **4,** 40–44.

Michel, S., Schmidt, R., Shroot, B., and Reichert, U. (1988b). Morphological and biochemical characterization of the cornified envelopes from human epidermal keratinocytes of different origin. *J. Invest. Dermatol.* **91,** 11–15.

Michel, S., Reichert, U., Isnard, J. L., Shroot, B., and Schmidt, R. (1989). Retinoic acid controls expression of epidermal transglutaminase at the pretranslational level. *FEBS (Fed. Eur. Biochem. Soc.) Lett.* **258,** 35–38.

Michel, S., Courseaux, A., Miquel, C., Bernardon, J. M., Schmidt, R., Shroot, B., Thacher,

S. M., and Reichert, U. (1991). Determination of retinoid activity by an enzyme-linked immunosorbent assay. *Anal. Biochem.* **192**, 232–236.

Nagae, S., Lichti, U., De Luca, L., and Yuspa, S. H. (1987). Effect of retinoic acid on cornified envelope formation: difference between spontaneous envelope formation *in vivo* or *in vitro* and expression of envelope competence. *J. Invest. Dermatol.* **89**, 51–58.

Negi, M., and Ogawa, H. (1981). Possible activation mechanisms of transglutaminase in epidermis. *Arch. Dermatol. Res.* **271**, 101–105.

Negi, M., Matsui, T., and Ogawa, H. (1981). Mechanism of regulation of human epidermal transglutaminase. *J. Invest. Dermatol.* **77**, 389–392.

Negi, M., Colbert, M. C., and Goldsmith, L. A. (1985). High-molecular-weight human epidermal transglutaminase. *J. Invest. Dermatol.* **85**, 75–78.

Negi, M., Park, J. K., and Ogawa, H. (1990). Alteration of human epidermal transglutaminase during its activation. *J. Dermatol. Sci.* **1**, 167–172.

Obinata, A., and Endo, H. (1977). Induction of epidermal transglutaminase by hydrocortisone in chick embryonic skin. *Nature* **270**, 440–441.

Obinata, A., and Endo, H. (1979). Induction of chick epidermal transglutaminase by hydrocortisone *in ovo* and *in vitro* with reference to the differentiation of epidermal cells. *J. Biol. Chem.* **254**, 8487–8490.

Ogawa, H., and Goldsmith, L. A. (1976). Human epidermal transglutaminase. Preparation and properties. *J. Biol. Chem.* **251**, 7281–7288.

Ogawa, H., and Goldsmith, L. A. (1977). Human epidermal transglutaminase. II. Immunologic properties. *J. Invest. Dermatol.* **68**, 32–35.

Parenteau, N. L., Pilato, A., and Rice, R. H. (1986). Induction of keratinocyte type-I transglutaminase in epithelial cells of the rat. *Differentiation* **33**, 130–141.

Parenteau, N. L., Eckert, R. L., and Rice, R. H. (1987). Primate involucrins: Antigenic relatedness and detection of multiple forms. *Proc. Natl. Acad. Sci. U.S.A.* **84**, 7571–7575.

Park, S. C., Kim, S. Y., Kim, H. C., Thacher, S., and Chung, S. I. (1988). Differential expression of transglutaminases in human foreskin and cultured keratinocytes (abstract). *J. Cell Biol.* **107**, 139a.

Peck, G. L. (1984). Synthetic retinoids in dermatology. In "The Retinoids" (Sporn, M. B. Roberts, A. B., Goodman, D. S., eds.), Vol. 2, pp. 209–287. Academic Press, Orlando.

Peterson, L. L., and Wuepper, K. D. (1984). Epidermal and hair follicle transglutaminases and crosslinking in skin. *Mol. Cell Biochem.* **58**, 99–111.

Phillips, M. A., Stewart, B. E., Qin, Q., Chakravarty, R., Floyd, E. E., Jetten, A. M., and Rice, R. H. (1990). Primary structure of keratinocyte transglutaminase. *Proc. Natl. Acad. Sci.* **87**, 9333–9337.

Phillips, S. B., Kubilus, J., Grassi, A. M., Goldaber, M., and Baden, H. P. (1990). The pancornulins: a group of basic low-molecular-weight proteins in mammalian epidermis and epithelium that may function as cornified envelope precursors. *Comp. Biochem. Physiol.* **95B**, 781–788.

Piacentini, M., Fesus, L., Sartori, C., and Ceru, M. P. (1988a). Retinoic acid-induced modulation of rat liver transglutaminase and total polyamines *in vivo*. *Biochem. J.* **253**, 33–38.

Piacentini, M., Martinet, N., Beninati, S., and Folk, J. E. (1988b). Free and protein-conjugated polyamines in mouse epidermal cells. *J. Biol. Chem.* **263**, 3790–3794.

Piacentini, M., Autuori, F., Dini, L., Farrace, M. G., Ghibelli, L., Piredda, L., and Fesus, L. (1991a). "Tissue" transglutaminase is specifically expressed in neonatal rat liver cells undergoing apoptosis upon epidermal growth factor-stimulation. *Cell Tissue Res.* **263**, 227–235.

Piacentini, M., Fesus, L., Farrace, M. G., Ghibelli, L., Piredda, L., and Melino, G. (1991b). The expression of "tissue" transglutaminase in two human cancer cell lines is related with the programmed cell death (apoptosis). *Eur. J. Cell. Biol.* **54**, 246–254.

Pillai, S., and Bikle, D. D. (1991). Role of intracellular-free calcium in the cornified envelope formation of keratinocytes: differences in the mode of action of extracellular calcium and 1,25-dihydroxyvitamin D_3. *J. Cell. Physiol.* **146,** 94–100.

Pisano, J. J., Finlayson, J. S., and Peyton, M. P. (1969). Chemical and enzymatic detection of protein crosslinks. Measurement of ϵ-(γ-glutamyl)lysine in fibrin polymerized by factor XIII. *Biochemistry* **8,** 871–876.

Polakowska, R., and Goldsmith, L. A. (1991). The cell envelope and transglutaminase. *In* "Physiology, Biochemistry and Molecular Biology of the Skin" (L. A. Goldsmith, ed.), pp. 168–201. Oxford University Press, New York.

Polakowska, R., Herting, E., and Goldsmith, L. A. (1991a). Isolation of cDNA for human epidermal type I transglutaminase. *J. Invest. Dermatol.* **96,** 285–288.

Polakowska, R. R., Eddy, R. L., Shows, T. B., and Goldsmith, L. A. (1991b). Epidermal type I transglutaminase (TGM1) is assigned to human chromosome 14. *Cytogenet. Cell. Genet.* **56,** 105–107.

Prasad, K. N. (1980). Butyric acid with diverse biological functions. *Life Sci.* **27,** 1351–1358.

Qutaishat, S. S., and Kumar, V. (1988). Cell envelopes of psoriatic scales bear antigen(s) of antigenic determinants different from callus. *J. Invest. Dermatol.* **90,** 600–600.

Reichert, U., and Fesus, L. (1991). Programmed cell death. *Retinoids Today and Tomorrow* **24,** 31–34.

Rice, R. H., and Green, H. (1977). The cornified envelope of terminally differentiated human epidermal keratinocytes consists of crosslinked protein. *Cell* **11,** 417–422.

Rice, R. H., and Green, H. (1978). Relation of protein synthesis and transglutaminase activity to formation of the crosslinked envelope during terminal differentiation of the cultured human epidermal keratinocyte. *J. Cell Biol.* **76,** 705–711.

Rice, R. H., and Green, H. (1979). Presence in human epidermal cells of a soluble protein precursor of the crosslinked envelope: Activation of the crosslinking by calcium ions. *Cell* **18,** 681–694.

Rice, R. H., Rong, X., and Chakravarty, R. (1990). Proteolytic release of keratinocyte transglutaminase. *Biochem. J.* **265,** 351–357.

Rice, R. H., Mehrpouyan, M., O'Callahan, W., Parenteau, N. L., and Rubin, A. L. (1992). Keratinocyte transglutaminase: differentiation marker and member of an extended family. *Epithelial Cell Biol.,* **1,** 128–137.

Richards, S., Scott, I. R., Harding, C. R., Liddell, J. E., Powell, G. M., and Curtis, C. G. (1988). Evidence for filaggrin as a component of the cell envelope of the newborn rat. *Biochem. J.* **253,** 153–160.

Roberts, A. B., and Sporn, M. B. (1984). Cellular biology and biochemistry of the retinoids. In "The Retinoids" (Sporn MB; Roberts AB; Goodman DS, eds.), Vol. 2, pp. 209–287. Academic Press, Orlando.

Rubin, A. I., and Rice, R. H. (1986). Differential regulation by retinoic acid and calcium of transglutaminases in cultured neoplastic and normal human keratinocytes. *Cancer Res.* **46,** 2356–2361.

Rubin, A. L., and Rice, R. H. (1988). Characterization of calcium sensitivity of differentiation in SCC-13 human squamous carcinoma cells. *In Vitro Cell. Dev. Biol.* **24,** 857–861.

Schaefer, H., and Reichert, U. (1990). Retinoids and their perspectives in dermatology. Les Rétinoïdes et leurs perspectives en dermatologie. *Nouv. Dermatol.* **9,** 3–6.

Schmidt, R., Reichert, U., Michel, S., Shroot, B., and Bouclier, M. (1985). Plasma membrane transglutaminase and cornified envelope competence in cultured human keratinocytes. *FEBS (Fed. Eur. Biochem. Soc.) Lett.* **186,** 201–204.

Schmidt, R., Reichert, U., Demarchez, M., Shroot, B., and Michel, S. (1987). *In vitro* and *in vivo* studies on epidermal transglutaminases. Noble Conference in Cellular and Molecular Biology. Transglutaminase and Protein Crosslinking Reactions. Miami, Florida.

Schmidt, R., Michel, S., Shroot, B., and Reichert, U. (1988a). Plasma membrane transglutaminase and cytosolic transglutaminase form distinct envelope-like structures in transformed human keratinocytes. *FEBS (Fed. Eur. Biochem. Soc.) Lett.* **229,** 193–196.

Schmidt, R., Michel, S., Shroot, B., and Reichert, U. (1988b). Transglutaminases in normal and transformed human keratinocytes. *J. Invest. Dermatol.* **90,** 475–479.

Schmidt, R., Cathelineau, C., Cavey, M. T., Dionisius, V., Michel, S., Shroot, B., and Reichert, U. (1989). Sodium butyrate selectively antagonizes the inhibitory effect of retinoids on cornified envelope formation in cultured human keratinocytes. *J. Cell. Physiol.* **140,** 281–287.

Schmidt, R., Parish, E. J., Dionisius, V., Cathelineau, C., Michel, S., Shroot, B., Rolland, A., Brzokewicz, A., and Reichert, U. (1991). Modulation of cellular cholesterol and its effect on cornified envelope formation in cultured human epidermal keratinocytes. *J. Invest. Dermatol.* **97,** 771–775.

Schwartz, M. L., Pizzo, S. V., Hill, R. L., and McKee, P. A. (1973). Human factor XIII from plasma and platelets. Molecular weights, subunit structures, proteolytic activation, and crosslinking of fibrinogen and fibrin. *J. Biol. Chem.* **248,** 1395–1407.

Scott, K. F. F., Meyskens, F. L., and Russel, D. H. (1982). Retinoids increase transglutaminase activity and inhibits ornithine decarboxylase activity in chinese hamster ovary cells and in melanoma cells stimulated to differentiate. *Proc. Natl. Acad. Sci. U.S.A.* **79,** 4093–4097.

Serre, G., Mils, V., Haftek, M., Vincent, C., Croute, F., Reano, A., Ouhayoun, J. P., Bettinger, S., and Soleilhavoup, J. P. (1991). Identification of late differentiation antigens of human cornified epithelia, expressed in reorganized desmosomes and bound to crosslinked envelope. *J. Invest. Dermatol.* **97,** 1061–1072.

Simon, M., and Green, H. (1984). Participation of membrane-associated proteins in the formation of the crosslinked envelope of the keratinocyte. *Cell* **36,** 827–834.

Simon, M., and Green, H. (1985). Enzymatic crosslinking of involucrin and other proteins by keratinocyte particulates in vitro. *Cell* **40,** 677–683.

Simon, M., and Green, H. (1988). The glutamine residues reactive in transglutaminase-catalyzed crosslinking of involucrin. *J. Biol. Chem.* **263,** 18093–18098.

Simon, M., and Green, H. (1989). Involucrin in the epidermal cells of subprimates. *J. Invest. Dermatol.* **92,** 721–724.

Slaughter, R. S., Smart, C. E., Wong, D.-.S., and Lever, J. E. (1982). Lysosomotropic agents and inhibitors of cellular transglutaminase stimulate dome formation, a differentiated characteristic of MDCK kidney epithelial cell cultures. *J. Cell. Physiol.* **112,** 141–147.

Staiano-Coico, L., Helm, R. E., McMahon, C. K., Pagan-Charry, I., LaBruna, A., Piraino, V., and Higgins, P. J. (1989). Sodium-*N*-butyrate induces cytoskeletal rearrangements and formation of cornified envelopes in cultured adult human keratinocytes. *Cell Tissue Kinet.* **22,** 361–375.

Steinert, P. M., and Idler, W. W. (1979). Postsynthetic modification of mammalian epidermal keratin. *Biochemistry* **18,** 5664–5669.

Steven, A. C., Bisher, M. E., Roop, D. R., and Steinert, P. M. (1990). Biosynthetic pathways of filaggrin and loricrin—Two major proteins expressed by terminally differentiated epidermal keratinocytes. *J. Struct. Biol.* **104,** 150–162.

Suedhoff, T., Birckbichler, P. J., Lee, K. N., Conway, E., and Patterson, M. K. (1990). Differential expression of transglutaminase in human erythroleukemia cells in response to retinoic acid. *Cancer Res.* **50,** 7830–7834.

Sun, T. T., and Green, H. (1976). Differentiation of the epidermal keratinocyte in cell culture: formation of the cornified envelope. *Cell* **9,** 511–521.

Sung, L. A., Chien, S., Chang, L. S., Lambert, K., Bliss, S. A., Bouhassira, E. E., Nagel, R. L., Schwartz, R. S., and Rybicki, A. C. (1990). Molecular cloning of human protein 4.2: a major component of the erythrocyte membrane. *Proc. Natl. Acad. Sci. U.S.A.* **87,** 955–959.

Swartzendruber, D. C., Wertz, P. W., Madison, K. C., and Downing, D. T. (1987). Evidence that the corneocyte has a chemically bound lipid envelope. *J. Invest. Dermatol.* **88,** 709–713.

Swartzendruber, D. C., Kitko, D. J., Wertz, P. W., Madison, K. C., and Downing, D. T. (1988). Isolation of corneocyte envelopes from porcine epidermis. *Arch. Dermatol. Res.* **280,** 424–429.

Szodoray, L. (1930). Beiträge zur Eiweiβstruktur des Hautepithels. *Arch. Dermatol. Syphilis* **159,** 605–610.

Takahashi, M., and Tezuka, T. (1989). Characterization of a 15 k-Da protein as a novel substrate of transglutaminase (abstract). *J. Invest. Dermatol.* **92,** 526.

Takahashi, N., Takahashi, Y., and Putnam, F. W. (1986). Primary structure of blood coagulation factor XIII*a* (fibrinoligase, transglutaminase) from human placenta. *Proc. Natl. Acad. Sci. U.S.A.* **83,** 8019–8023.

Taylor-Papadimitriou, J., Purkis, P., Lane, E. B., McKay, I. A., and Chang, S. E. (1982). Effect of SV40 transformation on the cytoskeleton and behavioural properties of human keratinocytes. *Cell Differ.* **11,** 169–180.

Tezuka, T., and Takahashi, M. (1987). The cystine-rich envelope protein from human epidermal stratum corneum cells. *J. Invest. Dermatol.* **88,** 47–51.

Thacher, S. M., and Rice, R. H. (1985). Keratinocyte-specific transglutaminase of cultured human epidermal cells: relation of crosslinked envelope formation and terminal differentiation. *Cell* **40,** 685–695.

Thacher, S. M., Coe, E. L., and Rice, R. H. (1985). Retinoid suppression of transglutaminase activity and envelope competence in cultured human epidermal carcinoma cells. *Differentiation* **29,** 82–87.

Thomazy, V., and Fesus, L. (1989). Differential expression of tissue transglutaminase in human cells. A immunohistochemical study. *Cell Tissue Res.* **255,** 215–224.

Tseng, H., and Green, H. (1988). Remodeling of the involucrin gene during primate evolution. *Cell* **54,** 491–496.

Tseng, H., and Green, H. (1990). The involucrin genes of pig and dog: comparison of their segments of repeats with those of prosimians and higher primates. *Mol. Biol. Evol.* **7,** 293–302.

Upchurch, H. F., Conway, E., Patterson Jr., M. K., Birckbichler, P. J., and Maxwell, M. D. (1987). Cellular transglutaminase has affinity for extracellular matrix. *In Vitro Cell. Dev. Biol.* **23,** 795–800.

Warhol, M. J., Roth, J., Lucocq, J. M., Pinkus, G. S., and Rice, R. H. (1985). Immunoultrastructural localization of involucrin in squamous epithelium and cultured keratinocytes. *J. Histochem. Cytochem.* **33,** 141–149.

Watt, F. M., and Green, H. (1981). Involucrin synthesis is correlated with cell size in human epidermal cultures. *J. Cell Biol.* **90,** 738–742.

Weisberg, L. J., Shiu, D. T., Conkling, P. R., and Shuman, M. A. (1987). Identification of normal peripheral blood monocytes and liver as sites of synthesis of coagulation factor XIII *a* chain. *Blood* **70,** 579–582.

Wertz, P. W., Swartzendruber, D. C., Abraham, W., Madison, K. C., and Downing, D. T. (1987). Essential fatty acids and epidermal integrity. *Arch. Dermatol.* **123,** 1381–1384.

Wertz, P. W., Madison, K. C., and Downing, D. T. (1989a). Covalently bound lipids of human stratum corneum. *J. Invest. Dermatol.* **92,** 109–111.

Wertz, P. W., Swartzendruber, D. C., Kitko, D. J., Madison, K. C., and Downing, D. T. (1989b). The role of the corneocyte lipid envelopes in cohesion of the stratum corneum. *J. Invest. Dermatol.* **93,** 169–172.

Wolpl, A., Lattke, H., Board, P. G., Arnold, R., Schmeiser, T., Kubanek, B., Robin-Winn, M., Pichelmayr, R., and Goldmann, S. F. (1987). Coagulation factor XIII *a* and *b* subunits in bone marrow and liver. *Transplantation* **43,** 151–153.

Wright, J. (1973). Morphology and growth rate changes in chinese hamster cells cultured in the presence of sodium butyrate. *Exp. Cell Res.* **78,** 456–460.

Yamanishi, K., Min Liew, F., Konishi, K., Yasuno, H., Doi, H., Hirano, J., and Fukushima, S. (1991). Molecular cloning of human epidermal transglutaminase cDNA from keratinocytes in culture. *Biochem. Biophys. Res. Commun.* **175,** 906–913.

Yuspa, S. H., Ben, T., and Steinert, P. (1982). Retinoic acid induces transglutaminase activity but inhibits cornification of cultured epidermal cells. *J. Biol. Chem.* **257,** 9906–9908.

Yuspa, S. H., Ben, T., and Lichti, U. (1983). Regulation of epidermal transglutaminase activity and terminal differentiation by retinoids and phorbol esters. *Cancer Res.* **43,** 5707–5712.

Zettergren, J. G., Peterson, L. L., and Wuepper, K. D. (1984). Keratolinin: the soluble substrate of epidermal transglutaminase from human and bovine tissue. *Proc. Natl. Acad. Sci. U.S.A.* **81,** 238–242.

Zhou, X. M., Idler, W. W., Steven, A. C., Roop, D. R., and Steinert, P. M. (1988). The complete sequence of the human intermediate filament chain keratin 10. *J. Biol. Chem.* **263,** 15584–15589.

5

Loricrin

Daniel Hohl and Dennis Roop

Introduction

Terminal differentiation of epidermal cells begins with the migration of cells from the basal layer, continues with the progression of cells through the spinous and granular layers, and terminates with the formation of mature epidermal cells (i.e., squames, scales, or corneocytes) of the stratum corneum. This process has been shown by morphological and biochemical studies to occur in stages (Matoltsy, 1975). Keratins K5 and K14 are major products of basal epidermal cells (Woodcock-Mitchell *et al.*, 1982). These proteins assemble into intermediate filaments and, with microtubules (tubulin) and microfilaments (actin), comprise the cytoskeleton of epidermal cells (Steinert and Roop, 1988). One of the earliest changes associated with the commitment to differentiation and migration into the spinous layer is the induction of a differentiation-specific pair of keratins, K1 and K10. Intermediate filaments containing K1 and K10 replace those containing K5 and

Molecular Biology of the Skin: The Keratinocyte

K14 as the major products of cells in the spinous layer (Fuchs and Green, 1980; Woodcock-Mitchell *et al.*, 1982; Roop *et al.*, 1983; Schweizer *et al.*, 1984). In the granular layer of the epidermis, a nonintermediate filament protein, filaggrin, is synthesized as a high-molecular-weight precursor, which after processing is thought to promote keratin filament aggregation and disulfide bond formation (Dale *et al.*, 1978; Steinert *et al.*, 1981; Harding and Scott, 1983).

At approximately the same time, a set of proteins is deposited on the inside of the plasma membrane, which collectively constitutes a structure 10–20 nm in thickness termed the cornified cell envelope (CE; synonymous with "crosslinked cell envelope," "marginal," or "peripheral band;" see Polakowska and Goldsmith, 1991; Hohl, 1990). In addition to these proteins, covalently bound lipids—such as ceramides—contribute to the formation of the "compound CE" which consists of an inner protein envelope and an outer lipid envelope and accounts for about 10% of the dry weight of the CE (Swartzendruber *et al.*, 1988; Wertz *et al.*, 1989). The final terminally differentiated cornified cell consists of a flattened squame possessing a highly resistant compound CE containing the keratin intermediate filament–filaggrin complex.

The Cornified Cell Envelope

The inner protein CE is the most insoluble component of stratified squamous epithelial cells and is usually defined as the remnant after exhaustive extraction with NaOH (Matoltsy and Matoltsy, 1970) or SDS–ME (Sun and Green, 1976). The insolubility was attributed initially to crosslinking by disulfide bonds (Matoltsy and Matoltsy, 1970) and later to the presence of Nε-(γ-glutamyl)lysine isodipeptide bonds as well (Rice and Green, 1977; Polakowska and Goldsmith, 1991) because this crosslink is resistant to cleavage by both conventional protein solvents and harsh chemical reagents (Chung, 1972; Folk, 1983). The formation of this crosslink is catalyzed by transglutaminases; three distinct such enzymes have been detected in the epidermis (Chung, 1972; Rice and Green, 1977; Folk, 1983; Lichti *et al.*, 1985; Thacher and Rice, 1985):

1. cytosoluble, epidermal TGase of about 50 to 56 kDa (Buxman and Wuepper, 1976; Goldsmith *et al.*, 1974; Ogawa and Goldsmith, 1976; Peterson and Buxman, 1981) or 50 and 72–77 kDa, also termed "TGase E," or "pro-TGase E" (Negi *et al.*, 1985; Park *et al.*, 1988; Kim *et al.*, 1991a);
2. particulate, membrane-bound keratinocyte-specific TGase of 92 kDa also termed "type I" (Rice *et al.*, 1988; Thacher and Rice, 1985; Thacher, 1989; Phillips *et al.*, 1990; Kim *et al.*, 1991; Polakowska *et*

al., 1991) and epidermal (Lichti *et al.*, 1985; Schmidt *et al.*, 1988) TGase or "TGase K" (Park *et al.*, 1988).

3. cytosoluble, tissue type TGase of about 80 kDa (Chung, 1972), also termed "type II" (Thacher and Rice, 1985) or "TGase C" (Park *et al.*, 1988).

It is generally agreed that the last enzyme has no function in crosslinkage of the CE in epidermis; is induced by retinoic acid—which is known to inhibit CE formation—and is distributed in a wide variety of tissues (Lichti *et al.*, 1985; Lichti and Yuspa, 1988; Thacher and Rice, 1985).

However, both the cytosoluble TGase E (Goldsmith *et al.*, 1974; Kim *et al.*, 1989) and/or the particulate, membrane-bound TGase K (Rice *et al.*, 1988; Thacher and Rice, 1985; Thacher, 1989; M. A. Phillips *et al.*, 1990; Polakowska *et al.*, 1991) are considered to be involved in the formation of the CE.

At the moment there is no consensus as to which type of enzyme, "keratinocyte" (membrane-bound) or "epidermal" (cytosoluble) is responsible for the formation of the CE *in vivo*. It is also possible that both enzymes are involved and process different substrates sequentially in a very orderly fashion. It is obvious that during crosslinkage of a structure as thick as the CE, the early membrane-bound TGase K becomes progressively more embedded and restricted. Therefore, the presence of a cytosoluble TGase at later stages of differentiation has to be postulated in order to explain the formation of a mature CE 15-nm in thickness (Thacher, 1989). One might hypothesize that at earlier stages of differentiation, TGase K crosslinks mainly involucrin giving raise to the immature, fragile CE. Later, TGase E could promote the formation of a mature, rigid CE using predominantly loricrin as substrate (Hohl, 1990).

In fact, several lines of evidence suggest that the CE is assembled in stages involving immature intermediate forms:

1. Electron microscopic observations demonstrate a gradual progression of envelope formation *in vivo* with the deposition of a moderately dense band next to the plasma membrane in the upper stratum granulosum, which at later stages becomes progressively more dense in the stratum corneum (Farbman, 1966; Steven *et al.*, 1990).
2. In Nomarski contrast microscopy studies, two different types of CEs have been distinguished. CEs recovered by tape stripping from the interface of stratum granulosum and stratum corneum look irregularly shaped and fragile, whereas CEs from upper layers appear polygonal and rigid, indicating that they represent two distinct stages of maturation (Michel *et al.*, 1988).
3. Human foreskin keratinocytes grown as submerged cultures in Dulbecco's medium containing 10% fetal calf serum (Rheinwald and Green, 1975) are only competent to produce CEs of the fragile type.

The amino acid composition of fragile envelopes (Rice and Green, 1979), induced by permeabilization of keratinocytes in such cultures, is very similar to that of regular plasma membranes (Ogawa *et al.*, 1983). However, the amino acid composition determined from mature (rigid) CEs isolated from newborn and adult mice (Mehrel *et al.*, 1990) or human foreskin (Hohl *et al.*, 1991b) keratinocytes *in vivo* differs remarkably and shows (contrary to plasma membranes or fragile CEs) high levels of glycine, serine, and cysteine as well as low levels of alanine, asparagine, aspartic acid, and leucine. These data suggest that formation of the CE occurs as a sequence of events involving the deposition of different substrates at different timepoints.

Precursor Proteins

It appears that the CE is a complex structure in terms of its protein composition. Several proteins have been implicated as precursors of the CE on the basis of their reactivity to antibodies produced against purified CEs (Baden *et al.*, 1987a,b; Kubilus *et al.*, 1987; Michel *et al.*, 1987); their ability to serve as substrates for transglutaminase (Buxman *et al.*, 1980; Hanigan and Goldsmith, 1978; Kubilus and Baden, 1983; Rice and Green, 1979; Richards *et al.*, 1988; Mehrel *et al.*, 1990; Simon and Green, 1984; Simon and Green, 1988; Takahashi and Tezuka, 1989) or to become crosslinked (Baden *et al.*, 1987b; Hohl *et al.*, 1991b; Michel *et al.*, 1987; Piacentini *et al.*, 1988; Simon and Green, 1985) and on the basis of reactivity of antibodies produced against specific proteins with the periphery of cells located in the granular layer and stratum corneum (Hohl *et al.*, 1991b; Lobitz and Buxman, 1982; Ma and Sun, 1986; Rice and Green, 1979; Mehrel *et al.*, 1990; Tezuka and Masae, 1987; Zettergren *et al.*, 1984). The proteins include—besides loricrin—involucrin, keratolinin, the 195-kDa protein, the so-called "cystein-rich envelope protein," sciellin, the "sprs" and the pancornulins and will be discussed elsewhere in this book. It must be stressed that many of these proteins were identified and defined using *in vitro* experiments and that staining of the cell periphery does not constitute proof of staining of the CE.

The plethora of potential CE precursors (see Table 1) and the gross differences of biochemical data from CEs isolated under different conditions has led to the "dustbin hypothesis" (Michel *et al.*, 1987). It proposes that during late stages of epidermal differentiation, waste proteins originating from destroyed intracellular organelles may be reused as building blocks of the CE (Michel *et al.*, 1987), or that crosslinked CEs are composed of those proteins available to the crosslinking transglutaminase at the moment when this enzyme becomes activated (Nagae *et al.*, 1987). Possibly, the dustbin hypothesis is a valuable model to reflect the situation *in vitro* and perhaps in dermatological disorders. However, the dustbin hypothesis is ambiguous as

TGases have been shown to exhibit differences in substrate specificity (Folk, 1983). In addition, there is considerable question as to whether many of the described proteins are CE precursors *in vivo* since it has been demonstrated that even fibroblasts can be induced to form microscopically visible and insoluble crosslinked fragments upon activation of transglutaminase *in vitro* (Simon and Green, 1984). *In vivo,* fibroblasts and fibrocytes do not form a CE. Therefore, proteins crosslinked in fibroblasts under *in vitro* conditions that induce CEs in keratinocytes might represent artifacts with respect to their definition as CE precursors. In addition, it should be explicitly stated that staining of the cell periphery does not constitute proof of staining of the CE. In fact, confirmation that a protein is a precursor for the CE *in vivo* is lacking in most cases. Ultimate proof that a given protein is a precursor of the CE would have to come from a comparison of the known amino acid sequence of this protein with peptides derived from purified mature CEs, preferably those peptides containing a crosslink. To our knowledge, loricrin is the only precursor of the CE for which such experiments have successfully been performed (Hohl *et al.*, 1991b).

The isolation and characterization of loricrin is discussed in more detail in the following sections.

Identification of a Novel Protein

In initial isolations of cDNA clones encoding major proteins expressed at different differentiation states in mouse epidermis, three classes were identified complementary to equally abundant mRNAs of 2.4, 2.0, and 1.6 kb (Roop *et al.*, 1983). On the basis of hybridization selection analysis (Roop *et al.*, 1983) and nucleotide sequence information (Steinert *et al.*, 1983; Steinert *et al.*, 1985), the clones complementary to the 2.4- and 2.0-kb mRNAs were determined to correspond to the differentiation-specific keratins K1 and K10, respectively. On the basis of preliminary hybridization selection data and partial nucleotide sequence analysis, the clones complementary to the 1.6-kb mRNA were presumed to encode a differentiation-specific 55-kDa keratin (Roop *et al.*, 1983). However, analysis of the complete nucleotide sequence of the original clones (pk321 and pk1005) (Roop *et al.*, 1983) and additional clones (Mehrel *et al.*, 1990) failed to reveal a coiled-coil region characteristic of all keratin intermediate filament proteins (Steinert and Roop, 1988). Instead, the sequences possess numerous inexact peptide repeats containing glycine, serine, cysteine, and tyrosine residues reminiscent of the amino- and carboxy-terminal end domains of the keratins K1 (Steinert *et al.*, 1985) and K10 (Steinert *et al.*, 1983). However, the exact peptide repeat sequences are different from the keratins and extend for much longer stretches. Curiously, these peptide repeats are interrupted in two locations by sequences enriched in glutamines and, in addition, the amino- and carboxy-terminal ends also contain clusters of glutamines.

TABLE 1 Proven and Putative Precursor Proteins to the CE

Current designation	Species	Tissue	Extraction buffer	MW	CE Criterium	Reference
	rat, human	epidermis	DTT $CaCl_2$	12–16 kDa high and very high	TGase substrate	Goldsmith, 1977; Hanigan and Goldsmith, 1978
Involucrin	human, primates	stratified squamous epithelia	SDS–ME	68 kDa (deduced from cDNA)	peripheral staining TGase substrate	Rice and Green, 1979; Watt and Green, 1981; Eckert and Green, 1986; Tseng and Green, 1988
	human	cultured keratinocytes	SDS–ME	330 kDa 210 kDa, ks[a] 195 kDa, ks	TGase substrates	Simon and Green, 1984
Involucrin				140 kDa, ks 100 kDa 95 kDa 70 kDa		
Keratolinin	bovine, human	epidermis	NH_4-acetate	36 kDa	TGase substrate	Lobitz and Buxman, 1982; Zettergren *et al.*, 1984
	human	foreskin, esophagus	Tris SDS–ME	195 kDa	peripheral staining	Ma and Sun, 1986

	human	stratified epithelia	Tris DTT or Tris Urea–ME	195 kDa	antibody to CE, peripheral staining, TGase substrate	Kubilus and Baden, 1983; Baden *et al.*, 1987; Kubilus *et al.*, 1987; Phillips *et al.*, 1990; Kvedar *et al.*, 1992
Involucrin	monkey, cow, sheep, dog, rat, guinea pig	epidermis only		143 kDa		
Sciellin				82 kDa		
Pancornulins[b]				25 kDa		
Pancornulins[b]				22 kDa		
Pancornulins[b]				17 kDa		
Pancornulins[b]				15 kDa		
Involucrin	human	cultured SV40 transformed foreskin	Tris SDS–ME Glycerol	140 kDa 90 kDa 61 kDa 53 kDa 36 kDa	cross-linkage, antibody to CE	Michel *et al.*, 1987
CREP	human	plantar epidermis	Tris	16 kDa	peripheral staining	Tezuka and Masae, 1987
Loricrin	mouse	stratified epithelia	SDS–ME (boiled)	38 kDa (deduced from cDNA)	peripheral staining, TGase substrate	Mehrel *et al.*, 1990
Loricrin	human	epidermis	SDS–ME	26 kDa (deduced from cDNA)	sequenced cross-links	Hohl *et al.*, 1991

[a]ks, Keratinocyte-specific.
[b]Identical with spr proteins or cornifins.

From the compiled sequence information, it was possible to calculate that this cDNA encodes a protein with a molecular weight of 37,828 and an estimated pI of 9.5. At this time, an epidermal protein with these characteristics was novel and not yet described. However, Nischt *et al.* (1988) reported a sequence of a partial cDNA clone from mouse footpad epidermis that was extremely GC rich. The authors did not identify the protein encoded by this cDNA, presumably because of disruptions in the open reading frame of their sequence, but it has high homology (>98%) with pk 321 (nucleotides 1031–1734), suggesting that they are identical.

To identify this protein, antibodies were produced against a synthetic peptide corresponding to 14 unique residues at the carboxyl terminus of its sequence. This antiserum was purified by affinity chromatography with the synthetic peptide and immunoprecipitated a prominent protein of approximately 45 kDa from a [^{35}S]cysteine-labeled total protein extract of newborn mouse epidermis. Most interestingly, the antiserum reacted by indirect immunofluorescence with discrete granules in the first layers of the granular compartment and even more abundantly with the periphery of cells in the upper granular and lower cornified layers in newborn mouse epidermis and benign mouse skin tumors (Mehrel *et al.*, 1990). This staining pattern was consistent with that predicted for proteins associated with the CE.

The following observations suggest that this protein is in fact a precursor protein to the CE.

1. Mature envelopes that had been purified from the stratum corneum of adult mouse epidermis fail to react with the antibodies. However, an intense staining was observed after these CEs had been fragmented by sonification. This indicated that the carboxy-terminal epitope was located on the interior surface of CEs (Mehrel *et al.*, 1990).
2. This observation was confirmed by immune electronic microscopy: (i) transverse thin sections of fragmented CEs labeled in solution with the antibodies revealed a heavy decoration with gold particles located > 90% on the inner surface of CEs (Mehrel *et al.*, 1990; (ii) in thin sections of newborn mouse epidermis, gold particles concentrated around the cell periphery throughout the stratum corneum (Steven *et al.*, 1990).
3. Initial experiments demonstrated that the amount of immunoprecipitable protein present after a 3-hr labeling period with [^{35}S]cysteine decreased after an 18-hr chase period. This decrease may have resulted from crosslinking the labeled protein into insoluble complexes by transglutaminase. Therefore, a pulse–chase experiment was carried out in the presence and absence of LTB-2, a specific irreversible inhibitor of epidermal translutaminase that blocks the formation of crosslinked envelopes (Killackey *et al.*, 1989). The inhibitor was able

to block substantially the decrease in immunoprecipitable labeled protein during the chase period and indicated this protein was a substrate for transglutaminase (Mehrel *et al.*, 1990).

On the basis of its presumed function, we have called the protein "loricrin" (from the latin *lorica*—a protective shell or cover). Further proof that the protein is in fact a CE precursor was obtained by the isolation of peptides from purified human CEs that contained recognizable loricrin sequences and which were crosslinked by the Nϵ-(γ-glutamyl)lysine isodipeptide bond (Hohl *et al.*, 1991b).

Loricrin

As A Major Component of the CE

A comparison of the amino acid composition of isolated mouse (Mehrel *et al.*, 1990) and human (Hohl *et al.*, 1991b) CEs with the deduced composition of the respective loricrin reveals a high degree of similarity. The deduced sequences have a high content of Gly (55.1% and 46.8%, respectively) and Ser (22.3 and 22.8%). These amino acids are also the most abundant in mouse and human CEs (Mehrel *et al.*, 1990; Hohl *et al.*, 1991b). Thus, loricrin appears to be a major envelope component. This argument is further strengthened by comparing data available for three other putative envelope components: involucrin [initially isolated from cultured human keratinocytes by Rice and Green (1979) and shown to have a high Glu/Gln content (46%)]; keratolinin [extracted from human and bovine tissue by Zettergren *et al.* (1984) and which is enriched in Glu/Gln (13.5%), Ser (10.4%) and Thr (9.2%)]; and a cysteine-rich envelope protein isolated from human epidermis by Tezuka and Masae (1987), which is predominantly composed of Gly (18.5%), Glu/Gln (12.7%), Ser (9.8%) and, to a lesser extent, Cys (4.3%) (see Table 2). Additional evidence that loricrin is a major, if not the most abundant, precursor to the CE, comes from visual judgment of Northern blots in comparison with the abundance of keratins 1, keratins 10, and filaggrin (Hohl *et al.*, 1991b), intensity of the *in situ* hybridization signal (Mehrel *et al.*, 1990; Hohl *et al.*, 1991b), as well as abundance of cDNA clones in epidermal cDNA libraries.

As the Sulfur-rich Component of Keratohyaline Granules

Results reported by several laboratories imply that cell envelopes contain a cysteine-rich component. The first amino acid analysis of cell envelopes was determined by Matoltsy and Matoltsy (1966) who reported a cysteine content of 4.9% for envelopes isolated from human plantar stratum corneum. Autoradiography performed by Fukuyama and Epstein (1975) 6 hr after

TABLE 2 Amino Acid Composition of CE Contents and Components[a]

	Human CE in culture (Rice and Green, 1979)	Human CE plantar (Ogawa *et al.*, 1983)	Human CE foreskin (Hohl *et al.*, 1991b)	Human involucrin (Rice and Green, 1979)	Bovine keratolinin (Lobitz and Buxman, 1982)	Human CREP (Tezuka and Masae, 1987)	Human loricrin (Hohl *et al.*, 1991b)
Asp + Asn	9.1[a]	2.7	1.73	2.8	8.71	8.25	0.32
Thr	4.8	3.1	2.08	1.6	9.22	5.35	2.22
Ser	7.2	13.6	19.40	1.6	10.43	9.77	22.78
Glu + Gln	15.9	11.1	8.85	45.8	13.46	12.68	4.43
Pro	7.8	13.3		5.7	4.50	Trace	2.85
Gly	9.2	19.2	33.94	6.7	8.80	18.54	46.84
Ala	7.0	3.5	3.44	1.5	8.44	5.33	0.95
Val	5.1	4.1	3.63	3.7	6.32	5.25	3.48
Cys	2.5	10.5	4.56	0.3	0.97	4.3	6.01
Met	1.5	0.9	0.27	0.9	0.63	Trace	—
Ile	3.5	1.4	1.97	0.4	3.13	3.66	1.58
Leu	7.9	2.3	1.54	14.6	8.19	6.78	—
Tyr	1.8	2.4	2.25	0.8	1.63	Trace	2.53
Phe	2.9	2.5	2.42	0.6	2.67	2.91	2.85
Lys	7.6	4.9	4.77	7.4	5.65	9.04	2.22
His	1.9	2.0	0.64	4.7	1.13	2.25	0.32
Arg	4.3	2.5	1.79	0.7	2.46	6.32	—
Trp	—	—	—	0.2	—	—	0.32
Citrulline	—	—	—	—	2.38	—	—

[a]Units, mol/100 mol.

injection of [^{3}H]cysteine showed heavy labeling near the cell envelope, and in addition, certain dense homogeneous deposits were preferentially labeled. These deposits are identical to the single granule components that Jessen (Jessen, 1970) described as being sulfur rich by X-ray microanalytical analysis. Jessen also suggested that these granules participate in formation of the cell envelope since they are preferentially localized along the periphery of granular cells and disappear as the envelope is formed in transitional cells (Jessen, 1973; Jessen *et al.*, 1976). Such data were confirmed by Goldsmith (1977) who found that the injection of radioactive cysteine into newborn rats resulted in significant incorporation into cell envelope fractions. Likewise, Tezuka and Hirai (1980) found that over 90% of incorporated [^{35}S]cysteine was present in an insoluble fraction, suggesting that the radioactivity was in the cell envelope. Furthermore, techniques that depend on free disulfide bonds, using fluorescein–isothiocyanate (Christophers and Braun-Falco, 1971) or 7-(*N*-dimethylamino-4-methyl-3-coumarine) (Ogawa et al., 1979; Hirotani *et al.*, 1981; Tezuka, 1982) stain the cell periphery of cornified epidermal cells.

Studies addressing the chemical composition of keratohyaline granules have given conflicting accounts (see Holbrook, 1989; Resing and Dale, 1991). A cysteine-rich protein of 19.5 kDa with a content of 9.2% cysteine was isolated from keratohyaline granules by Matoltsy; however, it was only partially characterized (Matoltsy, 1975; Matoltsy and Matoltsy, 1970). During fetal development of mouse epidermis, a cystine-rich protein of 66 kDa was found to correlate temporally and quantitatively well with the presence of keratohyaline granules (Balmain *et al.*, 1979). More recently, Tezuka and Takahashi have reported the extraction of a cystine-rich envelope protein (CREP) from the membrane region of stratum corneum cells from human sole epidermis (Tezuka and Masae, 1987). CREP has a mass of 16 kDa, contains 4.3% cysteine, and has been reported to be a substrate of transglutaminase (Takahashi and Tezuka, 1989; Tezuka and Masae, 1987). However, localization of this protein within granules was not reported, and its molecular weight and solubility in aqueous buffers clearly distinguish it from loricrin.

As mentioned above, initial experiments with antibodies against loricrin using indirect immunofluorescence reveal spotted deposits in the granular layer of newborn mouse epidermis (Mehrel *et al.*, 1990). Further evidence in supporting the notion that loricrin is in fact the sulfur-rich component of keratohyaline has been obtained by electron microscopic immunohistochemical studies. Steven *et al.* (1990) used the affinity purified carboxy-terminal peptide antiserum against mouse loricrin and a peptide antiserum against mouse filaggrin (Rothnagel *et al.*, 1987) to follow the biosynthetic pathway of keratohyaline granules in newborn mouse epidermis (Steven *et al.*, 1990). The immunolabeling results distinguish granules that contain filaggrin epitopes (F-granules) from those that contain loricrin epitopes (L-granules). The existence of a basic distinction between them is supported by pronounced differences in shape and size: L-granules are invariably round, in contrast to F-granules' elongated shapes and tufted outline; and on average L-granules are three times smaller. Moreover, the interiors of F-granules visualized in electron micrographs vary, in a way that is not shared by L-granules, both according to proximity to the stratum corneum and with fixation conditions, further suggesting a compositional difference between them (Steven *et al.*, 1990). Thus, mouse loricrin first accumulates in a particular class of cytoplasmic granules. When the cell matures from the granulosum state to the corneum state, loricrin is rapidly incorporated into the CE (Steven *et al.*, 1990). This precursor–product relationship between L-granules and the CE corroborates indeed the earlier suggestions of Jessen (Jessen, 1973; Jessen *et al.*, 1974; Jessen *et al.*, 1976). By conventional immunofluorescence microscopy, granular deposits labeled by our antiserum were also observed in the epidermis of other rodents such as hamster and rat, as well as in the epidermis of marsupials (opossum) and the tongue epithelium in lambs (Hohl *et al.*, 1991c).

The occurrence of heterogeneous keratohyaline granules, similar the ones reported by Jessen, have repeatedly been observed in distinct human tissues such as the acrosyringeal cells (Anton-Lamprecht, 1980), the cells of the fetal nail (Breathnach, 1971), the interfollicular epidermis in warts (Laurent *et al.*, 1978), in certain palmoplantar hyperkeratoses (Laurent *et al.*, 1985), and in the buccal mucosa of multiple fibroepithelial hyperplasia (Haneke, 1982). Moreover, Kastl and Anton-Lamprecht (1990) reported a distinct "bicomponent" keratohyaline in normal human ridged skin. Studies have been initiated to address the question whether some of these granules contain loricrin; to date L-granules have not been identified in human epidermis (Yoneda *et al.*, 1991), except for the acrosyringium of sweat gland ducts (A. Ishida-Yamamoto, R. Eady, and D. Hohl unpublished observations). It should be noted that another protein known to form granules, trichohyalin (Rothnagel and Rogers, 1986; Fietz *et al.*, 1990) has recently been demonstrated to occur in normal human interfollicular epidermis and thus could also account for the heterogeneity of epidermal granules. However, in plantar epidermis, trichohyalin was found to be absent (Hamilton *et al.*, 1991).

Solubility

The failure of previous attempts to identify this major envelope component can be attributed to several factors. First, most investigators have assumed that envelope components would initially be in the soluble (Buxman and Wuepper, 1976; Lobitz and Buxman, 1982; Zettergren *et al.*, 1984; Kubilus *et al.*, 1987; Tezuka and Masae, 1987) or membrane-bound fractions (Rice and Green, 1979; Simon and Green, 1985) and only become insoluble after crosslinking. Second, several investigators have attempted to isolate envelope components from keratinocyte cultures (Simon and Green, 1984; Simon and Green, 1985; Michel *et al.*, 1987). Loricrin is highly insoluble, even prior to crosslinking. It can only be isolated reproducibly by homogenization of tissue in 5% SDS–20% 2-mercaptoethanol followed by boiling (Mehrel *et al.*, 1990; Hohl *et al.*, 1991b).

We think that the extraordinary insolubility of loricrin is in part due to the presence of the high content of glycine and other hydrophobic residues and the formation of disulfide bonds. This implies that before incorporation into the CE, loricrin is crosslinked to itself or other structures by disulfide bonds. Even so, the amount of non-crosslinked loricrin released by such extraction procedures is quite small. It can only be detected readily by reaction with antibodies or after labeling with [^{35}S]cysteine followed by immunoprecipitation (Mehrel *et al.*, 1990). This suggests that the steady state concentration of monomeric loricrin in the epidermis is small. At first glance, these findings seem inconsistent with the apparent abundance of loricrin. However, as mentioned above, loricrin becomes completely insoluble within 6 hr after synthesis, even in the presence of reducing agents, so that it cannot penetrate an SDS–polyacrylamide gel (Mehrel *et al.*, 1990). Presumably loricrin monomers become quickly crosslinked by isodipeptide bonds by the action of epidermal transglutaminases into large aggregates. This view has been repeatedly confirmed by use of an inhibitor of transglutaminases that showed that the amount and thus solubility of loricrin was greatly increased in cells cultured under conditions that induce terminal differentiation (Mehrel *et al.*, 1990; Hohl *et al.*, 1991a).

Structure of Mouse and Human Loricrins

Thus far, loricrin cDNAs have been cloned in mouse and human. The murine protein of about 38 kDa with a predicted pI of about 9 is encoded by a mRNA of 1.75 kb (Mehrel *et al.*, 1990), whereas the markedly smaller human loricrin of about 26 kDa with a pI of about 12 is encoded by a mRNA species of 1.25 kb (Hohl *et al.*, 1991b). It should be mentioned that in addition to the mouse sequence of Nischt *et al.* (1988), a GC-rich human cDNA sequence called A8 with a very close homology to mouse and human loricrin in the 3′ prime portion has been reported (Magnaldo *et al.*, 1990). However, the first 600 bases in the 5′ prime part of A8 (Magnaldo *et al.*, 1990) and human loricrin differ completely (Hohl *et al.*, 1991b). Most probably A8 represents a hybrid clone that emerged as an artifact during cDNA library construction since genomic clones of human loricrin (Hohl *et al.*, 1989) show a complete match with the published cDNA sequence (Hohl *et al.*, 1991b). Cloning of the A8 gene should resolve this discrepancy.

Structurally, the loricrins can be subdivided into distinct domains. These are several glycine–serine-rich domains (5 in mouse and 4 in human loricrin) separated by lysines or glutamine-rich sequences and bounded by lysine–glutamine-rich amino- and carboxy-terminal end domains (Mehrel *et al.*, 1990; Hohl *et al.*, 1991b). Most probably, these lysines and glutamines serve as substrates for transglutaminases and are involved in N^{ϵ}-(γ-glutamyl)lysine crosslinks. Indeed, we have isolated four crosslinked peptides between K88 and Q153, K88 and Q219, K315 and Q157, as well as K315 and Q215 of human loricrin from CEs matured *in vivo* (Hohl *et al.*, 1991b). However, there is no final experimental proof whether these crosslinks involving loricrin are inter- or intramolecular, or both.

The striking feature of loricrin is its high content of glycine and serine, often configured in tandem quasi peptide repeats (Hohl *et al.*, 1991b). Interestingly, this pattern is present in the non-α-helical end domains of the two keratin proteins (keratin 1 and keratin 10) also coexpressed in terminally differentiating epidermal tissue (Johnson *et al.*, 1985; Zhou *et al.*, 1988). This high content of glycine renders the protein rather hydrophobic and thus insoluble in an aqueous environment in the absence of detergents. Moreover, it has been hypothesized that these glycine–serine-rich sequences adopt an Ω-loop conformation on the keratins (Zhou *et al.*, 1988). It is therefore possible that the related sequences of loricin might similarly form Ω-loops and that keratin intermediate filaments might dock at the cell periphery by interaction of the related Ω-loops of loricrin which is itself covalently bound into the cell envelope (Hohl, 1990). Interestingly, the cysteine residues are equally distributed over the entire molecule. One might argue that disulfide bonds might stabilize these Ω-loops at their bases, thereby exposing the hydrophobic glycine–serine-rich sequences to the outside of the molecule. This would explain the observed properties of solubility (Mehrel *et al.*, 1990; Hohl *et al.*, 1991b).

A comparison of mouse and human loricrins reveals that the overall amino acid composition is very similar, but the tandem peptide quasi repeats of the aliphatic form (glycine/serine/cysteine)$_n$ have not been conserved in either size or sequence between the mouse and human proteins (Hohl *et al.*, 1991b). Although these repeats are clearly related, they are not exact repeats and, as such, do not lend themselves to a simple evolutionary analysis as performed for involucrin (Eckert and Green, 1986; Tseng and Green, 1988; Djian and Green, 1989). In most cases, the aliphatic residue is tyrosine (mouse) and tyrosine, phenylalanine, or isoleucine (human). Most of the variations of the repeats between the two species involve insertions or deletions of individual glycine, serine, or cysteine residues. In general, the repeats are longer in the mouse protein, and the mouse protein is larger in part due to the insertion of several additional loops in the first glycine-rich domain (Mehrel *et al.*, 1990). These apparent insertions and deletions represent a classic example of the concept of "cryptic simplicity" (Tautz *et al.*,

1986); what appear to be major overt nucleic acid and amino acid sequence differences in the loricrins are in fact due to changes of a common sequence motif.

Loricrin Genes

Using a cDNA clone as a probe, we have isolated genomic DNA fragments that contain the full-length human loricrin gene. Sequence analysis reveals a simple structure with a single intron of 1186 bp in the 5′ untranslated region (Hohl *et al.,* 1989). Moreover, a potential epidermis-specific promoter sequence found in several genes expressed in the epidermis (Johnson *et al.,* 1985; Krieg *et al.,* 1985; Eckert and Green, 1986) (AANCCAAA, where N is a purine), a potential CAT-box, the TATA-box and the polyadenylation signal were identified (Hohl *et al.,* 1989).

The gene structure is remarkably similar to that of the involucrin (Eckert and Green, 1986) and profilaggrin (Gan *et al.,* 1990) genes, both coexpressed in the epidermis. All three genes consist in large part of quasi-repeating peptide repeats that have not been well conserved between different species during evolution; all contain conserved sequences at their termini and involucrin and loricrin contain a single intron in 5′ untranslated regions. Moreover, comparison of the human involucrin and loricrin intron sequences reveals about 15 fragments of 12-bp length and 20 fragments of 8-bp length which are completely identical while there is no overall close sequence homology (J. Carroll and D. Hohl, unpublished observation). Most interestingly the human loricrin gene maps to chromosome 1q21, which is very near to the location of the involucrin (Simon *et al.,* 1989), profilaggrin (McKinley-Grant *et al.,* 1989), trichohyalin and the epithelial S100 genes. By genetic linkage analysis, the loricrin and profilaggrin genes are closely linked with a $\theta = 0.015$ and a $Z = 0.29$ but separated by informative meioses (Yoneda *et al.,* 1991). Restriction analysis of total genomic human DNA and several genomic clones, sequence analysis of genomic fragments spanning over 4.3 kb (Hohl *et al.,* 1989), analysis of rodent–human somatic cell hybrids, and direct chromosomal *in situ* hybridization (Yoneda *et al.,* 1992) have not provided any evidence for loricrin pseudogenes thus far and suggest that loricrin occurs as a single locus gene. However it might well be that a distinct but related gene exists (Magnaldo *et al.,* 1990).

At present there is no final answer to why these important epidermal structure genes have a similar structure and why they are located in the same area on chromosome 1. However, several highly homologous genes encoding for proline-rich proteins expressed in keratinocytes (spr 1, 2, and 3) (Kartasova and Van de Putte, 1988; Kartasova *et al.,* 1988; Gibbs *et al.,* 1990) also map to chromosome 1q21 (Backendorf, 1990). They have an identical gene structure: small first exon, one single intron, and a second exon that contains the entire open reading frame. Moreover, a computed analysis of loricrin, involucrin, and spr proteins reveals a significant sequence homology of the amino- and carboxy-terminal peptide sequences while the internal sequences consist of numerous repeats of a common sequence motif that are not homologous among of these genes. This might indicate that these genes were created by gene duplication of a common ancestor and have diverged thereafter by evolving specific internal domains (Backendorf and Hohl, 1992). Such a view is consistent with the phylogenic analysis of involucrins that has shown that the ancestral elements reside in the amino- and carboxy-terminal sequences (Tseng and Green, 1988). One is even tempted to speculate that the *spr* genes might encode CE precursor proteins. In fact, the CE contains a fair amount of proline (Matoltsy and Matoltsy, 1966; Ogawa *et al.*, 1983) for which the known CE precursor proteins such as involucrin, loricrin, keratolinin, or CREP cannot account (see Table 2).

Preliminary results indicate that the organization of the mouse loricrin gene shows a high degree of similarity to the human gene (J. Rothnagel, M. Longley, D. Bundman and D. Roop, unpublished data).

Expression and Regulation in Vivo

By indirect immunofluorescence microscopy, loricrin is often detected as spotted deposits in the cytoplasm and regularly at the periphery of cells in the granular and lower cornified layers, while transcripts are found in the stratum granulosum of both mouse and human epidermis using hybridization *in situ* (Mehrel *et al.*, 1990; Hohl *et al.*, 1991b). This indicates that loricrin is expressed very late in epidermal differentiation (comparable to filaggrin) and furthermore suggests that the expression is regulated transcriptionally or posttranscriptionally in terms of mRNA stability. By immunoblotting and immunohistology, loricrin—or perhaps more prudently—the epitope used for antibody production has been identified in all mammals tested so far: mouse, hamster, rat, opossum, dog, jaguar, rabbit, lamb, cow, and human (Mehrel *et al.*, 1990; Hohl *et al.*, 1991b; Hohl *et al.*, 1991c).

Analysis of various normal epithelia from the mouse, hamster, and rat including whole body crosssections of newborn mice reveal that loricrin is basically expressed in all stratified orthokeratotic squamous epithelia of rodents (Mehrel *et al.*, 1990; Hohl *et al.*, 1991c). We have noticed an eminent variability of loricrin expression in rat epidermis depending on body site. For example, epidermis of the upper trunk shows less loricrin expression than epidermis from the pad where a very strong loricrin expression is found. In rat tail epidermis and mouse lingual papillae, areas of normally keratinizing epidermis are regularly interrupted by hyperparakeratotic portions. In these regions of disturbed keratinization, no loricrin is produced

(Hohl *et al.*, 1991c). In rabbits, loricrin is expressed in epidermis and in the mucosa from the dorsal side of the tongue. However, loricrin is not found in other stratified internal epithelia such as the ventral side of the tongue or the esophagus. Finally, in humans the epithelium lining, the hard palate, and parts of the gingiva as well as single cells of the upper layers of the lingual papillae and the vagina express loricrin. Otherwise, extra-epidermal expression of loricrin is not found in buccal mucosa, esophagus, cervix, or bladder, nor even thymic epithelial islets. Again, as in rats, epidermal loricrin expression is variable with high expression in the intraepidermic portions of follicle and sweat gland ducts as well as in outer leaf of foreskin and perianal skin (Hohl *et al.*, 1991c).

A comparison of human loricrin and involucrin expression reveals that

1. loricrin is expressed much later in epidermal differentiation in normal human interfollicular skin.
2. loricrin is basically restricted to the infundibular part of the follicle; involucrin is expressed additionally in the isthmus, the duct of the sebaceous glands, and some parts of the hair shaft, the inner, and the outer hair root sheet (Hashimoto *et al.*, 1987).
3. loricrin in humans is largely restricted to the epidermis whereas involucrin is found in all stratified epithelia including urothelium (Walts *et al.*, 1985).

Experiments have been performed to further define loricrin expression patterns in rodents. First, squamous metaplasia is induced by vitamin A depletion in various nonkeratinized epithelia of syrian hamsters (Hohl *et al.*, 1991c). Loricrin is expressed in tracheal squamous metaplasia in vitamin A deficient animals, while not in normal tracheal epithelium. Esophagus, oral mucosa, cheek pouch, and skin express loricrin physiologically, therefore, no apparent off–on type of regulation is found (D. Hohl and D. R. Roop, manuscript submitted).

Urinary bladder tumors have been chemically induced in mice by butylnitrosamine. Chemically induced tumors in mice usually exhibit a mixed phenotype with both squamous and transitional differentiation. In the areas of squamous differentiation of such tumors, loricrin was found to be heavily expressed (Hohl *et al.*, 1991c). However, normal urinary bladder epithelium in mice does not express loricrin.

Stratification and keratinization have been induced in the vaginal epithelium of ovarectomized rats by the administration of estradiol. Vaginal epithelium in rats is highly hormone dependent and can be easily changed from a mucous-secreting epithelium to a highly stratified squamous epithelium (Long and Evans, 1922). Loricrin transcripts appear after 24 and 48 hr of induction and correlate well with stratification and keratinization observed morphologically (Roop, 1987).

Expression and Regulation in Keratinocyte Culture

Mouse Loricrin

Using submerged cultures of murine keratinocytes on plastic dishes, loricrin expression has been found to be tightly linked to extracellular calcium concentration (Yuspa *et al.*, 1989). Loricrin transcripts and protein are detected maximally at 0.12 m*M* Ca^{2+} and are reduced at either higher or lower calcium concentrations. Expression of loricrin is minimal 24 hr after the calcium shift, but readily apparent after 48 hr. These results indicate that loricrin is expressed within a limited range of extracellular calcium (Yuspa *et al.*, 1989).

Human Loricrin

In submerged and serum-free cultured normal human keratinocytes, only Ca^{2+} concentrations above 0.1 m*M* permit expression of loricrin mRNA. Highly increased mRNA levels can be obtained at 0.35 m*M* Ca^{2+} and a critical cell density appears to be required for optimal accumulation of loricrin transcripts (Hohl *et al.*, 1991a). Retinoic acid at 10^{-7} to 10^{-9} *M* completely blocks Ca^{2+}-induced loricrin mRNA synthesis when applied simultaneous with increased Ca^{2+}. Furthermore, addition of RA to cultures already exposed to higher Ca^{2+} levels results in the complete loss of loricrin mRNA within about 48–72 hours (Hohl *et al.*, 1991a). Therefore, extracellular Ca^{2+} and retinoid concentration and cell density are important factors modulating transcriptional activity of the human loricrin gene *in vitro*. Loricrin thereby expands the growing list of markers of terminal epidermal differentiation such as the suprabasally expressed keratins K1 and K10, the pemphigus vulgaris antigen, or filaggrin known to be regulated by similar stimuli (Fuchs and Green, 1981; Stanley and Yuspa, 1983; Fleckman *et al.*, 1984; Thivolet *et al.*, 1984; Kipan *et al.*, 1987; Roop, 1987; Fleckman *et al.*, 1989; Ryle *et al.*, 1989; Yuspa *et al.*, 1989). In fact, a comparison of loricrin and filaggrin mRNA levels reveals that they change in parallel with response to the various culture conditions (Hohl *et al.*, 1991a). Thus, the transcriptional control of loricrin and filaggrin expression in the epidermis appears to be closely coordinated.

Using reconstituted human skin cultured at the air–liquid interface, an identical suppression of human loricrin expression by retinoic acid has been noted. In this system, retinoic acid at $10^{-5}M$ (Hendricks *et al.*, 1991) or $10^{-6}M$ (Magnaldo *et al.*, 1992) completely inhibits expression of loricrin but not of involucrin. This indicates differences in the regulation of these two components of the CE. For example, in human epidermal cell culture, involucrin expression *in vitro* is basically unaffected by retinoids (Asselineau *et al.*, 1989; Green and Watt, 1982) but synthesis begins immediately above the basal layer and is thought to be controlled primarily by changes in cell size (Watt and Green, 1981) or cell shape (Watt *et al.*, 1988). Therefore, the expression of different constituents of the CE are regulated by distinct mechanisms. Moreover, early studies have shown that, while the expression of involucrin is unaffected by the addition of retinyl acetate to culture medium, the number of CEs counted is drastically reduced (Green and Watt, 1982). This reduction is partially explained by the finding that RA diminishes particulate transglutaminase activity (Jetten and Shirley, 1986); (Lichti and Yuspa, 1988; Rubin and Rice, 1986; Thacher *et al.*, 1985). However, it has been shown that RA not only prevents formation of CEs by reduction of particulate transglutaminase activity but also by suppressing the synthesis of putative precursor proteins (Nagae *et al.*, 1987). This inhibition of mature CE formation by RA *in vitro* is also mediated through the inhibition of synthesis of loricrin.

Developmental Expression

In mouse skin, only keratin K14 is synthesized prior to stratification at days 12 to 14. While keratins K1 and K10 are expressed simultaneously with the appearance of stratification of the epidermis at day 15, loricrin is initially detected at day 16. Filaggrin expression is not observed until day 17 (Greer and Roop, 1991).

In a study of human epidermal development, loricrin has been found to be absent in the periderm but expressed in intermediate cells after 7–8 weeks. In the hair follicle, loricrin is apparently detected in the layer of Henle in the bulbous hair peg at 18 to 19 weeks (Holbrook *et al.*, 1991). This is interesting since we have not observed expression of loricrin in adult follicles except for the infundibulum as assessed by antibodies (D. Hohl and D. Roop, submitted) and *in situ* hybridization (Hohl *et al.*, 1991b).

CE and Loricrin in Skin Disorders

Little is known about the pathophysiological role of the CE and particularly of loricin in disorders of keratinization (DOK) for several reasons.

First, there is no known animal model for DOK involving abnormalities of CE formation. Second, most basic information on DOK comes from ultrastructural observations. Unfortunately, electron microscopy cannot discriminate the various components believed to contribute to the formation of the CE and reveals the CE as a complete structure only. However, it is feasible that the impairment of one of its constituents changes interactions with the structures connected to the CE (i.e., the lipids on the outer side and the keratins and other proteins possibly on the inner side). Third, such alterations not only could be due to changes of constituents but also to altered expression of one of the enzymes involved in CE formation. Indeed, a highly increased activity of membrane-bound transglutaminase is found in scales from patients with nonerythrodermic autosomal recessive lamellar ichthyosis (van Hooijdonk *et al.*, 1991). However, the finding of increased enzyme activity in scales in a recessively inherited disorder is most surprising. Possibly, these results reflect the situation in scales rather than the situation in the epidermis *in vivo*. Therefore, these results should be confirmed by measurement of transglutaminase activity in keratinocytes.

So far, distortion of the CE as observed by electron microscopy has been reported in hidrotic ectodermal dysplasia (Ando *et al.*, 1988) and in recessive ichthyosis congenita type II (Niemi *et al.*, 1991).

In addition, the CE can be judged by Normaski microscopy, which reveals altered and fragile appearing CEs in Darier's disease, ichthyoses vulgaris, ichthyosis brittle hair syndrome, and psoriasis (Michel *et al.*, 1988; Michel and Juhlin, 1990). This is consistent with the abnormal expression sequence of various differentiation markers in psoriasis where involucrin is expressed precociously and filaggrin and crosslinked CEs late or not at all (Bernard *et al.*, 1986; Bernard *et al.*, 1988). Indeed, immunohistology also demonstrates delayed and diminished expression of loricrin in psoriasis (T. Mehrel and D. Hohl, unpublished observation).

Preliminary results using over 120 biopsy specimens from various cutaneous disorders and tumors indicate that diminished and delayed expression of loricrin in psoriasis is nonspecific and simply represents a disturbance of late terminal epidermal differentiation. In fact, loricrin expression is highly diminished or absent in any agranulotic and parakeratotic epidermis studied so far and therefore does not parallel involucrin expression (D. Hohl, 1993). Several biopsies from patients with ichthyosis vulgaris show normal staining in the granular layers. In three cases of lamellar ichthyosis, a high expression of loricrin and involucrin with a peculiar and abnormal cytoplasmic staining concur with a diminished cytoplasmic staining of keratinocyte transglutaminase (Thacher and Rice, 1985). Interestingly, this pattern was absent in a collodion baby at birth but present 2 weeks later before a clinical phenotype of lamellar ichthyosis appeared. Therefore, immunohistology might serve as an early diagnostic and prognostic tool in the management of collodion babies (D. Hohl, *et al.*, in press). Our findings in lamellar ichthyosis could be explained by two mechanisms. For example, if involucrin serves as a scaffold in the formation of the compound CE (Swartzendruber *et al.*, 1989), defects in its proper placement or crosslinkage by mutation of keratinocyte transglutaminase could distort the arrangement of the intercellular lipid lamellae. In addition, the deposition of precursor proteins to the cytoplasmic side of the CE could be disturbed and thereby alter the properly controlled multistage process of mature CE formation (Farbman, 1966; Hohl *et al.*, 1991a). Alternatively, primary defects of fatty acid metabolism including palmitate and myristate acylation could cause both malformation of the compound CE by an altered keratinocyte transglutaminase membrane anchorage and by formation of lipid vacuoles as observed in ichthyosis congenita II (Niemi *et al.*, 1991). Further experiments such as immunoelectron microscopy, studies of keratinocyte transglutaminase acylation, protein analysis by immunoprecipitation, or sequencing of the amino-terminal sequences of keratinocyte transglutaminase in lamellar ichthyosis will help to test these hypotheses. Finally, it should be mentioned that the amount of protein detected by using antibodies does not necessarily reflect the absolute amount of loricrin synthesized by the cells, but a steady-state level representing the balance of synthesis and crosslinking of loricrin. Therefore, results obtained by immunohistology should be confirmed by hybridization *in situ* with a cDNA probe to loricrin.

Future Directions

Many studies have been performed using antibodies against putative precursors of the CE or against enzymes implicated in crosslinking CEs. These antibodies have proven to be helpful in the determination of stages or patterns of differentiation as well as tissue origin, particularly in cancer research and tumor diagnosis. Therefore, it is anticipated that these antibodies will be widely applied as markers for diagnostic purposes.

Immunohistology might serve as an early diagnostic and prognostic tool in the management of collodion babies. However, loricrin has never been found to be absent in any dermatosis or inherited disorder of keratinization studied thus far. This suggests that life is incompatible with a complete lack of cell envelopes and that severe mutations directly interfering with the formation of functional CE might be lethal. This also emphasizes the necessity of a proper protection of the body at its outermost layers by a highly resistant structure. Nevertheless, it is highly probable that less damaging mutations altering structural and regulatory sequences or factors regulating gene and protein expression or protein activity occur and result in disease. Certainly, the transgenic mouse model will be used to introduce mutated loricrin genes into the germline of mice, whose incorporation might alter envelope structure and/or function in a dominant manner (Palmiter and Brinster, 1986). Such experiments will help to analyze the complex machinery of CE assembly and might allow to identify phenotypes similar to human inherited disorders of keratinization. The feasibility of this approach has been demonstrated for osteogenesis imperfecta (Stacey *et al.*, 1988), epidermolysis bullosa simplex (Vassar *et al.*, 1991; Coulombe *et al.*, 1992) or bullous congenital ichthyosifom erythroderma (Rothnagel, *et al.*, 1992). Moreover, in the near future, homologous recombination might be widely applicable and now appears to be a highly attractive but technically difficult system. Using this method, targeted genes can be deleted and replaced by a chosen engineered mutated allele. For instance, it could be used in a crucial regulatory or structural sequence in order to alter transcriptional activities or functional protein domains (Mansour *et al.*, 1988; Zimmer and Gruss, 1989). Thereby both recessive and dominant mutations could be created. Finally, it would appear worthwhile to investigate inherited disorders by a direct approach using direct sequencing or restriction fragment length polymorphisms with probes located close to or within genes suspected to be involved in a given disorder. This procedure has been applied to the study of spondyloepiphyseal dysplasia (Lee *et al.*, 1989), epidermolysis bullosa simplex (Bonifas *et al.*, 1991; Coulombe *et al.*, 1991; Lane *et al.*, 1992), dominant dystrophic epidermolysis bullosa (Ryyänen *et al.*, 1982) and bullous congenital ichthyosifom erythroderma (Rothnagel, *et al.*, 1992).

Acknowledgment

This work was supported in part by grants to D. H. from the Swiss National Science Foundation (31-30323.90) and to D. R. from the NIH (AR 40240). The authors are gratefully indebted to Ulrike Lichti, Joseph Rothnagel, Peter Steinert, and Stuart Yuspa for numerous helpful discussions and for their continued interest and support.

References

Ando, Y., Tanaka, T., *et al.* (1988). Hidrotic ectodermal dysplasia: a clinical and ultrastructural observation. *Dermatologica* **176**, 205–211.

Anton-Lamprecht, I. (1980). Keratinization of the human sweat duct. *J. Cutaneous. Pathol.* **7**, 192–193.

Asselineau, D., Bernard, B. A. *et al.* (1989). Retinoic acid improves epidermal morphogenesis. *Dev. Biol.* **133**, 322–335.

Backendorf, C. (1990). *Mutagenesis* **5**: 77.

Backendorf, C. and Hohl, D. (1992). A common origin for cornified envelope precursor proteins? Nature Genetics, **2**, 91.

Baden, H. P., Kubilus, J. *et al.* (1987a). Characterization of monoclonal antibodies generated to the cornified envelope of human cultured keratinocytes. *J. Invest. Dermatol.* **89**(5), 454–459.

Baden, H. P., Kubilus, J. *et al.* (1987b). A new class of soluble basic protein precursors of the cornified envelope of mammalian epidermis. *Biochim. Biophys. Acta* **925**, 63–73.

Balmain, A., Loehren, D. *et al.* (1979). Proteins synthesis during fetal development of mouse epidermis. *Dev. Biol.* **73**, 338–344.

Bernard, B. A., Réano, A. *et al.* (1986). Precocious appearance of involucrin and epidermal transglutaminase during differentiation of psoriatic skin. *Br. J. Dermatol.* **114**, 279–283.

Bernard, B., Asselineau, A. D. *et al.* (1988). Abnormal sequence of expression of differentiation markers in psoriatic epidermis: Inversion of two steps in the differentiation program? *J. Invest. Dermatol.* **90**(6), 801–805.

Bonifas, J., Rothman, A. *et al.* (1991). Epidermolysis bullosa simplex: Evidence in two families for keratin gene abnormalities. *Science* **254**, 1202–1205.

Breathnach, S. (1971). "An atlas of the ultrastructure of human skin." London, Churchill.

Buxman, M. M. and Wuepper, K. D. (1976). Isolation, purification and characterization of bovine epidermal transglutaminase. *Biochim. Biophys. Acta* **452,** 356–369.

Buxman, M. M., Lobitz, C. J. *et al.* (1980). Epidermal transglutaminase. Identification and purification of a soluble substrate with studies of *in vitro* crosslinking. *J. Cell Biol.* **255,** 1200–1203.

Christophers, E. and Braun-Falco, O. (1971). Fluorescein–isothiocyanate as a stain for keratinizing epithelia. *Arch. Dermatol. Forsch.* **241,** 199–209.

Chung, S. I. (1972). Comparative studies on tissue transglutaminase and factor XIII. *Ann. N.Y. Acad. Sci. U.S.A.* **202,** 240–255.

Coulombe, P., Hutton, M. *et al.* (1991). Point mutations in human keratin 14 genes of epidermolysis bullosa simplex: Genetic and functional analysis. *Cell* **66,** 1301–1311.

Coulombe, P., Hutton, M. *et al.* (1992). A function of keratins and a common thread among different types of epidermolysis bullosa simplex diseases. *J. Cell Biol.* **115,** 1661–1674.

Dale, B. A., Holbrook, K. A. *et al.* (1978). Assembly of stratum corneum basic protein and keratin filaments in macrofibrils. *Nature (London)* **276,** 729–731.

Djian, P. and Green, H. (1989). Vectorial expansion of the involucrin gene and the relatedness of the hominoids. *Proc. Natl. Acad. Sci. U.S.A.* **86,** 8447–8451.

Eckert, R. L., and Green, H. (1986). Structure and evolution of the human involucrin gene. *Cell* **46,** 583–589.

Farbman, A. I. (1966). Plasma membrane changes during keratinization. *Anat. Rec.* **156,** 269–282.

Fietz, M. J., Presland, R. B. *et al.* (1990). The cDNA-deduced amino acid sequence of trichohyalin, a differentiation marker in the hair follicle, contains a 23 amino acid repeat. *J. Cell Biol.* **110,** 427–436.

Fleckman, P., Haydock, P. *et al.* (1984). Profilaggrin and the 67 kD keratin are coordinately expressed in cultured human epidermal keratinocytes. *J. Cell Biol.* **99,** 315.

Fleeckman, P., Haydock, P. *et al.* (1989). Differentiation of cultured human epidermal keratinocytes upon reaching confluence. *J. Invest. Dermatol.* **92,** 428.

Folk, J. E. (1983). Mechanisms and basis for specificity of transglutaminase-catalyzed ϵ-(γ-glutamyl)lysine bond formation. *Adv. Enzymol.* **54,** 1–54.

Fuchs, E. and Green H. (1980). Changes in keratin gene expression during terminal differentiation of the keratinocyte. *Cell* **19,** 1033–1042.

Fuchs, E. and Green, H. (1981). Regulation of terminal differentiation of cultured human keratinocytes by vitamin A. *Cell* **25,** 617–625.

Fukuyama, K. and Epstein, W. L. (1975a). A comparative autoradiographic study of keratohyaline granules containing cystine and histidine. *J. Ultrastruct. Res.* **51,** 314–325.

Fukuyama, K. and Epstein, W. L. (1975b). Heterogeneous proteins in keratohyaline granules studied by quantitative autoradiography. *J. Invest. Dermatol.* **65,** 113–117.

Gan, S. Q., McBride, W., *et al.* (1990). Organisation, structure, and polymorphisms of the human profilaggrin gene. *Biochemistry* **29,** 9432–9440.

Gibbs, S., Lohman, F. *et al.* (1990). Characterization of the human spr2 promoter: induction after UV irradiation or TPA treatment and regulation during differentiation of cultured primary keratinocytes. *Nucleic Acids Res.* **18,** 4401–4407.

Goldsmith, L. A. (1977). ϵ-(γ-Glutamyl)lysine crosslinks in proteins. *In* "Biochemistry of Cutaneous Epidermal Differentiation" pp. 398–418. University of Tokyo Press, Tokyo.

Goldsmith, L. A., Baden, H. P. *et al.* (1974). Vertebral epidermal transamidases. *Biochim. Biophys. Acta* **351,** 113–125.

Green, H. and F. Watt M. (1982). Regulation by vitamin A of envelope cross-linking in cultured keratinocytes derived from different human epithelia. **2**(9), 1115–1117.

Greer, J. and Roop, D. (1991). Loricrin, a major keratinocyte cell envelope protein is expressed late in development. *J. Invest. Dermatol.* **96,** 553.

Hamilton, E. H., Payne, R. E. *et al.* (1991). Trichohyalin: presence in the granular layer and stratum corneum of normal human epidermis. *J. Invest. Dermatol.* **96**(5), 666–672.

Haneke, E. (1982). Composite and heterogeneous keratohyaline in the human buccal mucosa. *Arch. Dermatol. Res.* **272,** 127–134.

Hanigan, H. and Goldsmith, L. A. (1978). Endogenous substrates for epidermal transglutaminase. *Biochim. Biophys. Acta* **522,** 589–601.

Harding, C. R. and Scott, I. R. (1983). Histidine-rich proteins (filaggrins): structural and functional heterogeneity during epidermal differentiation. *J. Mol. Biol.* **170,** 651–673.

Hashimoto, T., Inamoto, N. *et al.* (1987). Involucrin expression in skin appendage tumours. *Br. J. Dermatol.* **117,** 325–332.

Hendricks, L., Geesin, J. *et al.* (1991). Retinoid suppression of loricrin expression in reconstituted human skin cultured at the air–liquid interface. *J. Invest. Dermatol.* **96,** 548.

Hirotani, T., Manabe, M. *et al.* (1981). Characteristics of horny cell membrane isolated from stratum corneum. *J. Jpn. Dermatol.* **91,** 677–680.

Hohl, D. (1990). The cornified cell envelope. *Dermatologica* **180,** 201–211.

Hohl, D., Idler, W. *et al.* (1989). Loricrin—a major component of the cornified envelope. Isolation, restriction analysis and sequencing of genomic DNA clones. *Dermatologica* **179,** 235.

Hohl, D., Lichti, U. *et al.* (1991a). Transcription of the human loricrin gene *in vitro* is induced by calcium and cell density and suppressed by retinoic acid. *J. Invest. Dermatol.* **96,** 414–418.

Hohl, D., Mehrel, T. *et al.* (1991b). Characterization of human loricrin-Structure and function of a new class of epidermal cell envelope proteins. *J. Biol. Chem.* **266**(10), 6626–6636.

Hohl, D., Olano, B. Ruf *et al.* (1991c). Loricrin is a marker of squamous differentiation in rodents and a marker of epidermal differentiation in higher mammals. *J. Invest. Dermatol.* **96,** 1030.

Hohl, D. (1993). Expression patterns of loricrin in dermatological disorders. *Am. J. Dermatopathol.* **15(1)**, 20–27.

Hohl, D., Huber, M. and Frenk, E. (1993). Analysis of the cornified cell envelope in lamellar ichthyosis. *Arch. Dermatol.* in press.

Holbrook, K. A. (1989). Biologic structure and function: Perspectives on morphologic approaches to the study of the granular layer keratinocyte. *J. Invest. Dermatol.* **92**(4), 84s–104s.

Holbrook, K., Underwood, B. *et al.* (1991). Cornified cell envelope (CCE) in human fetal skin: Involucrin, keratolinin, loricrin and transglutaminase expression and activity. *J. Invest. Deramtol.* **96,** 542.

Jessen, H. (1970). Two types of keratohyalin granules. *J. Ultrastruct. Res.* **33,** 95–115.

Jessen, H. (1973). Electron cytochemical demonstration of sulfhydryl groups in keratohyalin granules and the peripheral envelope of cornified cells. *Histochemie* **33,** 15–29.

Jessen, H., Peters, P. D. *et al.* (1974). Sulphur in different types of keratohyaline granules: A quantitative assay by X-ray microanalysis. *J. Cell. Sci.* **15,** 359–377.

Jessen, H., Peters, P. D. *et al.* (1976). Sulphur in epidermal keratohyalin granules: A quantitative assay by X-ray microanalysis. *J. Cell Sci.* **22,** 161–171.

Jetten, A. M. and Shirley, J. E. (1986). Characterization of transglutaminase activity in rabbit tracheal cells. Regulation by retinoids. *J. Biol. Chem.* **261,** 15097–15101.

Johnson, L. D., Idler, W. W. *et al.* (1985). Structure of a gene for the human epidermal 67-kDa keratin. *Proc. Natl. Acad. Sci. U.S.A.* **82,** 1896–1900.

Kartasova, T. and Van de Putte, P. (1988). Isolation, characterization, and UV-stimulated expression of two families of genes encoding polypeptides of related structure in human epidermal keratinocytes. *Mol. Cell Biol.* **8,** 2195–2203.

Kartasova, T., Van Muijen, G. *et al.* (1988). Novel protein in human epidermal keratinocytes: Regulation of expression during differentiation. *Mol. Cell. Biol.* **8,** 2204–2210.

Kastl, I. and Anton-Lamprecht, I. (1990). Biocomponent keratohyaline in normal human ridged skin. *Arch. Dermatol. Res.* **282**, 71–75.

Killackey, J. J., Bonaventura, B. J. *et al.* (1989). A new class of mechanism-based inhibitors of transglutaminase enzymes inhibits the formation of crosslinked envelopes by human malignant keratinocytes. *Mol. Pharmacol.* **35**, 701–706.

Kim, H. C., Song, K. Y. *et al.* (1989). Immunological identity and distribution of epidermal and hair follicle transglutaminse of mouse, guinea pig, and human. *J. Invest. Dermatol.* **92**, 458.

Kim, H. C., Lewis, M. S. *et al.* (1991a). Protransglutaminase E from guinea pig skin. *J. Biol. Chem.* **265**, 21971–21978.

Kim, H. C., Idler, W. W. *et al.* (1991b). The complete amino acid sequence of the human transglutaminase K enzyme deduced from the nucleic acid sequences of cDNA clones. *J. Biol. Chem.* **266**, 536–539.

Kopan, R., Traska, G. *et al.* (1987). Retinoids as important regulators of terminal differentiation: Examining keratin expression in individual epidermal cells at various stages of keratinization. *J. Cell Biol.* **105**, 427–440.

Krieg, T. M., Schafer, M. P. *et al.* (1985). Organization of a type I keratin gene. *J. Biol. Chem.* **260**, 5867–5870.

Kubilus, J. and Baden, H. P. (1983). The role of cross-linking in epidermal differentation. *Curr. Probl. Dermatol.* **11**, 253–263.

Kubilus, J., Kvedar, J. *et al.* (1987). Identification of new components of the cornified envelope of human and bovine epidermis. *J. Invest. Dermatol.* **89**, 44–50.

Kvedar, Y. C., Manabe, M., Phillips, S. B., Ross, B. S., Baden, H. P. (1992). Characterization of sciellin, a precursor to the cornified envelope of human keratinocytes. *Differentiation* **49**, 195–204.

Lane, E., Rugg, E. *et al.* (1992). A mutation of the conserved helix termination peptide of keratin 5 in hereditary skin blistering. *Nature (London)* **356**, 244–246.

Laurent, R., Nicollier, M. *et al.* (1978). Heterogeneous keratohyaline formation in warts. *Arch. Dermatol. Res.* **262**, 83–96.

Laurent, R., Prost, O. *et al.* (1985). Composite keratohyaline granules found in palmoplantar keratoderma: An ultrastructural study. *Arch. Dermatol. Res.* **277**, 384–394.

Lee, B., Vissing, H. *et al.* (1989). Identification of the molecular defect in a family with spondyloepiphyseal dysplasia. *Science* **244**, 978–980.

Lichti, U. and Yuspa, S. H. (1988). Modulation of tissue and epidermal transglutaminases in mouse epidermal cells after treatment with 12-O-tetradecanoylphorbol-13-acetate and/or retinoic acid *in vivo* and *in vitro*. *Cancer Res.* **48**, 74–81.

Lichti, U., Ben, T. *et al.* (1985). Retinoic acid-induced transglutaminase in mouse epidermal cells is distinct from epidermal transglutaminase. *J. Biol. Chem.* **260**(3), 1422–1426.

Lobitz, C. J. and Buxman, M. M. (1982). Characterization and localization of bovine epidermal transglutaminase substrate. *J. Invest. Dermatol.* **78**(2), 150–154.

Long, J. A. and Evans, H. M. (1922). *Mem. Univ. Calif.* **6**, 1–148.

Ma, A. S.-M. and Sun, T.-T. (1986). Differentiation-dependent changes in the solubility of a 195-kD protein in human epidermal keratinocytes. *J. Cell Biol.* **103**(1), 41–48.

Magnaldo, M., Pommes, L. *et al.* (1990). Isolation of a GC-rich cDNA identifying mRNAs present in human epidermis and modulated by calcium and retinoic acid in cultured keratinocytes. Homology with murine loricrin mRNA. *Mol. Biol. Rep.* **14**, 237–246.

Magnaldo, T., Bernerd, F. *et al.* (1992). Expression of loricrin is negatively controlled by retinoic acid in human epidermis reconstructed *in vitro*. *Differentiation* **49**, 39–46.

Mansour, S. L., Thomas, K. R. *et al.* (1988). Disruption of the proto-oncogene *int*-2 in mouse embryo-derived stem cells: A strategy for targeting mutations to nonselectable genes. *Nature (London)* **336**, 348–352.

Matoltsy, A. G. (1975). Desmosomes, filaments and keratohyaline granules: Their role in the stabilization and keratinization of the epidermis. *J. Invest. Dermatol.* **65**, 127–142.

Matoltsy, A. G. and Matoltsy, M. N. (1966). The membrane protein of horny cells. *J. Invest. Dermatol.* **46**(1), 127–129.

Matoltsy, A. G. and Matoltsy, M. N. (1970). The chemical nature of keratohyalin granules of the epidermis. *J. Cell Biol.* **47**, 593–603.

McKinley-Grant, L. J., Idler, W. W. *et al.* (1989). Characterization of a cDNA clone encoding human filaggrin and localization of the gene to chromosome region 1q21. *Proc. Natl. Acad. Sci. U.S.A.* **86**, 4848–4852.

Mehrel, T., Hohl, D. *et al.* (1990). Identification of a major keratinocyte cell envelope protein, loricrin. *Cell* **61**, 1103–1112.

Michel, S. and Juhlin, L. (1990). Cornified envelopes in congenital disorders of keratinization. *Br. J. Dermatol.* **122**, 15–21.

Michel, S., Schmidt, R. *et al.* (1987). Identification and subcellular distribution of cornified envelope precuror proteins in the transformed human keratinocyte line SV-K14. *J. Invest. Dermatol.* **88**(3), 301–305.

Michel, S., Schmidt, R. *et al.* (1988). Morphological and biochemical characterization of the cornified envelopes from human epidermal keratinocytes of different origin. *J. Invest. Dermatol.* **91**(1), 11–15.

Nagae, S., Lichti, U. *et al.* (1987). Effect of retinoic acid on cornified envelope formation: Difference between spontaneous envelope formation *in vivo* or *in vitro* and expression of envelope competence. *J. Invest. Dermatol.* **89**(1), 51–58.

Negi, M., Colbert, M. *et al.* (1985). High molecular weight epidermal transglutaminase. *J. Invest. Dermatol.* **85**, 75–78.

Niemi, K.-M., Kanerva, L. *et al.* (1991). Recessive ichthyosis congenita type II. *Arch. Dermatol. Res.* **283**, 211–218.

Nischt, R., Rentrop, M. *et al.* (1988). Localization of a novel mRNA in keratinizing epithelia of the mouse: Evidence for the sequential activation of differentiation-specific genes. *Epithelia* **1**(2), 165–177.

Ogawa, H. and Goldsmith, L. A. (1976). Human epidermal transglutaminase. *J. Biol. Chem.* **251**, 7281–7288.

Ogawa, H., Taneda, A. *et al.* (1979). The histochemical distribution of protein bound sulfhydryl groups in human epidermis by the new staining method. *J. Histochem. Cytochem.* **27**, 942–946.

Ogawa, H., Manabe, M. *et al.* (1983). Comparative studies of the marginal band and plasma membrane of the epidermis. *Curr. Probl. Dermatol.* **11**, 265–276.

Palmiter, R. D. and Brinster, R. L.(1986). Germ-line transformation of mice. *Annu. Rev. Genet.* **20**, 465–499.

Park, S. C., Kim, S. Y. *et al.* (1988). Differential expression of transglutaminases in human foreskin and cultured keratinocytes. *J. Cell Biol.* **107**, 139a.

Peterson, L. L., and Buxman, M. M. (1981). Rat hair follicle and epidermal transglutaminases biochemical and immunochemical isoenzymes. *Biochim. Biophys. Acta* **657**, 268–276.

Phillips, M. A., Stewart, B. E. *et al.* (1990). Primary structure of keratinocyte transglutaminase. *Proc. Natl. Acad. Sci. U.S.A.* **87**, 9333–9337.

Phillips, S. B., Kubilus, J. *et al.* (1990). The pancornulins: a group of basic low molecular weight proteins in mammalian epidermis and epithelium that may function as cornified envelope precursors. *Comp. Biochem. Physiol.* **95B**(4), 781–788.

Piacentini, M., Martinet, N. *et al.* (1988). Free and protein-conjugated polyamines in mouse epidermal cells. *J. Biol. Chem.* **263**, 3790–3794.

Polakowska, R. and Goldsmith, L. (1991). the cell envelope and transglutaminases. *In* "Biochemistry, Physiology and Molecular Biology of the Skin" Vol. 5, pp. 168–201. Oxford University Press, New York.

Polakowska, R., Herting, E. *et al.* (1991). Isolation of cDNA for human epidermal type I transglutaminase. *J. Invest. Dermatol.* **96**, 285–289.

Resing, K. and Dale, B. (1991). The proteins of keratohyaline. *In* "Biochemistry, Physiology and Molecular Biology of the Skin. Vol. 4, pp. 148–167. Oxford University Press, New York.

Rheinwald, J. G. and Green H. (1975). Serial cultivation of strains of human epidermal keratinocytes: The formation of keratinizing colonies from single cells. *Cell* **6**, 331–344.

Rice, R. H. and Green, H. (1977). The cornified envelope of terminally differentiated human epidermal keratinocytes consists of cross-linked protein. *Cell* **11**, 417–422.

Rice, R. H. and Green H. (1979). Presence in human epidermal cells of a soluble protein precursor of the cross-linked envelope: Activation of the cross-linking by calcium ions. *Cell* **18**, 681–694.

Rice, R. H., Chakravarty, R. *et al.* (1988). Keratinocyte transglutaminase: Regulation and release. *In* "Posttranslational modifications of proteins and ageing." pp. 51–61. Plenum Press, New York.

Richards, S., Scott, I. R. *et al.* (1988). Filiaggrin composition of cell envelopes in newborn rat. *Biochem. J.* **253**, 153–160.

Roop, D. R. (1987). Regulation of keratin gene expression during differentiation of epidermal and vaginal epithelial cells. *In* Current Topics In Developmental Biology." Pp. 195–207. Academic Press, San Diego.

Roop, D. R., Hawley-Nelson, P. *et al.* (1983). Keratin gene expression in mouse epidermis and cultured epidermal cells. *Proc. Natl. Acad. Sci. U.S.A.* **80**, 716–720.

Rothnagel, J. A. and Rogers, G. E. (1986). Trichohyalin, an intermediate filament-associated protein of the hair follicle. *J. Cell Biol.* **102**, 1419–1429.

Rothnagel, J. A., Mehrel, T. *et al.* (1987). The gene for mouse epidermal filaggrin precursor. *J. Biol. Chem.* **262**(32), 15643–15648.

Rothnagel, T. A., Dominey, A. M., Dempsey, L. D., Longley, M. A., Greengalph, D. A., Gagne, T. A., Huber, M., Frenk, E., Hohl, D., and Roop, D. R. (1992). Mutations in the rod domains of keratin 1 and 10 in epidemolytic hyperkeratosis. *Science* **257**, 1128–1130.

Rubin, A. L. and Rice, R. H. (1986). Differential regulation by retinoic acid and calcium of transglutaminases in cultured neoplastic and normal human keratinocytes. *Cancer Res.* **46**, 2356–2361.

Ryle, C. M., Breitkreutz, D. *et al.* (1989). Density dependent modulation of keratin synthesis *in vitro* and dissociation of K1 and K10 expression by the human keratinocyte line Ha*cat,* and *ras*-transfected tumorigenic clones. *Differentiation* **40**, 42–54.

Ryyänen, M., Ryyänen, J. *et al.* (1982). Genetic linkage of type VII collagen (Col7A1) to dominant dystrophic epidermolysis bullosa in families with abnormal anchoring fibrils. *J. Clin. Invest.* **89**, 974–980.

Schmidt, R., Michel, S. *et al.* (1988). Transglutaminases in normal and transformed human keratinocytes. *J. Invest. Dermatol.* **90**, 475–479.

Schweizer, J., Kinjo, M. *et al.* (1984). Sequential expression of mRNA-encoded keratin sets in neonatal mouse epidermis: Basal cells with properties of terminally differentiating cells. *Cell* 37, 159–170.

Simon, M. and Green, H. (1984). Participation of membrane-associated proteins in the formation of the cross-linked envelope of the keratinocyte. *Cell* **36**, 827–834.

Simon, M. and Green, H. (1985). Enzymatic cross-linking of involucrin and other proteins by keratinocyte particulates *in vitro*. *Cell* **40**, 677–683.

Simon, M. and Green, H. (1988). The glutamine residues reactive in transglutaminase-catalyzed cross-linking of involucrin. *J. Biol. Chem.* **263**, 18093–18098.

Simon, M., Phillips, M. *et al.* (1989). Absence of a single repeat from the coding region of the human involucrin gene leading to RFLP. *Am. J. Hum. Genet.* **45**, 910–916.

Stacey, J. R., Bateman, J. *et al.* (1988). Perinatal lethal osteogenesis imperfecta in transgenic mice bearing an engineered mutant pro-α-(I) collagen gene. *Nature (London)* **332**, 131–132.

Stanley, J. R. and Yuspa, S. H. (1983). Specific epidermal protein markers are modulated during calcium-induced terminal differentiation. *J. Cell Biol.* **96**, 1809–1814.

Steinert, P. M. and Roop, D. R. (1988). Molecular and cellular biology of intermediate filaments. *Annu. Rev. Biochem.* **57**, 593–625.

Steinert, P. M., Cantieri, J. S. *et al.* (1981). Characterization of a class of cationic proteins that specifically interact with intermediate filaments. *Proc. Natl. Acad. Sci. U.S.A.* **78**, 4097–4101.

Steinert, P. M., Rice, R. H. *et al.* (1983). Complete amino acid sequence of a mouse epidermal keratin subunit: implications for the structure of intermediate filaments. *Nature (London)* **302**, 794–800.

Steinert, P. M., Parry, D. A. D. *et al.* (1985). Amino acid sequences of mouse and human epidermal type II keratins of M_r 67,000 provide a systematic basis for the structural and functional diversity of the end domains of keratin intermediate filament subunits. *J. Biol. Chem.* **260**(11), 7142–7149.

Steven, A. C., Bisher, M. E. *et al.* (1990). Biosynthetic pathways of filaggrin and loricrin—two major proteins expressed by terminally differentiated epidermal keratinocytes. *J. Struct. Biol.* **104**, 150–162.

Sun, T.-T. and Green, H. (1976). Differentiation of the epidermal keratinocyte in cell culture: Formation of the cornified envelope. *Cell* **9**, 522–521.

Swartzendruber, D. C., Kitko, D. J. *et al.* (1988). Isolation of corneocyte envelopes from porcine epidermis. *Arch. Dermatol. Res.* **280**, 424–429.

Swartzendruber, D. C., Wertz, P. W. *et al.* (1989). Molecular models of the intercellular lipid lamellae in mammalian stratum corneum. *J. Invest. Dermatol.* **92**, 251–257.

Takahashi, M. and Tezuka, T. (1989). Characterization of the 15 KDa protein as a novel substrate of transglutaminase. *J. Invest. Dermatol.* **92**, 526.

Tautz, D., Trick, M. *et al.* (1986). Cryptic simplicity in DNA is a major source of genetic variation. *Nature (London)* **322**, 652–656.

Tezuka, T. (1982). Dyskeratosic process of hyperkeratosis lenticularis perstans (Flegel). *Dematologica* **164**, 379–385.

Tezuka, T. and Hirai, R. (1980). The synthesis of the cystine-rich proteins in rat epidermis. I. Analysis by [^{35}S]cystine incorporation. *In* "Biochemistry of Normal and Abnormal Epidermal Differentation." University of Tokyo Press, Tokyo.

Tezuka, T. and Masae, T. (1987). The cystine-rich envelope protein from human epidermal stratum corneum cells. *J. Invest. Dermatol.* **88**(1), 47–51.

Thacher, S. M. (1989). Purification of keratinocyte transglutaminase and its expression during squamous differentiation. *J. Invest. Dermatol.* **92**, 578–584.

Thacher, S. M., Coe, E. L. *et al.* (1985a). Retinoid suppression of transglutaminase activity and envelope competence in cultured human epidermal carcinoma cells. *Differentiation* **29**, 82–87.

Thacher, S. M. and Rice, R. H. (1985b). Keratinocyte-specific transglutaminase of cultured human epidermal cells: Relation to cross-linked envelope formation and terminal differentiation. *Cell* **40**, 685–695.

Thivolet, C. H., Hintner, H. H. *et al.* (1984). The effect of retinoic acid on the expression of pemphigus and pemphigoid antigens in cultured human keratinocytes. *J. Invest. Dermatol.* **82**(4), 329–334.

Tseng, H. and Green, H. (1988). Remodeling of the involucrin gene during primate evolution. *Cell* **54**, 491–496.

van Hooijdonk, C. A., Steijlen, P. M. *et al.* (1991). Epidermal transglutaminases in ichthyosis. *J. Invest. Dermatol.* **96**, 1024.

Vassar, R., Copulombe, P. *et al.* (1991). Mutant keratin expression in transgenic mice causes marked abnormalities resembling a human genetic skin disease. *Cell* **64**, 365–380.

Walts, A. E., Said, J. W. *et al.* (1985). Involucrin, a marker of squamous and urothelial differentation: An immunohistochemical study on its distribution in normal and neoplastic tissues. *J. Pathol.* **145**, 329–340.

Watt, F. M. and Green, H. (1981). Involucrin synthesis is correlated with cell size in human epidermal cultures. *J. Cell Biol.* **90**, 738–742.

Watt, F. M., Jordan, P. W. *et al.* (1988). Cell shape controls terminal differentiation in human epidermal keratinocytes. *Proc. Natl. Acad. Sci. U.S.A.* **85**, 5576–5580.

Wertz, P. W., Madison, K. C. *et al.* (1989). Covalently bound lipids of human stratum corneum. *J. Invest. Dermatol.* **92**, 109–111.

Woodcock-Mitchell, J., Eichner, R. *et al.* (1982). Immunolocalization of keratin polypeptides in human epidermis using monoclonal antibodies. *J. Cell Biol.* **95**, 580–588.

Yoneda, K., and Hohl, D. (1992). The human loricrin gene. *J. Biol. Chem.* **267**(**25**), 18060–18066.

Yuspa, S. H., Kilkenny, A. E. *et al.* (1989). Expression of murine epidermal differentiation markers is tightly regulated by restricted extracellular calcium concentrations *in vitro*. *J. Cell Biol.* **109**, 1207–1217.

Zettergren, J. G., Peterson, L. L. *et al.* (1984). Keratolinin: The soluble substrate of epidermal transglutaminase from human and bovine tissue. *Proc. Natl. Acad. Sci. U.S.A.* **81**, 238–242.

Zhou, X. M., Idler, W. W. *et al.* (1988). The complete sequence of the human intermediate filament chain keratin 10. *J. Biol. Chem.* **263**, 15584–15589.

Zimmer, A. and Gruss, P. (1989). Production of chimeric mice containing embryonic stem (ES) cells carrying a homeobox Hox 1.1 allele mutated by homologous recombination. *Nature (London)* **338**, 150–153.

Note Added in Proof

Since the submission of this chapter, K. W. Marvin *et al.* (1992). provided biochemical evidence that cornifin, a proline-rich protein in rabbits is a precursor protein of the CE (*Proc. Natl. Acad. Sci. USA* **89**, 11026–11030). In fact, cornifin is the rabbit homologue of the human spr-1 protein (89% overall identity and 97% in the NH2-terminus), which is most likely identical to one of the pancornulins. These data support our computed analysis on spr proteins, loricrin, and involucrin, suggesting a common origin of CE precursor proteins (Backendorf and Hohl, 1992).

6

Retinoic Acid in Epithelial and Epidermal Differentiation

Michel Darmon and Miroslav Blumenberg

History

Since antiquity, it has been known that certain types of food contain a substance necessary for night vision and eye health. At the beginning of the century, Stepp (1909), McCollum and Davis (1913), and Osborne and Mendel (1913) reported the isolation of a fat-soluble factor ("fat soluble A") that was a necessary requirement in the diet of rats (i.e., that was a vitamin) (Drummond, 1920). Later, it was shown that this fat-soluble factor, named vitamin A, prevented night blindness and xerophthalmia and was thus the substance whose existence was postulated in ancient times (reviewed by Sporn *et al.*, 1984). What is remarkable in a historical perspective is that the study of the two main symptoms of the eye disease caused by vitamin A, night blindness and xerophthalmia, led investigators to the elucidation of two entirely different pathogenic mechanisms that turned out to cover the two pathways of vitamin A effects: vision and cell differentiation. It was discovered that vitamin A (retinol) was the precursor of two physiologically important molecules (reviewed by DeLuca, 1979; Frolik, 1984): (1) retinaldehyde, which was identified as the chromophore of the visual pigment (Wald, 1934; Morton, 1944; Wald and Hubbard, 1950; reviewed by Sporn *et al.*, 1984), and (2) retinoic acid, a hormone-like agent necessary for a correct differentiation of many tissues, but more specifically of epithelia, including those of the eye (reviewed by Lotan, 1980; Sporn *et al.*, 1984; Roberts and Sporn, 1984; Underwood, 1984; Shapiro, 1986).

That retinoic acid, which cannot be reduced into retinol (Dowling and Wald, 1960), can replace retinol for most physiological functions, with the exception of vision (reviewed by DeLuca, 1979; Underwood, 1984), and that its potency is higher than that of retinol in bioassays run *in vitro* (reviewed by Sporn and Roberts, 1984), support the idea that retinoic acid is the hormonal form of vitamin A responsible for its effects on cell differentiation. Retinol carried in the plasma has been shown to be converted into retinoic acid by specific dehydrogenases within target tissues (McCormick and Napoli, 1982; Siegenthaler *et al.*, 1990; reviewed by DeLuca, 1979 and Frolik, 1984). From a molecular point of view, retinoic acid and its natural and synthetic analogues (retinoids) must actually be considered hormones since they mediate their biological effects after binding to nuclear receptors (RARs and RXRs) that belong to the superfamily of hormone-dependent transcriptional activators (Dejean *et al.*, 1986; Daly and Redfern, 1987; Petkovich *et al.*, 1987; Giguère *et al.*, 1987; De Thé *et al.*, 1987; Benbrook *et al.*, 1988; Brand *et al.*, 1988; Zelent *et al.*, 1989; Krust *et al.*, 1989; Ragsdale *et al.*, 1989; Mangelsdorf, 1990. Reviewed by Green and Chambon, 1988; Gudas, 1990; and Darmon, 1990). Moreover, many developmental studies have shown that retinoic acid is not only a modulator of the differentiation of epithelia and other tissue types (reviewed by Lotan, 1980; Roberts and Sporn, 1984; Sherman, 1986), but also a major morphogen during embryonic development, particularly in the formation of mesenchymal and nervous structures (Maden, 1982; Tickle *et al.*, 1982; Summerbell, 1983; Maden, 1985; Thaller and Eichele, 1987; Durston *et al.*, 1989; Wagner *et al.*, 1990).

The first evidence for control of epithelial differentiation by retinoids was obtained by Mori (1922). He showed that vitamin-A-deficient rats develop an eye disease characterized by a keratinization of conjunctiva and cornea. Wolbach and Howe (1925) showed that in vitamin-A-deficient animals, many epithelia are transformed into a stratified keratinized tissue resembling epidermis (squamous metaplasia). The importance of these early studies was later highlighted when Fell and Mellanby (1953) made the converse observation that an excess of vitamin A not only inhibits keratinization, but even induces a mucous metaplasia of chick embryo skin. These discoveries lead to the idea that a critical concentration of vitamin A is required for normal differentiation of epithelia (including epidermis), while an excess (or a deficiency) in this vitamin alters the normal pathways of differentiation and produces metaplasias in opposite directions. Similar conclusions were drawn from the examination of the cutaneous symptoms of human hypovitaminosis A and hypervitaminosis A. Frazier and Hu (1931), examining Chinese soldiers who had suffered from vitamin A deficiency, noticed that in addition to the typical ocular symptoms, these subjects exhibited a dry, shriveled, scaly skin characterized at the histological level by a hyperkeratinization of epidermis and hair follicles associated with a squamous metaplasia of the sweat ducts. Hypervitaminosis A is very rare in humans, but some observations of this condition have been made in arctic explorers who had eaten polar bear liver, or after massive oral vitamin A therapy (Kamm *et al.*, 1984). Although the cutaneous symptoms observed such as desquamation and alopecia cannot be interpreted with certainty as a decreased keratinization, they nevertheless show that excess vitamin A impairs epidermis and hair follicles.

The chronology of these discoveries illustrates well the reason why epidermis became a model tissue to study how retinoids may balance the epithelial phenotype, but much attention has also focused on the effect of retinoids on skin because of their utility in the treatment of diseases such as acne and psoriasis (reviewed by Bollag and Geiger, 1984), skin ageing (Kligman *et al.*, 1986), and skin cancer prevention (reviewed by Lippman *et al.*, 1987). The effects of retinoids on the expression of a malignant phenotype by epithelial tissues were recognized very early. Wolbach and Howe (1925) in their initial study on vitamin A deficiency suggested that the metaplastic tissues had acquired neoplastic properties. Shortly thereafter, Fujimaki (1926) reported a higher incidence of gastric cancers in vitamin-A-deprived rats. Later, experimental carcinogenesis was found to be elevated in vitamin-A-deficient animals (Rowe and Gorlin, 1959), while retinoids were able to induce a regression of tumors induced by topical treatment of the skin with DMBA and tumor promoters (reviewed by Boutwell, 1983). In humans, retinoids have been shown to be effective in the chemoprevention of premalignant and malignant lesions such as actinic keratosis and basal cell carcinoma and in the treatment of premalignancies such as bronchial metaplasia and bladder disease (reviewed by Bollag and Geiger, 1984 and Lippman *et al.*, 1987). Very recently, retinoids have been shown to reduce the relapse of surgically excised squamous cell carcinomas (Hong *et al.*, 1990) and to induce remissions in acute promyelocytic leukemias (Chomienne *et al.*, 1989).

Molecular Mechanisms of Retinoid Action

Retinoic Acid Receptors

The most important recent development in our understanding of the effects of vitamin A came with the discovery of nuclear receptors of retinoic acid. Nuclear receptors are proteins that have three fundamental functions: (1)

they bind their ligand with high affinity and specificity; (2) they bind to DNA at specific sites called recognition elements (REs) in the vicinity of regulated genes; and (3) they interact with the transcriptional machinery to modulate the level of transcription of regulated genes (review by Evans, 1988 and Green and Chambon, 1988). We can now assume that retinoids belong to a category of hormones including steroid and thyroid hormones that mediate their effects by binding to nuclear receptors. By association, (1) keratinocytes would be expected to contain the retinoic acid receptor(s), and (2) the promoter regions of genes expressed during epidermal differentiation and sensitive to retinoic acid would contain retinoic acid-responsive elements (RAREs).

Retinoic acid receptors (RARs) are built according to a blueprint common to all proteins of the steroid hormone receptor family: they consist of six domains, designated A–F, with separable, largely independent functions (Fig. 1). The C domain is possibly the best understood one: it binds DNA with two adjacent zinc finger structures. The zinc fingers provide direct

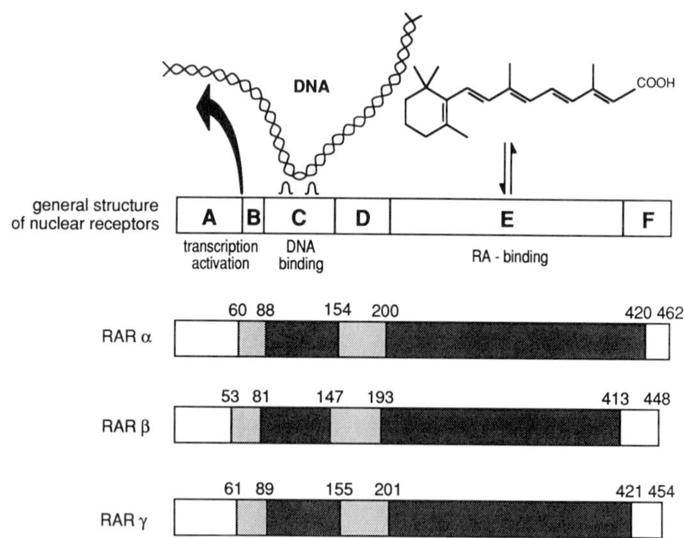

FIGURE 1 Schematic representation of the primary structure of human retinoic acid receptors RARα, RARβ, and RARγ. The primary structure of retinoic acid receptors can be divided into six regions A, B, C, D, E, F by analogy with the general structure of nuclear receptors (Green and Chambon, 1988). The A and B regions are involved in transcriptional activation, the C region in DNA binding, and the E region in RA binding. The sequence position of the last amino acid of each domain is indicated. Dark, light and no shading represent regions where similarities of the amino acid sequences are high (>80%), moderate (60–80%), or low (>25%), respectively. RARα (Petkovitch *et al.* 1987; Giguere *et al.* 1987; Brand *et al.* 1988), RARβ (De Thé *et al.* 1987; Benbrook *et al.* 1988; Brand *et al.* 1988), RARγ (Krust *et al.* 1989). (Reproduced by permission from Darmon, 1990.)

interaction with the bases in the major grove of DNA, and are thus responsible for the specificity of RE sequence recognition (Luisi *et al.*, 1991; Mader *et al.*, 1989). The function of the D domain is obscure and, while it is sometimes simply called the hinge, it has a role in down-regulation of transcription by those nuclear receptors that can both increase and decrease transcription (Adler *et al.*, 1988). The E domain binds ligand and is also responsible for the dimerization of the receptors (Forman and Samuels, 1990). In addition, the E domain interacts with the transcriptional machinery (i.e., it is a transcriptional activator) (Webster *et al.*, 1988). When linked to another nuclear regulatory protein, the E domain can confer ligand-dependent activation (Superti-Furga *et al.*, 1991). The amino-terminal domains A and B are also transcriptional activators, but their function is cell-type and target-gene dependent (Bocquel *et al.*, 1989).

Three receptors RARα, β, and γ mediate the effects of retinoic acid (Daly and Redfern, 1987; Petkovich *et al.*, 1987; Giguère *et al.*, 1987; Benbrook *et al.*, 1988; Brand *et al.*, 1988; Zelent *et al.*, 1989; Krust *et al.*, 1989), while another family of receptors, RXRs, may mediate the effects of other endogenous as yet unidentified retinoids (Mangelsdorf, 1990). The RXRs have almost three orders of magnitude weaker affinity for RA than the RARs, an observation that prompted the suggestion that their natural ligand is a retinoid which remains to be found ("retinoid X"),[1] and they were thus designated "RXR." Each receptor is encoded by its own gene, but multiple promoters and splicing variants combine to produce approximately seven different mRNA isoforms from each gene. Not all isoforms encode different proteins, however (Zelent *et al.*, 1991; Leroy *et al.*, 1991). The functions of multiple receptor isoforms are not understood but they seem to perform different regulatory activities and may inhibit the activity of other isoforms. For example, RARγ1 acts as a competitor of RARγ2 and RARβ2 in the transactivation of a RARE-*tk-cat* reporter (Husmann *et al.*, 1991). The three receptors can have different effects on transcription of certain genes (M. Blumenberg, unpublished) and different affinities for certain natural and synthetic retinoids (Delescluse *et al.*, 1991; Lehman *et al.*, 1991). The tissue distribution of RXRs is not yet known, but their transcriptional functions may be distinct from the functions of RARs (Rottman *et al.*, 1991; M. Blumenberg, unpublished).

The expression of RARs is tissue-specific. RARα is found, in relatively low levels, in many different tissues (Dollé *et al.*, 1989). Its function may be to provide a basal level of response to RA signal. In some tissues, RARβ is greatly induced by RA, which provides a feedback amplification of the RA signal. In F9 teratocarcinoma cells, for instance, RA-mediated induction of differentiation causes early induction of RARβ, which facilitates subsequent expression of differentiation markers (De Thé *et al.*, 1990a). Although some

[1]"Retinoid X" has been identified recently as 9-cis retinoic acid.

RARα may be expressed in epidermis, RARβ seems to be completely absent from this tissue. Contrary to other tissues that are the site of an autoregulatory loop of RAR expression (de Thé *et al.*, 1990a), RARβ is neither expressed nor induced by retinoic acid in keratinocytes (Elder *et al.*, 1991). Moreover, the responsive element involved in RARβ autoregulation (RARβ-RARE) does not function in keratinocytes (Miquel and M. Darmon *et al.*, 1993), probably because RARγ1, the major RAR isoform of this cell type (Kastner *et al.*, 1990) is unable to elicit a transactivation response after interacting with the RARβ/RARE (Husmann *et al.*, 1991).

The expression of RARγ is restricted to several tissues including skin, certain other stratified epithelia, and developing bone. RARγ is by far the predominant species expressed in epidermis (Krust *et al.*, 1989; Zelent *et al.*, 1989; Dollé *et al.*, 1989; Ruberte *et al.*, 1990). RARγ may thus have a tissue-specific function. An even more restricted tissue specificity seems to be conferred by alternative splicing: among seven possible RARγ mRNA isoforms (Giguère *et al.*, 1990; Kastner *et al.*, 1990), only one (RARγ1) is expressed in skin. RARγ1 mRNA differs from other RARγ isoforms both in its 5′ untranslated region and in the region coding for the N-terminal region of the receptor. These differences could confer both tissue-specific stability/translational efficiency and specificity of transcriptional function (Kastner *et al.*, 1990). Elegant *in situ* hybridization experiments (Dollé *et al.*, 1989; Ruberte *et al.*, 1990) have shown that in the skin of the late murine embryo and of the adult, RARγ is the predominant RAR specimen. Interestingly, this is also the case in the whisker follicles, in the oral cavity, in the esophagus, and in the forestomach, all of which are stratified epithelia known to be sites of abnormal keratinization in vitamin A deficiency. However, hybridization with the specific RARγ probe stops abruptly at the beginning of the glandular epithelium in the stomach.

Retinoic Acid Responsive Elements

Up to now, RAREs (or RXREs) have been identified in promoter regions of the genes that encode the following: RARβ (De Thé *et al.*, 1990a; Sucov *et al.*, 1990), laminin B1 (Vasios *et al.*, 1989), osteocalcin (Schüle *et al.*, 1990), CRBPI (Smith *et al.*, 1991), CRBPII (Mangelsdorf *et al.*, 1991), apoliprotein A1 (Rottman *et al.*, 1991), alcohol dehydrogenase ADH3 (Duester *et al.*, 1991), complement factor H (Munoz-Canoves, 1990), and phosphoenolpyruvate carboxykinase (Lucas *et al.*, 1991).

Both Rars and RXRs can recognize directly repeated consensus motifs
A T
G TCA. The same repeats can be recognized by the thyroid hormone and
G G
vitamin D3 receptors (Umesono *et al.*, 1991). The spacing between the half sites seems to determine the specificity of the effects of the receptors (Näär *et al.*, 1991; Umesono *et al.*, 1991). Palindromes and inverted palindromes of

the half sites are also functional. In keratin genes, the REs seem to consist of clusters of half sites (Tomic *et al.*, 1992). The specificity and level of activity of receptors is also modulated by accessory proteins that are as yet only partially characterized (Glass *et al.*, 1990).

Modulation of Tissular Retinoic Acid Concentration

The plasma carrier of vitamin A, retinol-binding protein (RBP), is able to diffuse into the intercellular space of epidermis, and is thus in contact with keratinocytes (Vahlquist *et al.*, 1985). Albumin, the carrier of retinoic acid, is also present in the intercellular space (Rabilloud *et al.*, 1989). Although retinol concentration in human plasma is in the micromolar range (Furr and Olson, 1988), retinoic acid concentration is a hundred-fold lower (Napoli *et al.*, 1985). Specific dehydrogenases enable keratinocytes and other cell types to transform retinol into retinoic acid (Williams and Napoli, 1984; Siegenthaler *et al.*, 1990). Thus, it must be assumed that the amount of biologically active retinoids synthesized *in situ* from retinol by keratinocytes is at least as important as the amount taken up from plasma (McCormick and Napoli, 1982). Interestingly, these dehydrogenases seem to be expressed only when keratinocytes differentiate (Siegenthaler *et al.*, 1990).

In addition to nuclear receptors, RA binds to cytosolic retinoic acid binding proteins CRABPs (Ong and Chytil, 1978; Jetten and Jetten, 1979; Giguère *et al.*, 1990), which do not appear to play an essential role in the control exerted by retinoids on specific cellular synthesis (Douer and Koeffler, 1982). Certain retinoic acid analogs, which are biologically active and induce the same panel of differentiation markers as retinoic acid, are unable to bind to CRABPs, but bind to retinoic acid receptors (Jetten *et al.*, 1987; Darmon *et al.*, 1988). Moreover, these analogs are able to modulate epidermal differentiation (Asselineau *et al.*, 1992) and embryonic limb morphogenesis (Maden *et al.*, 1991) in a manner similar to retinoic acid or those retinoid analogs able to bind to CRABPs. However, although CRABPs are not implicated in the transduction of the retinoid signal, they may be involved in the regulation of the intracellular concentration of active retinoic acid, since their dissociation constants for retinoic acid are close to those of the retinoic acid receptors. CRABPs can thus either sequester RA away from RARs, or deliver it to catabolizing enzymes (Fiorella and Napoli, 1991).

The tissue distribution of CRABPs is strictly controlled (Dollé *et al.*, 1989) and CRABPII is inducible by RA (Giguère *et al.*, 1990). Furthermore, the expression of cytoplasmic retinol binding protein (CRBP) is regulated by RARs via REs present in CRBPI and CRBPII genes (Smith *et al.*, 1991; Mangelsdorf *et al.*, 1991). Similarly, at least one of the genes encoding human alcohol dehydrogenase (ADH3) contains an RE regulated by RARs (Duester *et al.*, 1991). This alcohol dehydrogenase is capable of converting retinol to RA, thus potentially amplifying the retinoid signal. Therefore, we can distinguish three feedback loops by which biologically active retinoids might modulate their own concentration: (1) they can induce CRBPs and thus regulate the intracellular concentration of retinol, (2) they can up-regulate enzymes of retinoic acid biosynthesis, (3) they can up-regulate the levels of CRABPs and thus decrease the retinoic acid active concentration.

Retinoic Acid and Embryonic Development of Epithelia

The effects produced during embryonic development by the addition of retinoids are remarkable because, contrary to teratogens, which always produce defects, retinoids are able to provoke the formation of additional structures. The best examples are the duplications obtained in the limbs of the chick embryo or in limb regeneration in amphibians (Maden, 1982; Tickle *et al.*, 1982; Summerbell, 1983; reviewed by Tabin, 1991). These unique experimental effects and the presence of retinoids in developing tissues have lead to the conclusion that retinoids are the endogenous agents that specify directly or indirectly the morphogenesis of the limb bud (Thaller and Eichele, 1987). They are probably also involved in the development of certain structures of the central nervous system (Wagner *et al.*, 1990). Thus, the addition of exogenous retinoids may alter morphogenesis by changing the concentration of endogenous morphogens or by altering their spatiotemporal gradients.

It is not clear whether endogenous retinoids control the morphogenesis of epithelial organs, or whether they only modulate their differentiation. The effects produced by these agents during the development of these organs are nevertheless dramatic, particularly when the fate of the cells has not yet been determined. The importance of cell type determination is highlighted by the characteristics of the mucous metaplasia produced by retinoic acid in chick embryo skin (Fell and Mellanby, 1953). A treatment with high concentrations of vitamin A or retinoic acid arrests keratinization and provokes the appearance of mucin-containing cells in the skin of the embryo. This process is apparently reversible because a keratinized squamous epithelium reappears when the explant is transferred into a medium devoid of retinoic acid. However, the changes undergone by the metaplastic cells are irreversible because mucous epithelial cells are eliminated when retinoic acid is removed, and a new population, consisting of keratinocytes, develops from the basal cells (Peck *et al.*, 1977). These results indicate that in embryos of 7 to 14 days, the stem cells contained in epidermis are still bipotential. Hardy (1967) performed a kinetic study on explants of mouse embryo skin to demonstrate that the developmental stage (i.e., the potentiality) of the target cells conditions the effect produced by retinoids. When the skin of 14-day embryos is treated with vitamin A, there is an impairment of keratinization but no metaplasia, indicating that epidermal stem cells are irreversibly determined at that stage. However, this is not the case for vibrissae stem cells. Hardy (1968) also reported that when the skin of the snout area of 14-day embryos is cultured in the presence of vitamin A, the large vibrissae follicles are transformed into branching mucous-secreting glands. Another spectacular example of "transdetermination" of skin appendages in the chick was later described by Dhouailly *et al.* (1980). These authors obtained the formation of feathers on chick foot scales by injecting retinoic acid in the amniotic cavity of 10–12-day chick embryos.

The molecular mechanisms by which retinoids affect embryonic development are unknown, but interesting inroads into this question have been made by studying the effects of retinoic acid on regulation of expression of homeobox (*Hox*) genes. *Drosophila Hox* genes determine the insect body plan, and distribution of transcripts of mammalian *Hox* genes indicates that they may perform the same function (reviewed in Kessel and Gruss, 1990). Retinoic acid regulates the expression of human and murine *Hox* genes and can effect a "homeotic transformation" of the skeleton (Simeone *et al.*, 1990; Morriss-Kay *et al.*, 1991; Kessel and Gruss, 1991). These homeotic changes are dependent upon the timing of the retinoic acid addition, because, according to the developmental stage, different *Hox* genes can be regulated by retinoic acid (Kessel and Gruss, 1991).

As mentioned above, experiments performed with exogenous retinoids cannot predict the extent to which endogenous retinoids are involved in organ morphogenesis in the developing embryo. The study of effects on embryonic development produced by retinoid deficiency would probably provide an answer to this question; however, such experiments are not feasible because vitamin-A-deprived animals become sterile. Indirect approaches such as the study of embryos bearing targeted mutations in retinoic acid receptors, or the use of retinoic acid-antagonists (currently under investigation at this institution-M.D.) may provide future clues to the role of endogenous retinoids in development.

Retinoic Acid and Differentiation of Adult Epithelia

The study of the symptoms of vitamin A deprivation in deficiency diseases in humans, studied experimentally in animals, has contributed essential data on control of epithelial differentiation. They can be summarized as follows: (1) Simple epithelia such as liver, kidney, or intestine do not undergo metaplasia, but slight phenotypic changes may be observed (Manville, 1937; DeLuca *et al.*, 1969); (2) epidermis, a stratified keratinized epithelium, becomes hyperkeratinized (Frazier and Hu, 1931); (3) conjunctiva and cornea, which are stratified, but normally nonkeratinized, undergo a squamous metaplasia (Mori, 1922; Wolbach and Howe 1925) as do many columnar and transitional epithelia, such as lachrymal glands, trachea, bladder, and prostate (Wolbach and Howe, 1925; Moore, 1957; Lasnitzski, 1962; Sporn *et al.*, 1975; Reese and Friedman, 1978). Two interpretations of squamous metaplasia can be postulated. Either two different stem cells reside in these epithelia, one of which is dormant in the presence of retinoids or stem cells of these epithelia are still bipotential in the adult, and retinoic acid influences their determination.

The effects produced by systemic hypervitaminosis A in the adult cannot be interpreted as easily as those of deprivation. Epithelia are clearly affected as shown by abnormal desquamation and alopecia, but signs of general toxicity, weight loss, and bone defects predominate. To avoid the toxicity of high doses of vitamin A given systemically, the skin can be treated topically with retinoids and the effects on epidermis and appendages monitored. Such treatments certainly provoke an inhibition of keratinization, but the extent varies considerably from species to species. Although mucous-like material can be detected in epidermis (Schultz-Ehrenburg and Orfanos, 1981), a real mucous metaplasia is not seen. Christopher and Wolff (1975) treated guinea pig ear epidermis with retinoic acid. After five days, they observed an important hyperplasia and a complete block of terminal differentiation as evidenced by the persistence of nuclei in the most superficial layers. This absence of mature stratum corneum is called parakeratosis. Interestingly, although a few applications of the drug provoke a disappearance of keratohyaline granules, these reappear and form a thick granular layer when the drug is applied repeatedly. This implies that keratinocytes are able to adapt to the increased retinoid concentration. In mice and humans, topical treatment with retinoic acid produces hyperplasia and hypergranulosis, as well as a thinning and disorganization of the stratum corneum, but a complete parakeratosis is not observed (Elias, 1986).

Schweizer *et al.* (1987) treated tails of adult mice with retinoic acid and obtained paradoxical results: while the normal adult tail skin is formed by an alternation of orthokeratotic (i.e., with a mature anucleated stratum corneum) interscale epidermis and parakeratotic scale regions, the treated skin becomes uniformly orthokeratotic. This result contrasts with the decreased keratinization or even the parakeratosis obtained in other areas. These authors concluded from specific alterations in the keratin pattern that retinoic acid is able to provoke a complete renewal of the scaly parakeratotic epithelium by reprogramming basal cells toward an "orthokeratotic fate."

Effects of Retinoic Acid on Epidermal Differentiation

Studies performed with cultured keratinocytes, based on the knowledge obtained *in vivo*, have been aimed at (1) elucidating the mode of action of retinoids in epidermal differentiation and malignant transformation of keratinocytes, and (2) designing bioassays to evaluate the potency or the receptor selectivity of synthetic retinoids with the goal of developing new drugs.

Epidermal differentiation (for a review see Green, 1979; Rheinwald, 1979; Watt, 1989) is a continuous process leading to the formation of superficial cornified cells that form the protective stratum corneum and are eventually sloughed off. In the steady state, a number of cells equal to that of the desquamated cells is produced in the germinative compartment formed by the innermost layers. The successive steps of the keratinocyte differentiation program are coupled to the outward migration of the differentiated cells, which thus form a stratified epithelium with layers of different phenotypes.

Basal cells, attached to the basement membrane, contain a stem cell population, are highly proliferative, and display an undifferentiated keratinocyte phenotype characterized by specific antigens and the "basal" K5/K14 keratin couple. Phenotypic changes occur as soon as keratinocytes start detaching from the basement membrane. The most evident difference is the synthesis of a new set of keratins, the "suprabasal" K1/K10 couple (Fuchs and Green, 1980, 1981; reviewed by Fuchs, 1988). The density of desmosomes also increases, and the resulting morphology is said to be "spinous" (hence the name of the first suprabasal layers). Above these lie the granular layers. At this stage, the cytoplasm of the cells contains membrane coating granules (MCG) and keratohyaline granules. The latter structure is rich in filaggrin, a protein that plays a role in aggregation of keratin filaments. Granular cells also synthesize the precursors of crosslinked envelopes: involucrin, keratolinin, loricrin, and a membrane-bound transglutaminase, the enzyme responsible for the crosslinking process. Ultimately, crosslinking occurs leading to the formation of the resistant cornified envelopes. The nucleus is digested, the lipid content of the MCGs excreted, and the protective impermeable stratum corneum built.

The first observation made that retinoids suppress keratinization in tissue culture was reported by Yuspa and Harris (1974) who showed that retinyl acetate blocks the morphological differentiation of cultured mouse keratinocytes. They also showed that secretory activity and glycoprotein synthesis were increased, an observation similar to the mucous metaplasia observed in the chick embryo. These observations were confirmed and extended by DeLuca and Yuspa (1974) and others (reviewed by Shapiro, 1986). Fuchs and Green (1981) studied in more detail the effects of vitamin A on the biochemical and morphological differentiation of keratinocytes. They made the following intriguing observation: when human keratinocytes are cultured in the presence of nondelipidized serum (i.e., containing vitamin A), epidermal differentiation does not occur. But, delipidization of the serum is sufficient to induce differentiation. Readdition of vitamin A to the medium blocks this process, as shown by the suppression of K1 keratin synthesis.

Fuchs and Green (1981) reported that a concentration of vitamin A as low as 2% of that found in human plasma is sufficient to block epidermal differentiation. Then how do keratinocytes escape this inhibition *in vivo*? Experiments performed by growing the keratinocytes on a dermal substrate in an emerged position relative to the culture medium (Asselineau *et al.*, 1989) seem to solve the above paradox. In fact, it has been observed that, under these conditions which mimic the *in vivo* situation, keratinocytes are able to stratify and differentiate normally (orthokeratosis) in the presence of normal nondelipidized serum (i.e., in the presence of physiological concentrations of vitamin A). In contrast, when the keratinocytes are grown on the dermal substrate, but submerged by the culture medium, differentiation and stratification only occur when delipidized serum is used. To reconcile these results with the observation of Fuchs and Green, we propose that retinoids are filtered through the dermal substrate and less bioavailable to keratinocytes separated from the medium when emerged *in vitro* or separated from the plasma *in vivo*. Interestingly, when the emerged culture system is used with delipidized serum, the epidermis is hyperkeratotic, as it is *in vivo* in vitamin A deficiency (Frazier and Hu, 1931). By contrast, when retinoic acid is added to the emerged cultures grown in delipidized serum, a dose-dependent (from $10^{-9}M$ to $10^{-6}M$) reduction of keratinization is observed (see Fig. 2). At the lower concentration of retinoic acid, hyperkeratosis disappears and the epithelium becomes orthokeratotic, while at higher concentrations it gradually becomes parakeratotic, although stratification is not impaired. Parallel to the morphological effect, several differentiation markers progressively disappear when retinoic acid concentration is increased. These experiments demonstrated that isolated keratinocytes contain all the information necessary to undergo a complex morphogenetic process (the formation of a stratified epithelium) and that they do so in a way that is strictly dependent on the surrounding retinoic acid concentration. This approximates the situation *in vivo* with epidermal keratinocytes when vitamin A intake is not in the normal physiological range.

Since the pioneering work of Fuchs and Green (1981) that showed that the synthesis of K1 keratin is inhibited by retinoic acid, many other differentiation markers have been shown to be down-regulated. Retinoic acid not only inhibits the synthesis of the K1/K10 suprabasal keratin pair (Kopan *et al.*, 1987), but also the synthesis of profilaggrin (Fleckman *et al.*, 1984) and its conversion into filaggrin (Asselineau *et al.*, 1990). These effects correlate well with the inhibition of keratohyaline formation observed by Chopra and Flaxman in 1975. The synthesis of the cornified envelope precursor loricrin (Magnaldo *et al.*, 1992) and of the crosslinking membrane-bound transglutaminase (Thacher *et al.*, 1985; Rubin and Rice, 1986; Floyd and Jetten 1989; Michel *et al.*, 1988) are also inhibited by retinoic acid, explaining the global inhibition of cornified envelope formation (Green and Watt, 1982; Nagae *et al.*, 1987). Several bioassays of retinoids are based on these inhibi-

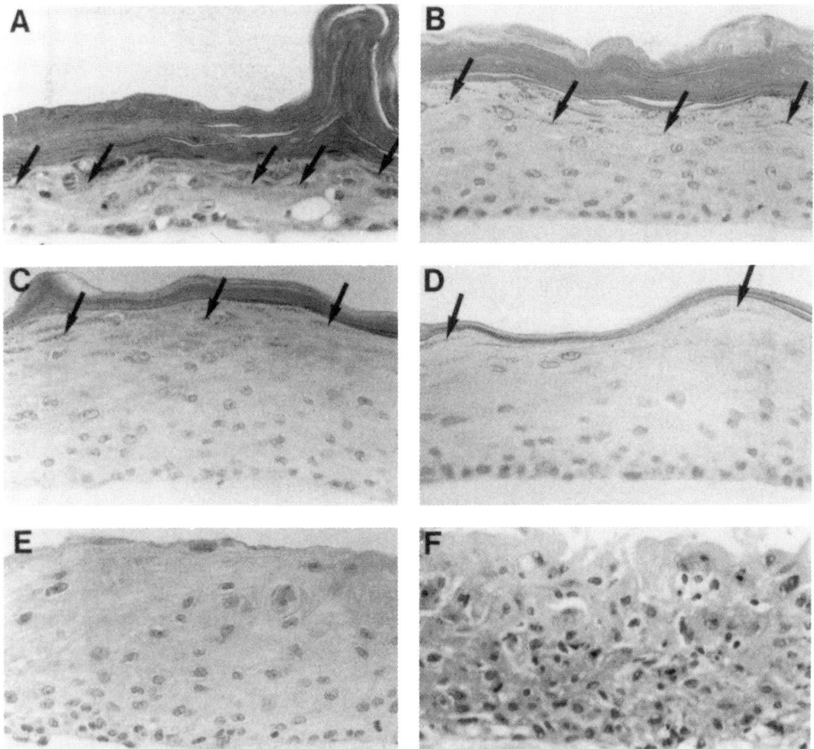

FIGURE 2 Morphology of the epithelium formed by human keratinocytes grown for 2 weeks on dermal equivalent emerged the second week in absence of retinoic acid (*A*), or in the presence of $10^{-10}M$ (*B*), $10^{-9}M$ (*C*), $10^{-8}M$ (*D*), $10^{-7}M$ (*E*), and $10^{-6}M$ (*F*) retinoic acid. Granular (*arrows*) and horny layers were focused. Bar, 30μ*M*. Note the gradual disappearance of the stratum corneum and stratum granulosum (From Asselineau, Bernard, Bailly, and Darmon, *Dev. Biol.* **133**, 322–335, 1989 with permission.)

tory effects on specific proteins or on cornified envelope formation (Régnier *et al.*, 1989; Michel *et al.*, 1991).

The inhibitory effects of retinoids on differentiation markers of keratinocytes and other epithelial cell types have been shown to occur at the transcriptional level, as demonstrated by northern blot analysis, RNAse protection assays, and run-off transcription of isolated nuclei (Eckert and Green, 1984; Gilfix and Eckert, 1985; Kim *et al.*, 1987; Floyd and Jetten, 1989, Stellmach and Fuchs, 1989; Magnaldo *et al.*, 1992). Although RARs and RXRs are probably involved in the downregulation of epidermal markers, negative RAREs have not yet been characterized in the corresponding genes except the gene encoding the EGF receptor (Hudson *et al.*, 1990) and possibly the K14 gene (Tomic *et al.*, 1992).

Some epidermal differentiation markers exhibit a higher sensitivity to retinoids than others (Gilfix and Eckert, 1985; Régnier and Darmon, 1988; Asselineau *et al.*, 1989). It has been observed that the markers most sensitive to retinoic acid are expressed later in differentiation (thus "higher" in the epithelium) than the less sensitive ones that are expressed "deeper" (Asselineau *et al.*, 1989). By manipulating the concentration of retinoic acid in the medium, it is possible to change the "depth" at which the synthesis of a given marker starts. These observations suggest that, as postulated in the case of the limb bud, a gradient of retinoic acid might span the epidermis with the concentration decreasing from the basal to external layers. Such a gradient associated with a differential gene sensitivity toward retinoic acid could explain the stepwise distribution of the markers in epidermis. One could imagine, for instance, that negative RAREs in the promoter regions of the corresponding genes would exhibit varying affinities for the RARs. In this context, a given marker would start to be expressed in a layer in which the retinoic acid concentration is below a certain threshold. If the affinity for the RARs of the RARE of marker A is lower than that of the RARE of marker B, the consequence would be a "deeper" expression in the epithelium of marker A than of marker B.

Some markers of terminal differentiation are not down-regulated by retinoic acid. An example is the case of one of the envelope precursors, involucrin (Green and Watt, 1982; Asselineau *et al.*, 1989). A quantitative estimate of the relative importance of markers belonging to the retinoid-controlled differentiation program versus the nonretinoid controlled program can be obtained by computerized comparison of two-dimensional gels of proteins in studies performed in the presence and absence of retinoids under conditions favoring or inhibiting terminal differentiation. From one of these studies it appears that only one third of the differentiation markers are repressed by retinoic acid (Rabilloud *et al.*, 1989). No differentiation marker was found to be induced by retinoic acid, but some markers of the undifferentiated stage were induced. This might explain why markers normally restricted to the basal layer, such as the fibronectin receptor, extend to several layers of cells in cultured epidermis treated with retinoic acid (Asselineau *et al.*, 1989). However, this result does not mean that the parakeratotic epithelium obtained *in vitro* by exposure to retinoic acid is formed by several layers of cells of "basal" phenotype. For instance, involucrin, a differentiation marker not sensitive to retinoic acid, appears in the suprabasal layers of retinoic-acid-treated parakeratotic epidermis as it does in normal epidermis. This indicates that the formation of two-cell compartments in epidermis: the basal and the suprabasal, is not controlled by retinoic acid. The existence of two compartments, independent of the retinoic acid concentration, is also demonstrated by the fact that suprabasal markers sensitive to retinoic acid, such as K1/K10 keratins, are not ectopically expressed in the basal layer when epidermis is cultured in the absence of retinoids (Asselineau *et al.*, 1989).

From the above results, it is clear that a simple model of epidermal differentiation that asserts that retinoic acid induces markers of the undifferentiated state and represses those of the differentiated state is unlikely. In fact, some markers of the basal keratinocytes are repressed (although only partially) by retinoic acid. This is true for mRNAs of the basal K5/K14 couple (Gilfix and Eckert, 1985). The case of the K14 keratin gene is particularly interesting since its promoter might harbor a negative responsive element (Tomic *et al.,* 1990) that is also sensitive to thyroid hormone (RARE/TRE). In view of this and other examples of hormone responsive-elements activated by more than one class of receptors (Umesono *et al.,* 1988; Glass *et al.,* 1989; Schüle *et al.,* 1990) the suggestion can be made that the actual modulation of transcription may depend upon the interplay of several receptors and hormones (Umesono *et al.,* 1988; Glass *et al.,* 1989; Graupner *et al.,* 1989). The promoter of the EGF receptor gene also contains a RARE/TRE that behaves as a negative responsive element in CV1 cells (Hudson *et al.,* 1990). Paradoxically, in keratinocytes, the EGF receptor down-regulation, which normally occurs during epidermal differentiation, is abolished by retinoids (Ponec *et al.,* 1987). This discrepancy suggests that, as in the case of the glucocorticoid receptor, the positive or negative behavior of a responsive element might depend not only upon interactions between hormone receptors, but also between hormone receptors and classical transcriptional activators such as AP_1 (Diamond *et al.,* 1990). Interaction between RARs and AP_1 might thus be responsible for the negative effect of RA on the transcription of the collagenase gene (Schüle *et al.,* 1991).

Some markers such as K19 keratin usually present in simple, but not in stratified epithelia, and K13 keratin usually present in wet stratified epithelia, are induced by retinoids in epidermal cells (Fuchs and Green, 1981; Eckert and Green, 1984; Gilfix and Eckert, 1985; Kopan *et al.,* 1987). The ectopic expression of these keratins can be interpreted as a sign of metaplasia, reminiscent of the early studies describing the acquisition of mucous features by retinoic acid-treated epidermis (Fell and Mellanby, 1953). The repression of the basal K5/K14 couple can also be interpreted as a global depression of the keratinocyte phenotype.

The molecular mechanisms by which RA controls epithelial differentiation have been the subject of intensive study. RA regulates transcription of both differentiation markers (see above) and transcription factors that may constitute steps in the cascade of regulation of differentiation, such as *Oct*3, AP_2 and *zif*268, (Lüscher *et al.,* 1989; Suva *et al.,* 1991; Okazawa *et al.,* 1991).

Another interesting effect of retinoic acid on epidermis is that it seems to abolish the polarity of the basal keratinocytes. The basal domain of the membrane of basal keratinocytes is characterized by the presence of the bullous pemphigoid (BP) antigen and certain integrins, while the lateral domains are characterized by different antigens and integrins (Darmon *et al.,* 1987; DeLuca *et al.,* 1990; Carter *et al.,* 1990). Retinoic acid is able to provoke the redistribution of the BP antigen into all domains of the membrane (Thivolet *et al.*, 1984; Régnier and Darmon, 1988), whereas the antigens characteristic of the lateral faces of the basal keratinocytes are induced in several suprabasal layers (Asselineau *et al.*, 1989). Moreover, laminin, one of the major components of the dermal–epidermal basement membrane, whose gene is known to be activated by retinoic acid and contains a RARE (Vasios *et al.*, 1989), is expressed ectopically in suprabasal cells of epidermis cultured in the presence of retinoic acid (D. Asselineau and M. Darmon, unpublished results). In conclusion, the parakeratotic epidermis produced by culturing keratinocytes in high retinoic acid concentrations is partially (but not totally) undifferentiated, has lost its polarity, and is slightly metaplastic.

The molecular approaches described above are aimed toward understanding the mechanism by which retinoid signals are interpreted by keratinocytes. It is equally important to understand the role played by epidermal cells in the generation of the "retinoid signals." In epidermis, 3,4-dehydroretinol can be synthesized from retinol (Vahlquist *et al.*, 1985) and possibly also 3,4-dehydroretinoic acid. This phenomenon of epidermis is interesting because in the embryo, 3,4-dehydroretinoic acid is probably an endogenous morphogen as important as retinoic acid (Thaller and Eichele, 1990). The presence of two hormonal forms derived from vitamin A in epidermis might have a functional significance if their synthesis/degradation rates differ and if their binding spectra to retinoid receptors (RARα, β, and γ and RXRs) and/or the CRABPs are different. This interpretation is compatible with the fact that the levels of CRABP- and retinoic-acid-synthesizing enzymes both increase when keratinocytes differentiate (Siegenthaler *et al.*, 1988; 1990).

To our knowledge, there has been no study of the enzymes that inactivate retinoic acid within epidermis, but these degradative enzymes may play an equally important role as that of synthesizing enzymes in the control of active retinoid levels (Williams and Napoli, 1984). The isolation, characterization, and cloning of the enzymes of retinoid metabolism will be a prerequisite for the understanding of how levels of biologically active retinoids are regulated in skin.

Effects of Retinoic Acid on Keratinocyte Proliferation and Tumorigenesis

In certain experimental systems, particularly those using murine keratinocytes, retinoids reduce proliferation. Moreover, several studies claim that the proliferation of tumor cells is more sensitive to retinoic acid than normal ones (Kim *et al.*, 1984; Rubin and Rice, 1986). In that regard, the induction of TGFβ2 and the repression of ornithine decarboxylase (ODC) by retinoic acid are particularly interesting (Glick *et al.*, 1989). In primary mouse keratinocytes, retinoic acid provokes an increase in the levels of TGFβ2 and its mRNA (probably via a posttranscriptional mechanism). Moreover, an anti-TGFβ2 antibody is able to partially block the inhibition of DNA synthesis caused by retinoic acid. Thus, the effects of retinoids on cell proliferation (reviewed by Shapiro, 1986), which vary much more according to the species and the cellular system used than the effects on differentiation, might be dependent upon the synthesis of other hormonal agents.

Olsen *et al.* (1990) showed that retinoids reduce the levels of ODC and of its mRNA by a transcriptional, but probably indirect mechanism. Since ODC levels reflect the proliferative state of cells, this example, and that of TGFβ2, indicate that retinoids might exhibit an anti-proliferative effect on normal and neoplastic cells. By controlling the differentiated phenotype and hence changing the cell repertoire of oncogenes, anti-oncogenes, growth factors, and growth factor receptors, retinoids may have indirect effects on the proliferation of premalignant tissues and tumors. Moreover, as in acute promyelocytic leukemia (De Thé *et al.*, 1990b; Kakizuka *et al.*, 1991), and perhaps in some hepatocarcinomas (Dejean *et al.*, 1986), RAR genes may be altered or rearranged in malignancies.

In epidermis, the negative effect of retinoic acid on collagenase (Bailly *et al.*, 1990) might also contribute to reduce the tumor phenotype as well as being involved in the antiinflammatory and antiageing effects of retinoids. The molecular basis of the control exerted by retinoic acid on the collagenase gene has been investigated recently (Lafyatis et al., 1990). Interestingly, it gives clues about the well-known antagonism between retinoic acid and tumor promoters such as TPA (reviewed by Boutwell, 1983). The collagenase gene can be induced by TPA via an AP_1 site in its promoter region. This stimulation of transcription is probably due to an induction of c-*jun* and c-*fos* by TPA. Retinoic acid has been shown not only to block the TPA induction of collagenase, but also to antagonize the TPA induction of c-*fos*. The inhibition of c-*fos* by retinoic acid may thus account for the inhibitory effect of this compound on collagenase transcription and its antagonism of TPA effects. Other mechanisms might converge towards an inhibition of collagenase transcription by retinoic acid. For instance, the RARs might interact directly with c-fos/c-jun proteins (Schüle *et al.*, 1990) as shown in the case of the glucocorticoid receptor (Diamond *et al.*, 1990; Jonat *et al.*, 1990).

Perspectives

In a field moving as fast as that of retinoid research, major new developments must be expected. The recognition of the roles of retinoic and didehydroretinoic acid as morphogens in vertebrates has brought life to notions such as morphogenetic gradients and positional information, which, although applying directly to invertebrate developmental models, had thus far remained speculative for vertebrates. As with research on the morphogenetic role of growth factors, retinoid research has allowed developmental biologists to "think in hormonal terms." In this context, it will be particularly important to understand how retinoids (and perhaps other hormones) control the expression of homeobox genes.

The possibility of producing genetically engineered embryos or animals that affect a gain or loss of function, or contain regulatory mutations, and the ability to address receptors to ectopic sites and to trace expression by appropriate promoters and responsive elements linked to reporter genes will represent a fantastic source of new information. Data concerning enzymes of retinoid metabolism will complete the information given by the distribution and activity of receptors and will tell us more about the generation and characteristics of retinoid signals. The discovery of new genes regulated by retinoids as ensembles coordinately or hierarchically expressed will eventually permit us to define which genetic elements and transactivators are responsible for a given phenotypic change. Information concerning epithelia will be only a part of the whole picture. Pharmacology and therapeutics will certainly benefit from these discoveries. One can hope that the synthesis of retinoid agonists and antagonists selective for receptor subclasses will help to improve the treatment of leukemias, cancers, and dermatological diseases, and to reduce adverse effects and teratogenic risks.

References

Adler, S., Waterman, M. L., He, X., and Rosenfeld, M. G. (1988). Steroid receptor-mediated inhibition of rat prolactin gene expression does not require the receptor DNA-binding domain. *Cell* **52**, 685–695.

Asselineau, D., Bernard, B. A., Bailly, C., and Darmon, M. (1989). Retinoic acid improves epidermal morphogenesis. *Dev. Biol.* **133**, 322–335.

Asselineau, D., Dale, B. A., and Bernard, B. A. (1990). Filaggrin production by cultured human epidermal keratinocytes and its regulation by retinoic acid. *Differentiation* **45**, 221–229.

Asselineau, D., Cavey, M. T., Shroot, B., and Darmon, M. (1992). Control of epidermal differentiation by a retinoid analogue unable to bind to cytosolic retinoic acid binding proteins (CRABPs). *J. Invest. Dermatol.* **98**, 128–134.

Bailly, C., Dreze, M. S., Asselineau, D., Nusgens, B., Lapière, C. M., and Darmon, M. (1990). Retinoic acid inhibits the production of collagenase by human epidermal keratinocytes. *J. Invest. Dermatol.* **94**, 47–51.

Benbrook, D., Lernhardt, E., and Pfahl, M. (1988). A new retinoic acid receptor identified from a hepatocellular carcinoma. *Nature (London)* **333**, 679–672.

Bocquel, M. T., Kumar, V., Stricker, C., Chambon, P., and Gronemeyer, H. (1989). The contribution of the N- and C-terminal regions of steroid receptors to activation of transcription is both receptor and cell-specific. *Nucleic Acids Res.* **17**, 2581–2595.

Bollag, W. and Geiger, J.-M. (1984). The development of retinoids in dermatology. *In* "Retinoid Therapy" (W. J. Cunliffe and A. J. Miller eds. pp. 1–7), MTP Press Limited. Lancaster, U.K.

Boutwell, R. K. (1983). The role of retinoids as protective agents in experimental carcinogenesis. *In* "Protective Agents in Cancer" (D. C. H. McBrien and T. F. Slater, eds.), pp. 279–297 Academic Press, New York.

Brand, N., Petkovich, M., Krust, A., Chambon, P., De Thé, H., Marchio, A., Tiollais, P., and Dejean, A. (1988). Identification of a second retinoic acid receptor. *Nature* **332,** 850–853.

Carter, W. C., Kaur, P., Gil, S. G., Gahr, P. J., and Wayner, E. A. (1990). Distinct functions for integrins $\alpha 3\beta 1$ in focal adhesions and $\alpha 6\beta 4$/bullous pemphigoid antigen in a new stable anchoring contact (SAC) of keratinocytes: relation to hemidesmosomes. *J. Cell. Biol.* **111,** 3141–3154.

Chomienne, C., Bellerini, P., Balitrand, N., Amar, M., Bernard, J. F., Bovin, P., Daniel, M. T., Berger, R., Castaigne, S., and Degos, L. (1989). Retinoic acid therapy for promyelocytic leukaemia. *Lancet* **8665,** 746–747.

Chopra, D. P., and Flaxman, B. A. (1975). The effect of vitamin A on growth and differentiation of human keratinocytes *in vitro. J. Invest. Dermatol.* **64,** 19–22.

Christopher, E. and Wolff, H. H. (1975). Effects of vitamin A acid in skin: *In vivo* and in vitro studies. *Acta Derm. Venereol.* **74** (Suppl.), 42–49.

Daly, A. K. and Redfern, C. P. F. (1987). Characterization of a retinoic acid binding component from F9 embryonal carcinoma cell nuclei. *Eur. J. Biochem.* **168,** 133–19.

Darmon, M., Schaffar, L., Bernard, B. A., Régnier, M., Asselineau, D., Verschoore, M., Lamaud, E., and Schalla, W. (1987). Polarity of basal keratinocytes, basement membrane and epidermal permeability. *In* "Pharmacology and the Skin" (H. Schaefer and B. Shroot, eds.), Vol. 1. P. 1021. Skin pharmacokinetics. Karger, Basel.

Darmon, M. (1990). The nuclear receptors of retinoic acid. *J. Lipid Mediators* **2,** 247–256.

Darmon, M., Rocher, M., Cavey, M. T., Martin, B., Rabilloud, T., Delescluse, C., and Shroot, B. (1988). Biological activity of retinoids correlates with affinity for nuclear receptors but not for cytosolic binding protein. *Skin Pharmacol.* **1,** 161–175.

De Thé, H., Marchio, A., Tiollais, P., and Dejean, A. (1987). A novel steroid thyroid hormone receptor-related gene inappropriately expressed in human hepatocellular carcinoma. *Nature* **330,** 667–670.

De Thé, H., del Mar Vivanco-Ruiz, M., Tiollais, P., Stunnenberg, H., and Dejean, A. (1990a). Identification of a retinoic acid responsive element in the retinoic acid receptor beta gene. *Nature* **343,** 177–180.

De Thé, H., Chomienne, C., Lanotte, M., Degos, L., and Dejean, A. (1990b). The *t*(15;17) translocation of acute promyelocytic leukaemia fuses the retinoic acid receptor alpha gene to a novel transcribed locus. *Nature* **347,** 558–561.

Dejean, A., Bougueleret, L., Grzeschik, K-H., and Tiollais, P. (1986). Hepatitis B virus DNA integration in a sequence homologous to *v-erb-A* and steroid receptor genes in a hepatocellular carcinoma. *Nature* **322,** 70–72.

Delescluse, C., Cavey, M. T., Martin, B., Bernard, B., Reichert, U., Maignon, J., Darmon, M., and Shroot, B. (1991). Selective high affinity RAR alpha or beta-RAR gamma retinoic acid receptor ligands. *Mol. Pharmacol.* **40,** 556–562.

DeLuca, H. F. (1979). Retinoic acid metabolism. *Federation Proc.* **38,** 2519–2523.

DeLuca, L. and Yuspa, S. H. (1974). Altered glycoprotein synthesis in mouse epidermal cells treated with retinyl acetate. *Exp. Cell Res.* **86,** 106–110.

DeLuca, L., Little, P. E., and Wolf, G. (1969). Vitamin A and protein synthesis by rat intestinal mucosa. *J. Biol. Chem.* **244,** 701–708.

DeLuca, M., Ramura, R. N., Kajiji, S., Bondanza, S., Rosino, P., Cancedda, R., Marchisio, P. C., and Quaranta, V. (1990). Polarized integrin mediates human keratinocyte adhesion to basal lamina. *Proc. Natl. Acad. Sci. U.S.A.* **87,** 6888–6892.

Dhouailly, D., Hardy, M. H., and Sengel, P. (1980). Formation of feathers on chick foot scales: a stage-dependent morphogenetic response to retinoic acid. *J. Embryol. Exp. Morphol.* **58,** 63–78.

Diamond, M. I., Miner, J. N., Yoshinaga, S. K., and Yamamoto, K. R. (1990). Transcription factor interactions: Selectors of positive or negative regulation from a single DNA element. *Science,* **249,** 1266–1272.

Dollé, P., Ruberte, E., Kastner, P., Petkovich, M., Stoner, C. M., Gudas, L. J., and Chambon, P. (1989). Differential expression of genes encoding alpha, beta, and gamma retinoic acid receptors and CRABP in the developing limbs of the mouse. *Nature,* **342,** 702–705.

Douer, D. and Koeffler, H. P. (1982). Retinoic acid. Inhibition of the clonal growth of human myeloid leukemia cells. *J. Clin. Invest.* **69,** 277–283.

Dowling, J. E. and Wald, G. (1960). The biological function of vitamin A acid. *Proc. Natl. Acad. Sci. U.S.A.* **46,** 587–608.

Drummond, J. C. (1920). Nomenclature of the so-called accessory food factors (vitamines). *Biochem. J.* **14,** 660.

Duester, G., Shean, M. L., McBride, M. S., Stewart, M. J. (1991). Retinoic acid response element in the human alcohol dehydrogenase gene ADH_3: Implications for regulation of retinoic acid synthesis. *Mol. Cell. Biol.* **11,** 1638–1646.

Durston, A. J., Timmermans, J. P. M., Hage, W. J., Hendriks, H. F. J., de Vries, N. J., Heideveld, M., and Nieuwkoop, P. D. (1989). Retinoic acid causes an anteroposterior transformation in the developing central nervous system. *Nature (London)* **340,** 140–144.

Eckert, R. L. and Green, H. (1984). Cloning of cDNAs specifying vitamin A-responsive human keratins. *Proc. Natl. Acad. Sci. U.S.A.* **81,** 4321–4325.

Elder, J. T., Fisher, G. J., Zhang, Q.-Y., Eisen, D., Krust, A., Kastner, P., Chambon, P., and Voorhees, J. (1991). Retinoic acid receptor gene expression in human skin. *J. Invest. Dermatol.* **96,** 425–433.

Elias, P. M. (1986). Epidermal effects of retinoids: Supramolecular observations and clinical implications. *J. Am. Acad. Dermatol.* **15,** 797–809.

Evans, R. M. (1988). The steroid and thyroid hormone receptor superfamily. *Science* **240,** 889–895.

Fell, H. B. and Mellanby, E. (1953). Metaplasia produced in cultures of chick ectoderm by vitamin A. *J. Physiol.* **119,** 470–488.

Fiorella, P. D. and Napoli, J. L. (1991). Expression of cellular retinoic acid binding protein (CRABP) in *Escherichia coli.* Characterization and evidence that holo-CRABP is a substrate in retinoic acid metabolism. *J. Biol. Chem.* **266,** 16572–16579.

Fleckman, P., Haydock, P., Blomquist, C., and Dale, B. A. (1984). Profilaggrin and the 67-kD keratin are coordinately expressed in cultured human epidermal keratinocytes. *J. Cell Biol.* **99,** 315A.

Floyd, E. E. and A. M. Jetten. (1989). Regulation of type I (epidermal) transglutaminase mRNA levels during squamous differentiation: Down regulation by retinoids. *Mol. Cell. Biol.* **9,** 4846–4851.

Forman, B. M. and Samuels, H. H. (1990). Interactions among a subfamily of nuclear hormone receptors: The regulatory zipper model. *Mol. Endocrin.* **4,** 1923–1301.

Frazier, C. N. and Hu, C. K. (1931). Cutaneous lesions associated with a deficiency in vitamin A in man. *Arch. Intern. Med.* **48,** 507–514.

Frolik, C. A. (1984). Metabolism of retinoids. *In* "The Retinoids" (eds. M. B. Sporn, A. B. Roberts, and D. S. Goodman, eds.), Vol. 2 pp. 177–208. Academic Press, Orlando, Florida.

Fuchs, E. (1988). Keratins as biochemical markers of epithelial differentiation. *Trends Genet.* **4,** 277–281.

Fuchs, E. and Green, H. (1981). Regulation of terminal differentiation of cultured human keratinocytes by vitamin A. *Cell* **25,** 617–625.

Fuchs, E. V. and Green, H. (1980). Changes in keratin gene expression during terminal differentiation of the keratinocyte. *Cell* **19,** 1033–1042.

Fujimaki, Y. (1926). Formation of gastric carcinoma in albino rats fed on deficient diets. *J. Cancer Res.* **10,** 469–477.

Furr, H. C. and Olson, J. A. (1988). A direct microassay for serum retinol (Vitamin A alcohol) by using size-exclusion high-pressure liquid chromatography with fluorescence detection. *Anal. Biochem.* **171,** 360–365.

Giguère, V., Ong, E. S., Segui, P., and Evans, R. (1987). Identification of a receptor for the morphogen retinoic acid. *Nature (London)* **330**, 624–629.

Giguère, V., Shago, M., Zirngibl, R., Tate, P., Rossant, J., and Varmuza, S. (1990). Identification of a novel isoform of the retinoic acid receptor gamma expressed in the mouse embryo. *Mol. Cell. Biol.* **10**, 2335–2340.

Gilfix, B. M. and Eckert, R. L. (1985). Coordinate control by vitamin A of keratin gene expression in human keratinocytes. *J. Biol. Chem.* **260**, 14026–14029.

Glass, C. K., Lipkin, S. M., Devary, O. V., and Rosenfeld, M. G. (1989). Positive and negative regulation of gene transcription by a retinoic acid–thyroid hormone receptor heterodimer. *Cell* **59**, 697–708.

Glass, C. K., Devary, R. V., and Rosenfeld, M. G. (1990). Multiple cell type-specific proteins differentially regulate target sequence recognition by the alpha retinoic acid receptor. *Cell* **63**, 729–738.

Glick, A. B., Flanders, K. C., Danielpour, D., Yuspa, S. H., and Sporn, M. B. (1989). Retinoic acid induces transforming growth factor-β_2 in cultured keratinocytes and mouse epidermis. *Cell Regulation* **1**, 87–97.

Graupner, G., Wills, K. N., Tzukerman, M., Zhang, X-K., and Pfahl, M. (1989). Dual regulatory role for thyroid-hormone receptors allows control of retinoic-acid receptor activity. *Nature (London)* **340**, 653–656.

Green, H. (1979). The keratinocyte as differentiated cell type. Harvey Lectures, Ser. **74**, 101–139.

Green, S. and Chambon, P. (1988). Nuclear receptors enhance our understanding of transcription regulation. *Trends Genet.* **4**, 309–314.

Green, H. and Watt, F. M. (1982). Regulation by vitamin A of envelope cross-linking in cultured keratinocytes derived from different human epithelia. *Mol. Cell Biol.* **2**, 1115–1117.

Gudas, L. J. (1990). Molecular mechanisms of retinoid action. *Am. J. Respir. Cell Mol. Biol.* **2**, 319–320.

Hardy, M. H. (1967). Responses in embryonic mouse skin to excess vitamin A in organotypic cultures from the trunk, upper lip, and lower jaw. *Exp. Cell Res.* **46**, 367–384.

Hardy, M. H. (1968). Glandular metaplasia of hair follicles and other responses to vitamin A excess in cultures of rodent skin. *J. Embryol. Exp. Morphol.* **19**, 157–180.

Hong, W. K., Lippman, S. M., Itri, L. M., Karp, D. D., Lee, J. S., Byers, R. M., Schantz, S. P., Kramer, A. M., Lotan, R., Peters, L. J., Dimery, I. W., Brown, B. W., and Goepfert, H. (1990). Prevention of second primary tumors with isotretinoin in squamous-cell carcinoma of the head and neck. *N. Engl. J. Med.* **323**, 795–801.

Hudson, L. G., Santon, J. B., Glass, C. K., and Gill, G. N. (1990). Ligand-activated thyroid hormone and retinoic acid receptors inhibit growth factor receptor promoter expression. *Cell* **62**, 1165–1175.

Husmann, M., Lehmann, J., Hoffmann, B., Hermann, T., Tzukerman, M., and Pfahl, M. (1991). Antagonism between retinoic acid receptors. *Mol. Cell. Biol.* **11**, 4097–4103.

Jetten, A. M. and Jetten, M. E. R. (1979). Possible role of retinoic acid binding protein in retinoic stimulation of embryonal carcinoma cell differentiation. *Nature (London)* **278**, 180–182.

Jetten, A. M., Anderson, K., Deas, M. A., Kagechika, H., Lotan, R., Rearick, J. I., and Shudo, K. (1987). New benzoic acid derivatives with retinoic activity: Lack of direct correlation between biological activity and binding to cellular retinoic acid binding protein. *Cancer Res.* **47**, 3523–3527.

Jonat, C., Rahmsdorf, H. J., Park, K-K., Cato, A. C. B., Gebel, S., Ponta, H., and Herrlich, P. (1990). Antitumor promotion and antiinflammation: Down-modulation of AP-$_1$ (*fos/jun*) activity by glucocorticoid hormone. *Cell*, **62**, 1189–1204.

Kakizuka, A., Miller, W. H. Jr., Umesono, K., Warrell, R. P. Jr., Frankel, S. R., Murty, V. V. V. S., Dmitrovsky, E., and Evans, R. M. (1991). Chromosomal translocation *t*(15;17) in human acute promyelocytic leukemia fuses RARalpha with a novel putative transcription factor, PML. *Cell* **66**, 663–674.

Kamm, J. J., Ashenfelter, K. O., and Ehrmann, C. W. (1984). Preclinical and clinical toxicology of selected retinoids. *In* "The Retinoids" (M. B. Sporn, A. B. Roberts, and D. S. Goodman, eds.), Vol. 2 pp. 288–326. Academic Press, Orlando, Florida.

Kastner, Ph., Krust, A., Mendelsohn, C., Garnier, J. M., Zelent, A., Leroy, P., and Staub, A. (1990). Murine isoforms of retinoic acid receptor gamma with specific patterns of expression. *Proc. Natl. Acad. Sci. U.S.A.* **87**, 2700–2704.

Kessel, M. and Gruss, P. (1990). Murine developmental control genes. *Science* **249**, 374–379.

Kessel, M. and Gruss, P. (1991). Homeotic transformations of murine vertebrae and concomitant alteration of hox codes induced by retinoic acid. *Cell* **67**, 89–104.

Kim, K. H., Schwartz, F., and Fuchs, E. (1984). Differences in keratin synthesis between normal epithelial cells and squamous cell carcinomas are mediated by vitamin A. *Proc. Natl. Acad. Sci. U.S.A.* **81**, 4280–4284.

Kim, K. H., Stellmach, V., Javors, J., and Fuchs, E. (1987). Regulation of human mesothelial cell differentiation: Opposing roles of retinoids and epidermal growth factor in the expression of intermediate filament proteins. *J. Cell. Biol.* **105**, 3039–3051.

Kligman, A. M., Grove, G. L., Hirose, R., and Leyden, J. J. (1986). Topical tretinoin for photoaged skin. *J. Am. Acad. Dermatol.* **15**, 836–859.

Kopan, R., Traska, G., and Fuchs, E. (1987). Retinoids as important regulators of terminal differentiation: Examining keratin expression in individual epidermal cells at various stages of keratinization. *J. Cell. Biol.* **105**, 427–440.

Krust, A., Kastner, P., Petkovich, M., Zelent, A., and Chambon, P. (1989). A third human retinoic acid receptor hRARγ. *Proc. Natl. Acad. Sci. U.S.A.* **86**, 5310–5314.

Lafyatis, R., Kim, S-J., Angel, P., Roberts, A. B., Sporn, B., Karin, M., and Wilder, R. L. (1990). Interleukin-l stimulates and all-*trans*-retinoic acid inhibits collagenase gene expression through its 5′ activator protein-l binding site. *Mol. Endocrinol.* **4**, 973–980.

Lasnitzki, I. (1962). Hypovitaminosis A in the mouse prostate gland cultured in chemically defined medium. *Exp. Cell Res.* **28**, 40–51.

Lehman, J. M., Dawson, M. I., Hobbs, P. D., Husmann, M., and Pfahl, M. (1991). Identification of retinoids with nuclear receptor subtype-selective activities. *Cancer Res.* **51**, 4804–4809.

Leroy, P., Krust, A., Zelent, A., Mendelsohn, C., Garnier, J.-M., Kastner, P., Dierich, A., and Chambon, P. (1991). Multiple isoforms of the mouse retinoic acid receptor alpha are generated by alternative splicing and differential induction by retinoic acid. *EMBO J.* **10**, 59–69.

Lippman, S. M., Kessler, J. F., and Meyskens, F. L. Jr. (1987). Retinoids as preventive and therapeutic anticancer agents (Part II). *Cancer Treatment Rep.* **71**, 493–515.

Lotan, R. (1980). Effects of vitamin A and its analogs (retinoids) on normal and neoplastic cells. *Biochim. Biophys. Acta* **605**, 33–91.

Lucas, P. C., O'Brien, R. M., Mitchell, J. A., Davis, C. M., Imai, E., Forman, B. M., Samuels, H. H., and Granner, D. K. (1991). A retinoic acid response element is part of a pleiotropic domain in the phosphoenolpyruvate carboxykinase gene. *Proc. Natl. Acad. Sci. U.S.A.* **88**, 2184–2188.

Luisi, B. F., Xu, W. X., Otwinowski, Z., Freedman, L. P., Yamamoto, K. R., and Sigler, P. B. (1991). Crystallographic analysis of the interaction of the glucocorticoid receptor with DNA. *Nature (London)* **352**, 497–505.

Lüscher, B., Mitchell, P. J., Williams, T., and Tjian, R. (1989). Regulation of transcription factor AP-2 by the morphogen retinoic acid and by second messengers. *Genes & Dev.* **3**, 1507–1517.

Maden, M. (1982). Vitamin A and pattern formation in the regenerating limb. *Nature (London)* **295**, 672–675.

Maden, M. (1985). Retinoids and the control of pattern in limb development and regeneration. *Trends Genet.* **1**, 103–107.

Maden, M., Summerbell, D., Maignan, T., Darmon, M., and Shroot, B. (1991). The respecification of limb pattern by new synthetic retinoids and their interaction with CRABP. *Differentiation* **47**, 49–55.

Mader, S., Kumar, V., de Verneuil, H., and Chambon, P. (1989). Three amino acids of the estrogen receptor are essential to its ability to distinguish an oestrogen from a glucocorticoid-responsive element. *Nature (London)* **338**, 271–274.

Magnaldo, M., Pommès, L., Asselineau, D., and Darmon, M. (1990). Isolation of a GC-rich cDNA identifying mRNAs present in human epidermis and modulated by calcium and retinoic acid in cultured keratinocytes. Homology with murine loricrin mRNA. *Mol. Biol. Rep.* **14**, 237–246.

Magnaldo, T., Bernerd, F., Asselineau, D., and Darmon, M. (1992). Expression of loricrin is negatively controlled by retinoic acid in human epidermis reconstructed in vitro. *Differentiation* **49**, 39–46.

Mangelsdorf, D. J., Ong, E. S., Dyck, J. A., and Evans, R. M. (1990). Nuclear receptor that identifies a novel retinoic acid response pathway. *Nature (London),* **345**, 224–229.

Mangelsdorf, D. J., Umesono, K., Kliewer, S. A., Borgmeyer, U., Ong, E. S., and Evans, R. M. (1991). A direct repeat in the cellular retinol-binding protein type II gene confers differential regulation by RXR and RAR. *Cell* **66**, 555–561.

Manville, I. A. (1937). The relationship of vitamin A and glucuronic acid in mucine metabolism. *Science* **85**, 44–45.

McCollum, E. V. and Davis, M. (1913). The necessity of certain lipids in the diet during growth. *J. Biol. Chem.* **15**, 167–175.

McCormick, A. M. and Napoli, J. L. (1982). Identification of 5,6-epoxyretinoic acid as an endogenous retinol metabolite. *J. Biol. Chem.* **257**, 1730–1733.

Michel, S., Reichert, U., Isnard, J. L., Shroot, B., and Schmidt, R. (1988). Retinoic acid controls expression of epidermal transglutaminase at the pretranslatational level. *FEBS Lett.* **258**, 35–38.

Michel, S., Courseaux, A., Miquel, C., Bernardon, J. M., Schmidt, R., Shroot, B., Thacher, S. M., and Reichert, U. (1991). Determination of retinoid activity by an enzyme-linked immunosorbent assay. *Anal. Biochem.* **192**, 232–236.

Miguel, C., Clusel, C., Semat, A., Gerst, C., and Darmon, M. (1993). Retinoic acid receptor RAR gamma 1: an antagonist of the transactivation of the RAR beta RARE in epithelial cell lines and normal human keratinocytes. *Mol. Biol. Rep.* **17**, 35–45.

Moore, T. (1957). Xerophthalmia and other epithelial lesions. *In* "Vitamin A" (T. Moore, ed.), pp. 301–313. Elsevier, Amsterdam.

Mori, S. (1922). The changes in the paraocular glands which follow the administration of diets low in fat-soluble vitamin A; with notes of the effect of the same diets on salivary glands and the muco of the larynx and trachea. *Bull. Johns Hopkins Hosp.* **33**, 357–359.

Morriss-Kay, G. M., Murphy, P., Hill, R. E., and Davidson, D. R. (1991). Effects of retinoic acid excess on expression of *Hox*-2.9 and *Krox*-20 and on morphological segmentation in the hindbrain of mouse embryos. *EMBO J.* **10**, 2985–2995.

Morton, R. A. (1944). Chemical aspects of the visual process. *Nature (London)* **153**, 69–71.

Munoz-Canoves, P., Vik, D. P., and Tack, B. F. (1990). Mapping of a retinoic acid-responsive element in the promoter region of the complement factor H gene. *J. Biol. Chem.* **265**, 20065–20068.

Näär, A. M., Boutin, J.-M., Lipkin, S. M., Yu, V. C., Holloway, J. M., Glass, C. K., and Rosenfeld, M. G. (1991). The orientation and spacing of core DNA-binding motifs dictate selective transcriptional responses to three nuclear receptors. *Cell* **65**, 1267–1279.

Nagae, S., Lichti, U., De Luca, L., and Yuspa, S. H. (1987). Effect of retinoic acid on cornified envelope formation: difference between spontaneous envelope formation *in vivo* or *in vitro* and expression of envelope competence. *J. Invest. Dermatol.* **89**, 51–58.

Napoli, J. L., Pramanik, B. C., Williams, J. B., Dawson, M. I., and Hobbs, P. D. (1985). Quantification of retinoic acid by gas–liquid chromatography–mass spectrometry: Total versus all-*trans*-retinoic acid in human plasma. *J. Lipid Res.* **26**, 387–392.

Okazawa, H., Okamoto, K., Ishino, F., Ishino-Kaneko, T., Takeda, S., Toyoda, Y., Muramatsu, M., and Hamada, H. (1991). The *oct*[3] gene, a gene for an embryonic transcription factor, is controlled by a retinoic acid repressible enhancer. *EMBO J.* **10**, 2997–3005.

Olsen, D. R., Hickok, N. J., Uitto, J. (1990). Suppression of ornithine decarboxylase gene expression by retinoids in cultured human keratinocytes. *J. Invest. Dermatol.* **94**, 33–36.

Ong, D. E. and Chytil, F. (1978). Cellular retinoic-acid binding protein from rat testis. *J. Biol. Chem.* **253**, 4551–4554.

Osborne, T. B. and Mendel, L. B. (1913). The relation of growth to the chemical constituents of the diet. *J. Biol. Chem.* **15**, 311–326.

Peck, G. C., Elias, P. M., and Wetzel, B. W. (1977). Effects of retinoic acid on embryonic chick skin. *J. Invest. Dermatol.* **69**, 463–476.

Petkovich, M., Brand, N. J., Krust, A., and Chambon, P. (1987). A human retinoic acid receptor which belongs to the family of nuclear receptors. *Nature (London)* **330**, 444–450.

Ponec, M., Weerheim, A., Havekes, L., and Boonstra, J. (1987). Effects of retinoids on differentiation, lipid metabolism, epidermal growth factor, and low-density lipoprotein binding in squamous carcinoma cells. *Exp. Cell Res.* **171**, 426–435.

Rabilloud, T., Asselineau, D., Bailly, C., Miquel, C., Tarroux, P., and Darmon, M. (1989). Study of the human epidermal differentiation by two-dimensional electrophoresis. *In* "Electrophoresis '88" (C. Schaefer-Nielsen, ed.), pp. 95–101. Verlag Chemie, Munich.

Ragsdale, C. W. Jr., Petkovich, M., Gates, P. B., Chambon, P., and Brockes, J. P. (1989). Identification of a novel retinoic acid receptor in regenerative tissues of the newt. *Nature (London)* **341**, 654–657.

Reese, D. H. and Friedman, R. D. (1978). Effect of retinoic acid on oxygen-cultured transitional epithelium. *Invest. Urol.* **16**, 39–45.

Régnier, M. and Darmon M. (1989). Human epidermis reconstructed *in vitro:* A model to study keratinocyte differentiation and its modulation by retinoic acid. *In Vitro Cell. & Devel. Biol.* **25**, 1000–1008.

Régnier, M., Eustache, J., Shroot, B., and Darmon, M. (1989). Bioassays for retinoic acid-like substances using cultured human keratinocytes. *Skin Pharmacol.* **2**, 1–9.

Rheinwald, J. G. (1979). The role of terminal differentiation in the finite culture lifetime of the human epidermal keratinocyte. *Int. Rev. Cytol.,* Suppl. 10, 25–33.

Roberts, A. B. and Sporn, M. B. (1984). Cellular biology and biochemistry of the retinoids. *In* "The Retinoids" (M. B. Sporn, A. B. Roberts, and D. S. Goodman, eds.), Vol. 2 pp. 209–286. Academic Press, Orlando, Florida.

Rottman, J. N., Widon, R. L., Nadal-Ginard, B., Mahdavi, V., and Karathanasis, S. K. (1991). A retinoic acid-responsive element in the apolipoprotein AI gene distinguishes between two different retinoic acid response pathways. *Mol. Cell. Biol.* **11**, 3814–3820.

Rowe, N. H. and Gorlin, R. J. (1959). The effect of vitamin A deficiency upon experimental oral carcinogenesis. *J. Dent. Res.* **38**, 72–83.

Ruberte, E., Dollé, P., Krust, A., Zelent, A., Morriss-Kay, G., and Chambon, P. (1990). Specific spatial and temporal distribution of retinoic acid receptor gamma transcripts during mouse embryogenesis. *Development* **108**, 213–222.

Rubin, A. L. and Rice, R. H. (1986). Differential regulation by retinoic acid and calcium of transglutaminases in cultured neoplastic and normal human keratinocytes. *Cancer Res.* **46**, 2356–2361.

Schüle, R., Umesono, K., Mangelsdorf, D. J., Bolado, J., Wesley Pike, J., and Evans, R. M. (1990). *Jun-fos* and receptors for vitamins A and D recognize a common response element in the human osteocalcin gene. *Cell* **61**, 497–504.

Schüle, R., Rangarajan, P., Yang, N., Kliewer, S., Ransone, L. J., Bolado, J., Verma, I. M., and

Evans, R. M. (1991). Retinoic acid is a negative regulator of AP-1-responsive genes. *Proc. Natl. Acad. Sci. U.S.A.* **88,** 6092–6096.

Schultz-Ehrenburg, U. and Orfanos, C. E. (1981). Light and electron microscopic changes of human epidermis under oral retinoid treatment. *In* "Retinoids" (C. E. Orfanos, ed.), pp. 85–92. Springer Verlag.

Schweizer, J., Furstenberger, G., and Winter, H. (1987). Selective suppression of two postnatally acquired 70kD and 65kD keratin proteins during continuous treatment of adult mouse tail epidermis with vitamin A. *J. Invest. Dermatol.* **89,** 125–131.

Shapiro, S. S. (1986). Retinoids and epithelial differentiation. *In* "Retinoids and Cell Differentiation" (M. I. Sherman, ed.), pp. 29–60 CRC Press, Boca Raton, Florida.

Sherman, M. I. (1986). Retinoids and cell differentiation (M. I. Sherman, ed.), CRC Press, Boca Raton, Florida.

Siegenthaler, G., Saurat, J. H., and Ponec, M. (1988). Terminal differentiation in cultured human keratinocytes is associated with increased levels of cellular retinoic acid-binding protein. *Exp. Cell Res.* **178,** 114–126.

Siegenthaler, G., Saurat, J. H., and Ponec, M. (1990). Retinol and retinal metabolism. *Biochem. J.* **268,** 371–378.

Simeone, A., Acampora, D., Arcioni, L., Andrews, P. W., Boncinelli, E., and Mavilio, F. (1990). Sequential activation of *Hox2* homeobox genes by retinoic acid in human embryonal carcinoma cells. *Nature (London)* **346,** 763–766.

Smith, W. C., Nakshatri, H., Leroy, P., Rees, J., and Chambon, P. (1991). A retinoic acid response element is present in the mouse cellular retinol binding protein I (mCRBPI) promoter. *EMBO J.* **10,** 2223–2230.

Sporn, M. B. and Roberts, A. B. (1984). Biological methods of analysis and assay of retinoids—relationships between structure and activity. *In* "The Retinoids" (M. B. Sporn, A. B. Roberts, and D. S. Goodman, eds.), Vol. I pp. 236–279. Academic Press, Orlando, Florida.

Sporn, M. B., Clamon, G. H., Dunlop, N. M., Newton, D. L., Smith, J. M., and Saffiotti, U. (1975). Activity of vitamin A analogues in cultures of mouse epidermis organ cultures of hamster trachea. *Nature (London)* **253,** 47–50.

Stellmach, V. M. and Fuchs, E. (1989). Exploring the mechanisms underlying cell type-specific and retinoid-mediated expression of keratins. *The New Biologist* **1,** 305–317.

Stepp, W. (1909). Experiments on feeding with lipid-free food. *Biochem. Z.* **22,** 452–460.

Sucov, H. M., Murakami, K. K., and Evans, R. M. (1990). Characterization of an autoregulated response element in the mouse retinoic acid receptor type beta gene. *Proc. Natl. Acad. Sci. U.S.A.* **87,** 5392–5396.

Summerbell, D. (1983). The effect of local application of retinoic acid to the anterior margin of the developing chick limb. *J. Embryol. Exp. Morphol.* **78,** 269–289.

Superti-Furga, G., Bergers, G., Picard, D., and Busslinger, M. (1991). Hormone-dependent transcriptional regulation and cellular transformation by *fos*-steroid receptor fusion proteins. *Proc. Natl. Acad. Sci. U.S.A.* **88,** 5114–5118.

Suva, L. J., Ernst, M., and Rodan, G. A. (1991). Retinoic acid increases *zif*268 early gene expression in rat preosteoblastic cells. *Mol. Cell. Biol.* **11,** 2503–2510.

Tabin, C. J. (1991). Retinoids, homeoboxes, and growth factors: Toward molecular models of limb development. *Cell* **66,** 199–218.

Thacher, S. M., Coe, E. L., and Rice, R. H. (1985). Retinoid suppression of transglutaminase activity and envelope competence in cultured human epidermal carcinoma cells. *Differentiation* **29,** 82–87.

Thaller, C. and Eichele, G. (1987). Identification and spatial distribution of retinoids in the developing chick limb bud. *Nature (London)* **327,** 625–628.

Thaller, C. and Eichele, G. (1990). Isolation of 3,4-didehydroretinoic acid, a novel morphogenetic signal in the chick wing bud. *Nature (London)* **345,** 815–819.

Thivolet, C. H., Hintner, H. H., and Stanley, J. R. (1984). The effect of retinoic acid on the expression of pemphigus and pemphigoid antigens in cultured human keratinocytes. *J. Invest. Dermatol.* **82,** 329–334.

Tickle, C., Alberts, B., Wolpert, L., and Lee, J. (1982). Local application of retinoic acid to the limb bud mimics the action of the polarizing region. *Nature (London)* **296,** 564–566.

Tomic, M., Jiang, C.-K., Epstein, H. S., Freedberg, I. M., Samuels, H. H., and Blumenberg, M. (1990). Nuclear receptors for retinoic acid and thyroid hormone regulate transcription of keratin genes. *Cell Regulation* **1,** 965–978.

Tomic-Canic, M., Sunjevaric, I., Freedberg, I., and Blumenberg, M. (1992). Identification of the retinoic acid and thyroid hormone receptor-responsive element in the human K14 keratin gene *J. invest. dermatol.* **99,** 842–847.

Umesono, K., Giguère, V., Glass, C. K., Rosenfeld, M. G., Evans, R. M. (1988). Retinoic acid and thyroid hormone induce gene expression through a common responsive element. *Nature (London)* **336,** 262–265.

Umesono, K., Murakami, K. K., Thompson, C. C., and Evans, R. M. (1991). Direct repeats as selective response elements for the thyroid hormone, retinoic acid, and vitamin D_3 receptors. *Cell* **65,** 1255–1266.

Underwood, B. A. (1984). Vitamin A in animal and human nutrition. *In* "The Retinoids" (M. B. Sporn, A. B. Roberts, and D. S. Goodman, eds.), Vol. I pp. 281–392. Academic Press, Orlando, Florida.

Vahlquist, A., Törma, H., Rollman, O., and Berne, B. (1985). Distribution of natural and synthetic retinoids in the skin. *In* "Retinoids: New Trends in Research and Therapy" (J. H. Saurat, ed.) Retinoid Symp. Geneva, pp. 159–167. Karger, Basel.

Vasios, G. W., Gold, J. D., Petkovich, M., Chambon, P., and Gudas, L. J. (1989). A retinoic acid-responsive element is present in the 5′ flanking region of the laminin B_1 gene. *Proc. Natl. Acad. Sci. U.S.A.* **86,** 9099–9103.

Wagner, M., Thaller, C., Jessell, T., and Eichele, G. (1990). Polarizing activity and retinoid synthesis in the floor plate of the neural tube. *Nature (London)* **345,** 819–822.

Wald, G. (1934). Carotenoids and the vitamin A cycle in vision. *Nature (London)* **134,** 65.

Wald, G. and Hubbard, R. (1950). The synthesis of rhodopsin from vitamin A. *Proc. Natl. Acad. Sci. U.S.A.* **36,** 92–102.

Watt, F. M. (1989). Terminal differentiation of epidermal keratinocytes. *Curr. Opinion Cell Biol.* **1,** 1107–1115.

Webster, N. J. G., Green, S., Jin, J. R., and Chambon, P. (1988). The hormone-binding domains of the estrogen and glucocorticoid receptors contain an inducible transcription activation function. *Cell* **54,** 199–207.

Williams, J. B. and Napoli, J. L. (1984). Metabolism of retinol and retinoic acid during differentiation of F9 embryonal carcinoma cells. *FASEB J.* **43,** 788.

Wolbach, S. B. and Howe, P. R. (1925). Tissue changes following deprivation of fat soluble A vitamin. *J. Exp. Med.* **42,** 753–777.

Yuspa, S. H. and Harris, C. C. (1974). Altered differentiation of mouse epidermal cells treated with retinyl acetate *in vitro*. *Exp. Cell Res.* **86,** 95–105.

Zelent, A., Krust, A., Petkovich, M., Kastner, P., Chambon, P. (1989). Cloning of murine alpha and beta retinoic acid receptors and a novel receptor gamma predominantly expressed in skin. *Nature (London)* **339,** 714–717.

Zelent, A., Mendelsohn, C., Kastner, P., Krust, A., Garnier, J.-M. Ruffenach, F., Leroy, P., and Chambon, P. (1991). Differentially expressed isoforms of the mouse retinoic acid receptor beta are generated by usage of two promoters and alternative splicing. *EMBO J.* **10,** 71–81.

7

Human Papillomavirus and Malignant Transformation

Bruno A. Bernard

Introduction

Pathology

In spite of the abundant literature dealing with the biology and biochemistry of papillomaviruses (for reviews see Gissmann, 1984; Pfister, 1984; Kirchner, 1986; Giri and Danos, 1986; McCance, 1986; Ward *et al.*, 1989; zur Hausen, 1989), recent progress in the understanding of molecular events involved in epithelial cell transformation have made it necessary to update our knowledge of this rapidly growing field.

Papillomaviruses comprise a group of species-specific viruses—members of the family Papovaviridae—that infect humans and a wide range of animals. Infection of cutaneous or mucosal epithelium by these viruses results in the development of benign tumors that contain a variable amount of infectious virus. These tumors are called papillomas, derived from *papilla,* meaning pimple or pustule, and the suffix *-oma,* which denotes a tumor or a neoplasm. In humans, common warts and plantar warts are the most recognizable cutaneous lesions, but it is now clear that papillomaviruses are associated with a variety of lesions on many squamous epithelial surfaces including the cervix, vaginal wall, vulva, penis (genital warts are also described as condylomas), larynx, tongue, oral mucosa, and conjunctiva, as well as the skin (for review see McCance, 1986).

Studies on cottontail rabbit papillomavirus (CRPV)-induced conversion of benign papillomas into malign epithelial carcinomas (for review see Kreider, 1980) and of a rare human disease (epidermodysplasia verruciformis) (for review see Orth, 1986) had pointed to the oncogenic potential of some papillomaviruses and led to the hypothesis that papillomavirus infection might be a key event in the development of a number of epithelial neoplasms in humans. Nevertheless little initial interest was generated in these viruses mainly because they had not been grown in tissue culture and they were not then linked to any major disease process. However, since 1983 (Dürst *et al.,* 1983), the close association of particular human papillomaviruses with genital cancers has been widely documented and the study of these viruses boosted. Indeed, cervical cancer is the second cause of cancer in females worldwide (Parkin *et al.,* 1984) and is the primary cause of cancer-linked mortality in several countries of South America and Africa.

About sixty different human papillomaviruses have thus far been identified on the basis of a homology of their DNA of less than 50% (Table 1). Among these, HPV6, 11, 16, 18, 31, 33, 34 and 35 are associated with squamous cell tumors of the lower genital tract; HPV6 and 11 are generally associated with benign tumors (Table 2). Because of the higher frequency of HPV16 and 18 in genital carcinomas, most recent studies have focused on these two HPV types as will this review.

Structure of Papillomavirus

The papillomavirus genome is a double-stranded circular DNA of about 8000 base pairs, encapsidated into an icosahedral protein shell (52–55 nm in diameter) consisting of 72 capsomers. The genome organization of the various papillomaviruses is highly conserved, with two long functional domains each containing a series of open reading frames (ORF) that code for viral proteins (Fig. 1). Only one strand of the papillomavirus genome is transcribed. The E (early) region contains up to seven ORFs that code for the proteins involved in replication, transcription control, and transforming activity. The L (late) region contains two large ORFs coding for the structural proteins of the virus capsid (Giri and Danos, 1986). A 900 base-pair

TABLE 1 **HPV Types and Associated Clinical Lesions**

HPV type	Associated clinical lesion
1	Verruca plantaris
2, 4, 26, 27, 29, 57	Verruca vulgaris
3, 10, 28, 49	Verruca plana
5, 8, 9, 12, 14, 15, 17, 19, 20, 21, 22, 23, 24, 25, 36, 46, 47, 50	Epidermodysplasia verruciformis
7	Butcher's wart
6, 11, 54	Condyloma acuminatum
13, 32	Focal epithelial hyperplasia
39, 55	Bowenoid papulosis
16, 18, 30, 31, 33, 34, 35, 39, 40, 42, 43, 44, 45, 51, 52, 56, 57, 58	Cervical intraepithelial neoplasia
16, 18, 31, 33, 35, 39, 45, 51, 52, 56	Cervical carcinoma
41, 48	Cutaneous squamous cell carcinoma
6, 11	Laryngeal papilloma
37	Keratoacanthoma
38	Malignant melanoma
53	Normal cervical mucosa

noncoding region lies between the stop codon of L1 ORF and the start codon of E6 ORF, and is identified as the regulatory region (long control region, LCR). In the absence of an *in vitro* culture system for HPV (Taichman *et al.,* 1984), our understanding of the regulation of transcription, replication, and the mechanisms of pathogenesis has relied almost exclusively on the cloning, sequencing, and expression of the different open reading frames (ORF).

Papillomaviruses and Malignant Transformation

In Vivo Transformation

Physical State of the Virus

As seen in Table 2, certain HPV types are associated with dysplasias and malignancies of the lower genital tract. These dysplasias are usually graded as mild, moderate, or severe, and termed cervical intraepithelial neoplasia (CIN) grade I, II, and III, respectively. Dysplasias of the vulva are classified as VIN I to III, and those of the penis PIN I to III. Interestingly, HPV6 and 11 are most commonly seen in cervical, vulvar, perianal, and penile condylomata acuminata (genital warts) which are benign and very rarely become malignant. On the other hand, HPV16 and 18 are consistently associated with the majority of malignant carcinomas of the cervix.

The papillomavirus genome is in a free and unintegrated state in benign condylomata and in premalignant lesions of the cervix, vulva, and penis.

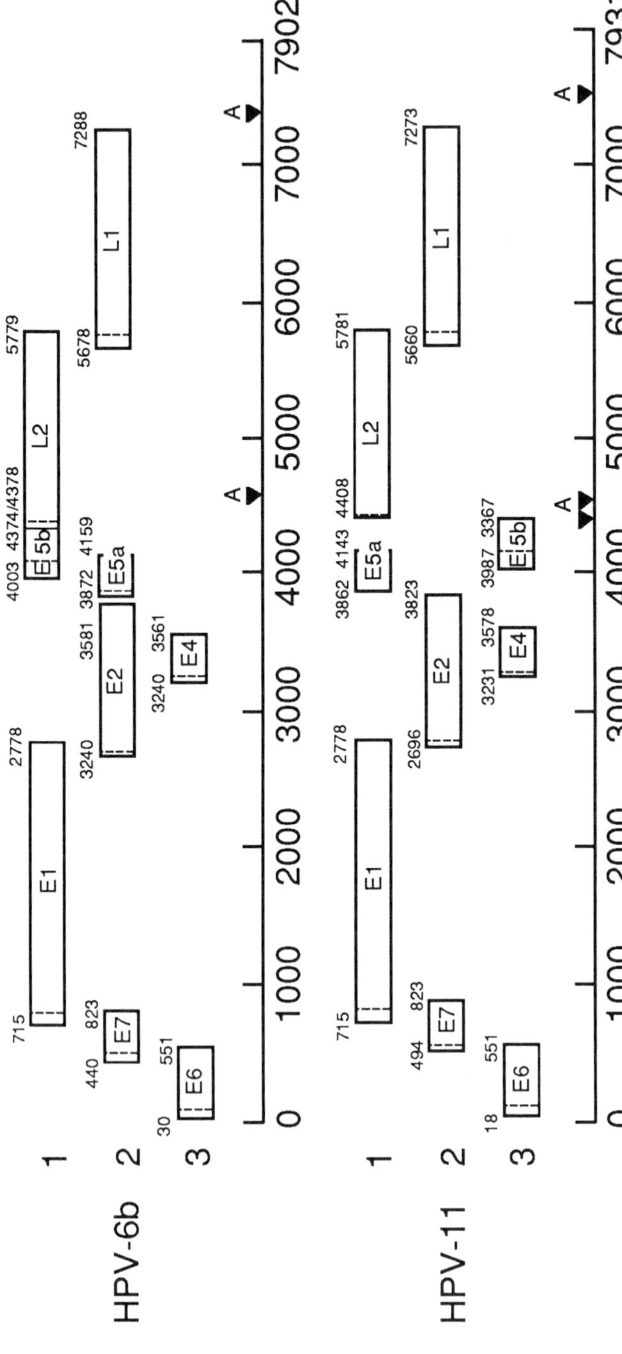

FIGURE 1 Genomic structure of HPV type 6b, 11, 16, 18, 31, and 33. Note the conserved general organization and the presence of the early region with 7 open reading frames (ORF), the late region with 2 ORFs and the unique long control region (LCR) lying between nt 7000 and 7900.

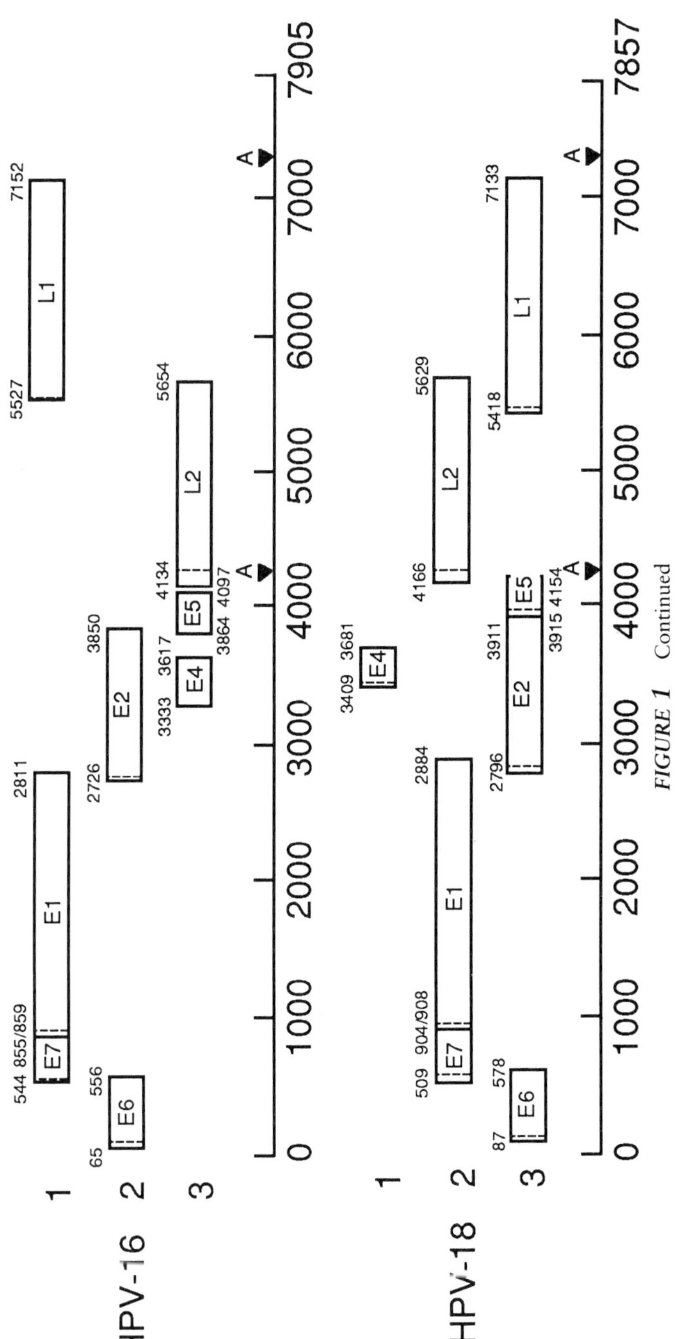

FIGURE 1 Continued

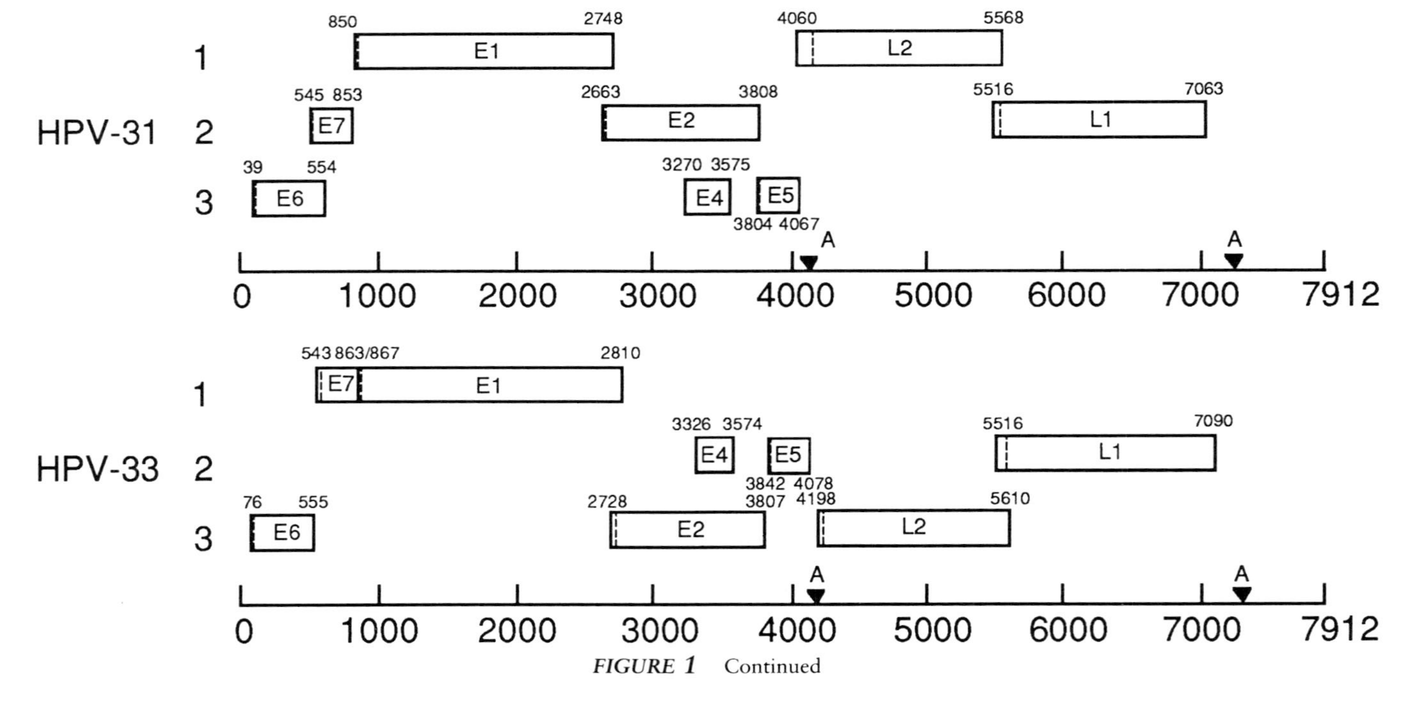

FIGURE 1 Continued

TABLE 2 **HPV Types in Squamous Cell Tumors of the Lower Genital Tract**

	HPV Type					
Clinital manifestation	6	11	16	18	31	33, 34, 35
Condyloma acuminatum (cervix, vagina, vulva, penis)	•	•	•			
Giant condyloma/Buske-Löwenstein's tumor (vulva, penis)	•	•				
Flat condyloma						
cervix (CIN I–III)	•	•	•		•	•
vagina		•				
Flat condylomatous lesion						
vulva (VIN I–II)	•	•	•			
penis (PIN I–II)	•	•	•			
Pigmented papular lesion						
vulva (VIN I–II)	•	•	•			
penis (PIN I–II)	•	•	•			
Pigmented bowenoid papulosis						
vulva (VIN III)			•			
penis (PIN III)			•			•
Bowen's disease						
vulva (VIN III)			•			
penis			•			
Carcinoma						
cervix uteri			•	•	•	•
vagina			•			
vulva			•			
penis			•	•		
Adenocarcinoma (cervix)			•			

Infectious virions are then produced and can be transmitted; this later feature is an epidemiologic characteristic of a sexually transmitted disease. In malignant lesions and in cell lines derived from them, however, the HPV16 and -18 genome appears to be integrated in the host genome in the majority of cases (Boshart *et al.*, 1984; Dürst *et al.*, 1985; Pater and Pater, 1985; Schwarz *et al.*, 1985; Yee *et al.*, 1985). Indeed, junction fragments containing HPV16 and human DNA sequences have been cloned (Dürst *et al.*,

1985), indicating that the viral DNA is integrated in the human chromosome. Two different patterns of integration have been observed: the HPV DNA is either repeated in a head-to-tail arrangement or inserted as a single copy linked to the cellular DNA. In some cell lines a single copy of HPV DNA is amplified together with the adjacent cellular sequences (Choo *et al.*, 1988; Lazo *et al.*, 1989; Couturier *et al.*, 1991). The physical state of the viral DNA—integrated versus extrachromosomal—may influence viral gene regulation (Dürst *et al.*, 1991). In fact, there is a remarkable specificity of integration into the viral genome: the viral integration sites always map within the E1 and E2 ORFs, thus resulting in the disruption of the early region (see Fig. 1). Moreover the LCR and the E6/E7 ORFs are invariably retained (Pater and Pater, 1985; El Awady *et al.*, 1987; Choo *et al.*, 1987; Schneider-Maunoury *et al.*, 1987; Shirasawa *et al.*, 1987; Choo *et al.*, 1988; Le and Defendi, 1988). The potential significance of such a disruption will be discussed later in the role of E2 protein.

Although integration sites in the host chromosomes initially seemed random or at least not as specific as that of the viral genome, it also appeared that integration sites could correspond to a heritable fragile site in host DNA, thus facilitating viral DNA integration (Popescu *et al.*, 1987). Moreover, in several carcinoma cell lines, papillomavirus sequences integrate near cellular oncogenes (Dürst *et al.*, 1987a). For example, in a primary cervical carcinoma, HPV16 sequences are found integrated in chromosome regions containing the protooncogenes *c-src*-1 and *c-raf*-1. In the SW756 carcinoma cell line, the HPV18 integration site localizes on chromosome 12 (Popescu *et al.*, 1987; Dürst *et al.*, 1987a) far from the Ki-ras2 gene locus, but in HeLa and C4-1 carcinoma cell lines, HPV18 DNA is integrated in locus 8q24.1 of chromosome 8, 5′ of the *c-myc* gene (Dürst et al., 1987a). Additionally, steady-state levels of *c-myc* mRNA is elevated in HeLa and C4-1 cells relative to other cervical carcinoma cell lines suggesting that in some genital tumors, *cis* activation of cellular oncogenes by HPV may be involved in malignant transformation of cervical cells. Indeed, constitutive expression of *c-myc* oncogene confers hormone independence and enhanced growth-factor responsiveness on cells transformed by HPV16 (Crook *et al.*, 1989a). The susceptibility of 8q24.1 locus to HPV integration is confirmed by the recent report of Couturier *et al.* (1991), in which another integration site is also described, namely locus 2p24 containing the *N-myc* gene. In three of the four invasive genital cancers studied (Couturier *et al.*, 1991), the *myc* protooncogene located near integrated viral sequences was found to be amplified and/or overexpressed. These results support the hypothesis that HPV integration may play a role in tumor progression via an activation of cellular oncogenes.

Nature of the Transcripts

HPV 16 and 18 sequences are generally found integrated in the genome of carcinoma cell lines, with the LCR/E6/E7 region uninterrupted and a disruption of E2 ORF. In a HPV16-associated cervical cancer and a cancer-derived cell line, the major viral transcript corresponds to the E7 ORF (Smotkin and Wettstein, 1987). Comparison of the three different HPV18-positive human cervical carcinoma cell lines HeLa, C4-1, and SW756, reveals common features in transcription of viral DNA (Schneider-Gädicke and Schwartz, 1986). Indeed, transcription of the HPV18 DNA in the three cell lines is confined to the E6–E7–E1 part of the early region. The cDNA clones from all three lines are in fact derived from virus–cell fusion transcripts in which 3′-terminal host cell sequences (different for each cell line) are differentially spliced to 5′-terminal exon sequences from the HPV18 E6–E7–E1 region. The corresponding RNAs may direct the synthesis of HPV18-specific proteins E6, E6*, E7, and (in HeLa) possibly E1. The mapping of the 5′-ends of the virus–cell fusion transcripts indicates that transcription is initiated at a viral promoter (Schneider-Gädicke and Schwartz, 1986; Inagaki *et al.*, 1988). From the cDNA structure, it has been predicted that E6 protein (158 amino acids, molecular weight about 19000) and E6* protein (57 amino acids, molecular weight about 6500) are identical in their amino-terminal 43 amino acids, and that E7 protein has 105 amino acids and molecular weight about 12000. In fact transcription analysis of carcinoma cell lines containing either HPV18 or HPV16 sequences reveals that viral transcription is always primarily derived from the E6 and E7 ORFs (Takebe *et al.*, 1987; Baker *et al.*, 1987; Seedorf *et al.*, 1987; Wilczynski *et al.*, 1988). In the case of HPV16, E6 and E7 proteins have molecular weights about 11 kDa and 15 kDa, respectively (Seedorf *et al.*, 1987). The presence of an interrupted E6* gene, however, may correlate with the oncogenic potential of genital HPV types, since E6* splice sites are conserved in other HPV types associated with malignant tumors but are absent in HPV6 and 11 DNAs (Schneider-Gädicke and Schwartz, 1986).

In Vitro Immortalization and Transformation

NIH 3T3 Cells

Although HPV16 and -18 DNA is present in cervical carcinoma cell lines and specific transcripts are detected, the biological significance of the association of HPV DNA with cancer remains to be determined. A first clue came from the work of Yasumoto *et al.* (1986), who showed that transfection into NIH 3T3 cells of a head-to-tail dimer of the full-length HPV16 genome resulted in malignant transformation of these cells with virus-specific RNA expression in the transformants. Transformation was assessed by foci formation and tumorigenicity in nude mice. This experiment clearly suggested that HPV16 and HPV18 must be considered not only as etiological agents but also as causative agents of cervical cancers. Tsunokawa *et al.* (1986) confirmed the transforming activity of HPV sequences by transfecting into NIH 3T3 cells a genomic DNA sample from cervical cancer tissue containing integrated HPV16 sequences. In fact, the HPV16 viral genes responsible for transformation are the E6 and E7 ORFs (Yutsudo *et al.*, 1988), in agreement with the persistence of these ORFs as integrated DNA sequences found in cervical carcinoma cells. Similarly, the E6–E7 region of HPV18 is sufficient for transformation of NIH 3T3 and Rat-1 cells (Bedell *et al.*, 1987).

Since NIH 3T3 cells are already immortalized cells, one could wonder about the potential effect of HPV DNA on primary cells. In baby rat kidney cells, HPV16 DNA has transforming activity only in the presence of an activated *EJ-ras* gene (Matlashewski *et al.*, 1987); this cooperating activity resides in a protein or proteins derived from the E6/E7 region of the HPV16 genome. Only DNA sequences of those HPV types most commonly found in carcinomas—types 16, 18, 31, and 33 (Table 2)—are capable of cooperating with *ras* to transform primary cells, while those types most commonly found in benign lesions—types 6 and 11 (Table 2)—are not. It turns out that the E7 ORF of HPV16 (Storey *et al.*, 1988) and HPV18 (Bedell *et al.*, 1989) is sufficient by itself to cooperate with activated *ras* to transform primary rat kidney epithelial cells.

Human Keratinocytes

In contrast to NIH 3T3 cells or rat kidney cells, transfection of oligomerized HPV16 DNA into human foreskin keratinocytes results only in immortalization of these cells, with one or possibly two integration sites for HPV16 DNA (Dürst *et al.*, 1987a). Similarly, when Pirisi *et al.* (1988) transfected the head-to-tail construct of HPV16 DNA—previously used by Yasumoto *et al.* (1986)—into human foreskin keratinocytes and fibroblasts (instead of NIH 3T3 cells) they obtained "transformed"—in fact immortalized—cells with an extended (fibroblasts) or indefinite (keratinocytes) lifespan compared to controls. However, none of the immortalized keratinocyte lines formed tumors in nude mice (Pirisi *et al.*, 1988). Similar effects were observed when HPV18 DNA was used instead of HPV16 (Kaur and McDougall, 1988, 1989). This clearly demonstrates that, even though the establishment of continuous cell lines is a direct consequence of the presence of viral sequences and expression of E6 and E7 proteins (Pirisi *et al.*, 1987; Woodworth *et al.*, 1988; Kaur and McDougall, 1989), additional events must take place for a progression to malignancy to occur.

Indeed, in terms of immortalization, mutational analysis and transfection studies using subfragments of HPV16 and HPV18 DNA clearly show that the E6 and E7 genes together are necessary and sufficient for immortalization of primary human keratinocytes (Hawley-Nelson *et al.*, 1989; Münger *et al.*, 1989a; Hudson *et al.*, 1990) as well as human embryonic fibroblasts (Watanabe *et al.*, 1989). The E7 gene of HPV16, expressed from a retroviral vector, has been shown to be sufficient to immortalize primary human keratinocytes (Halbert *et al.*, 1991). This supports the hypothesis that the observed maintenance and expression of E6 and E7 ORF in human cervical carcinomas has pathologic significance.

In terms of differentiation, human keratinocytes harboring integrated HPV16 or 18 sequences are characterized by a lack of stratification in culture (Dürst *et al.*, 1987b) and an absence of response to differentiation stimuli such as serum addition, increase of calcium concentration (up to 1.2 m*M*) or maintenance in suspension (Kaur and McDougall, 1988, 1989; Pirisi *et al.*, 1988; Schlegel *et al.*, 1988; Woodworth *et al.*, 1988). It is noteworthy that HPV18 DNA appears 3- to 4-fold more efficient at inducing altered keratinocyte differentiation than HPV16 DNA (Schlegel *et al.*, 1988). These features probably represent a phenotype that accompanies cellular immortalization since they have also been observed in SVK14 cells, an SV40-transformed keratinocyte cell line (Bernard *et al.*, 1985). In contrast, keratin patterns remain generally unaltered (McCance *et al.*, 1988; Pirisi *et al.*, 1988; Woodworth *et al.*, 1988) except in selected subclones that express the keratin pattern typical of simple epithelia (Pirisi *et al.*, 1988; Woodworth *et al.*, 1988). Here again the parallel with SV40-transformed keratinocytes is striking since, depending upon the cell line studied, the keratin pattern remains unchanged (Banks-Schlegel and Howley, 1983; Bernard *et al.*, 1985) or typical of simple epithelia (Bernard *et al.*, 1985). These differences in keratin expression might somehow reflect progression in the transformation process of keratinocytes. Such a progression is also suggested by the response of HPV16-immortalized keratinocytes to TGFb. Cells become resistant to the inhibitory effects of TGFb1 on cell growth and E6/E7 expression after prolonged cultivation *in vitro* or after malignant transformation (Braun *et al.*, 1990; Woodworth *et al.*, 1990a). In fact, such a progression has been observed when keratinocytes are immortalized by either the complete genome of HPV16 (McCance *et al.*, 1988; Hudson *et al.*, 1990) or the only HPV16-E6 and E7 ORFs (Hudson *et al.*, 1990), and then grown in the raft system that allows optimal epidermal differentiation (Asselineau *et al.*, 1985). Ultimately, the histology of the epithelium reconstructed from late passage HPV16-immortalized keratinocytes (e.g., from the 17th subculture after transfection) resembles that of cervical intraepithelial neoplasia grade III (McCance *et al.*, 1988). Interestingly, when confluent cultures of HPV16- or HPV18-immortalized keratinocytes (Woodworth *et al.*, 1988) are grafted onto nude mice, a dysplastic differentiation pattern is observed 2 to 3 weeks after grafting (Woodworth *et al.*, 1990b). Morphological alterations are accompanied by delayed commitment to terminal differentiation, alterations in the pattern of involucrin expression, and reduction in levels of involucrin and K1 keratin RNAs. As previously observed *in vitro* by Schlegel *et al.* (1988), HPV18-immortalized cells develop *in vivo* dysplastic changes more rapidly than cells immortalized by HPV16 (Woodworth *et al.*, 1990b). Recently, the progression of human epithelial cells to malignancy has been achieved using the FEP-1811 HPV18-immortalized human keratinocyte cell line (Hurlin *et al.*, 1991). FEP-1811 cells are nontumorigenic in nude mice through the 12th passage in culture, but after 32 passages, produce tumors that regress. After 62 passages, they produce invasive squamous cell carcinomas. The progression to malignancy is associated with an increase in the efficiency of forming colonies in soft agar and with altered differentiation properties. In the raft system, FEP-1811 cells at passages 12 and 32 exhibit features typical of premalignant intraepithelial neoplasia *in vivo,* and cells at passage 68 exhibit features consistent with squamous cell carcinomas. Interestingly, no change in the copy number of the HPV18 genome or in the level of expression of E6 and E7 proteins is detected between tumorigenic and nontumorigenic cells. On the other hand, chromosomal abnormalities segregate with the tumorigenic population. If this underlines the multistep character of progression to malignancy, it implies that synthesis of E6 and E7 proteins is one of the obligatory early steps of this process. This holds true for type 16, 18, 31, and 33 HPVs but not for HPVs type 6 and 11 with low oncogenic potential for cervical cancer, since in these latter cases, neither the complete DNA sequence nor the E6/E7 ORFs are able to induce keratinocyte immortalization (Pecoraro *et al.,* 1989; Woodworth *et al.,* 1989; Barbosa *et al.,* 1991).

E6 Gene Product

Structure and Properties of E6 Protein

The isolation and cloning of ORFs coding for putative E6 proteins have allowed the production of fusion proteins in bacteria and the subsequent preparation of specific antibodies (Androphy *et al.,* 1985; Banks *et al.,* 1987; Schneider-Gädicke *et al.,* 1988). The HPV16-E6 protein was then immunoprecipitated from human cervical carcinoma cells SiHa and CaSki and was identified as a 18-kDa protein (Androphy *et al.,* 1987). In SiHa cells, the half-life of this protein may be bimodal, with one component having a short half-life (less than 30 min) and the other half-life of about 4 hr. The putative E6* protein, deduced from cDNA analysis, was not detected in this study, perhaps because of poorly defined antibodies or low levels of synthesis.

The HPV18-E6 protein has been immunoprecipitated from HeLa cells and identified as a protein of 16.5 kDa (Banks *et al.,* 1987) even though, in HeLa cells, E6 protein is expressed at very low levels. Again, E6* protein was not detected. When expressed from a Baculovirus vector, HPV18-E6 has a molecular weight of 17 kDa and, when associated with the nuclear matrix, a half-life of 4 hr (Grossman *et al.,* 1989), similar to that ascribed to HPV16-E6 protein (Banks *et al.,* 1987). The slightly lower molecular weight of HPV18-E6 protein compared to the HPV16-E6 protein is predictable from their respective DNA sequences (Seedorf *et al.,* 1985). The immunoprecipitation of another 31-kDa protein from bacteria producing the HPV18-E6 protein (Banks *et al.,* 1987) suggests that E6 can form a dimer, probably through interchain disulfide bonds. Indeed the 31-kDa protein is sensitive to reducing agents and is quantitatively decreased when synthesis occurs in the presence of alkylating agents.

It is striking, from the work of Banks *et al.* (1987) and Androphy *et al.* (1987) that the HPV18-E6* protein cannot be immunoprecipitated from human carcinoma cells grown *in vitro*. As mentioned above, this might be due to poor antibodies or a very low level of synthesis. To answer this question, Schneider-Gädicke *et al.* (1988) expressed the HPV18-E6* protein in bacteria as a fusion protein and generated monoclonal antibodies reacting with E6* plus E6 proteins. As previously observed, it was impossible to detect E6* protein in human carcinoma cell lines grown *in vitro*. However, a 6.5 kDa E6* polypeptide was detected from a cervical carcinoma grown in nude mice. This suggests that E6* indeed exists, and that translation of the viral mRNA is different in culture cells *in vitro* and tumor cells *in vivo*. It is noteworthy that E6 and E6* proteins were isolated from the 300 m*M* NaCl nuclear protein fraction.

In fact, sequence analysis of putative proteins encoded by the E6 ORFs identifies these proteins as potential nucleic acid binding proteins. These proteins contain a high percentage of basic residues and a fourfold repetition with similar spacing for the tetrapeptide sequence Cys-X-X-Cys (Cole and Danos, 1987). This type of repeat is supposed to mediate zinc binding which stabilizes intramolecular protein loops now identified as "zinc fingers" (Klug and Rhodes, 1987) involved in nucleic acid recognition. In fact, HPV18-E6 protein binds zinc with high affinity through cysteine residues (Barbosa *et al.*, 1989) and also binds to double-stranded DNA with high affinity (Barbosa *et al.*, 1989; Grossman *et al.*, 1989), but with no sequence specificity for HPV18-DNA. Although the HPV16-E6 protein expressed in *E. coli* (Imai *et al.*, 1989) exhibits a higher specificity for a subfragment of HPV16-LCR—from nucleotides 7524 to 7756—the biological significance and the role of this binding activity in the transforming activity of E6 protein remains uncertain.

Interaction with p53 Antioncogene

The oncogenic potential of the E6 protein encoded by HPV16 and HPV18 has been revealed in transformation studies with primary human cells (Münger *et al.*, 1989a; Hawley-Nelson *et al.*, 1989; Watanabe *et al.*, 1989). Among possible mechanisms, and because of the absence of specificity in its DNA-binding activity (discussed earlier), its transforming properties may result from an ability to form complexes with and potentially modulate the activity of critical proteins that regulate cellular growth and differentiation. By analogy with the large T antigen of SV40 that can form a complex with the p53 protein (Lane and Crawford, 1979; Linzer and Levine, 1979), the possibility that E6 could bind p53 protein has been investigated. Werness *et al.* (1990) demonstrated, by *in vitro* coimmunoprecipitation assays, that p53 protein, from both mouse and human origin, and HPV16-E6 form a complex. While the E6 proteins of both HPV16 and 18 can associate with human wildtype p53 *in vitro*, no such association has been detected with the E6 proteins of HPV6 and 11. In these assays, mouse wildtype p53 was recovered from F9 cell lysate; human wildtype p53 and HPV16-E6 were produced by *in vitro* translation in reticulocyte lysate.

What could be the biological consequence of such an interaction? In fact, p53 is a tumor suppressor gene (for review see Levine *et al.*, 1991). p53 Protein is a 375 amino acid nuclear phosphoprotein originally discovered in extracts of transformed cells associated with large T antigen of SV40. It is only recently that p53 has been recognized as a tumor suppressor gene (Finlay *et al.*, 1989) that negatively regulates the cell cycle and requires loss-of-function mutations for tumor formation. Through its DNA-binding ability (Bargonetti *et al.*, 1991; Kern *et al.*, 1991) p53 could regulate the assembly or function of the DNA replication–initiation complex (Kern *et al.*, 1991). Alternatively, through its strong activation domain (Fields and Jang, 1990; Raycroft *et al.*, 1990) it could also act as a transactivator of gene transcription, either promoting or repressing mRNA synthesis of a set of genes that affect the passage from late G1 to S phase of the cell cycle.

Binding of SV40 large T antigen to wildtype p53 results in inhibition of p53 binding to DNA (Bargonetti *et al.*, 1991) and an increased half-life of p53 protein from 6 to 20 min or to several hours (Reich *et al.*, 1983). Both events are somehow involved in the process of inactivation of p53 normal function as a negative regulator of cell growth. HPV16- and HPV18-E6 proteins could act in a similar fashion, since they are also able to interact with p53 protein. However, in contrast to SV40-transformed cells, it is impossible to detect p53 protein in HeLa cells and other carcinoma cell lines despite the presence of translatable mRNA. Furthermore, the level of p53 in human keratinocytes transformed by HPV16 is low when compared to the level in primary keratinocytes or in SV40-transformed keratinocytes (Werness and Howley, unpublished results). This raises the possibility that E6 may preferably facilitate p53 degradation. In fact, Scheffner *et al.* (1990) demonstrated that the HPV16- and HPV18-E6 proteins promote the degradation of p53 protein by an ATP-dependent process involving the ubiquitin-dependent protease system (Ciechanover and Schwartz, 1989). This selective degradation of a tumor suppressor gene product involved in the control of cell growth might thus represent one of the key events necessary for the establishment of the transformed state.

E7 Gene Product

Structure and Properties of E7 Protein

As for the E6 gene product, the putative E7 protein contains regularly spaced Cys-X-X-Cys doublets. Computer analysis of ten available sequences suggests that both types of proteins may have evolved from a common 33-amino acid module, with successive duplications and drifts to form the E6 and E7 gene products (Danos and Yaniv, 1987). The cloning and expression of HPV16- and HPV18-E7 ORFs in bacteria has allowed the preparation of specific antibodies and the subsequent immunoprecipitation from cell lines harboring HPV16 or HPV18 DNA sequences (Oltersdorf *et al.*, 1987; Seedorf *et al.*, 1987). HPV16-E7 protein has been immunoprecipitated as a 15-kDa protein from CaSki and SiHa cell extracts; HPV18-E7 has been identified as a 12-kDa protein from HeLa cells. HPV16-E7 is 98 amino acid residues in length and consists of three regions. Regions 1 (AAs 1 to 20) and 2 (AAs 21 to 40) show homologies to the sequences of conserved domains 1 (AAs 37 to 49) and 2 (AAs 116 to 137), respectively, of adenovirus E1A, as well as to portions of the large T antigens of papovaviruses. Region 3 (AAs 41 to 98) contains the two zinc-binding motifs Cys-X-X-Cys (Watanabe *et al.*, 1990). Mutational analysis suggests that regions 1 and 3 are required for function while the zinc-binding domains of region 3 are required to maintain a stable or functional structure of the E7 protein (Watanabe *et al.*, 1990). In CaSki cells, the E7 protein has been shown to be located in the cytoplasm and not glycosylated; its levels were estimated to be less than 0.01% of total protein (Oltersdorf *et al.*, 1987). Smotkin and Wettstein (1987) also found that the E7 protein is located in the soluble cytoplasmic fraction and has a half-life of about 1 hr. The protein is phosphorylated at serine residues and sedimentation studies suggest either an oligomer formation or association with cellular protein. When the HPV18-E7 protein is expressed from a cassette vector, it is also localized by immunocytochemistry as a cytoplasmic component (Bernard *et al.*, 1987). However, when HPV16-E7 is expressed in monkey COS-1 cells, it is found in the nucleus by immunofluorescence and in the cytoplasmic fraction upon subcellular fractionating (Sato *et al.*, 1989; Watanabe *et al.*, 1990). Consequently E7 protein appears to be a small nuclear oncoprotein.

The role of the E6/E7 region of HPV16 and HPV18 genomes in the transformation process had been previously delineated. However, in contrast to E6 protein, E7 protein is sufficient to either immortalize primary human keratinocytes (Halbert *et al.*, 1991) or, in some cases, to transform cells (Tanaka *et al.*, 1989). Furthermore, the fact that the E7 gene is the most conserved region of HPV16 in cervical carcinomas suggests that it may play a role not only in the establishment but also in the maintenance of the malignant state (Wilczynski *et al.*, 1988). This hypothesis is supported by the requirement of continuous expression of HPV16-E7 protein for maintenance of the transformed phenotype of baby rat kidney cells cotransformed by HPV16 plus *EJ-ras* oncogene (Crook *et al.*, 1989b). The mechanism by which E7 exerts its action remains unclear. However, E7 gene encodes trans-activation and transformation functions similar to those of adenovirus E1A (AdE1A) and can *trans*-activate the adenovirus E2 promoter (Phelps *et al.*, 1988). This suggests that one effect of E7 may be at the level of transcription, with E7 protein being required for activation of cellular genes needed for cell division. Striking sequence similarities between HPV16-E7 and the conserved domains 1 and 2 of AdE1A, as well as similarity of their target in the Ad-E2 promoter (Phelps *et al.*, 1988) suggest possible similarities in the mode of action of HPV16-E7 and AdE1A proteins. Transactivation by E1A does not involve direct binding to promoter sequences but rather is mediated through interaction with cellular factors (Ferguson *et al.*, 1985). One such factor is E2F: E7 might affect E2F either by increasing the levels of the protein or possibly by modifying the affinity of E2F for its cognate DNA sequences.

Interaction with p105RB Antioncogene

A new insight into E7 function has come from the demonstration that HPV16-E7 protein is able to bind to the retinoblastoma gene product p105RB (Dyson *et al.*, 1989; Münger *et al.*, 1989b). The protein encoded by the RB locus is a nuclear protein that is found in phosphorylated and unphosphorylated states (Lee *et al.*, 1988). It was later identified as the 105-kDa protein that is complexed to the AdE1A transforming protein (Whyte *et al.*, 1988a), SV40 large T antigen (DeCaprio *et al.*, 1988), and large T antigens of many polyomaviruses (Dyson *et al.*, 1990). Significantly, binding of p105RB to E1A requires regions 1 and 2 of E1A, both being involved in transformation (Whyte *et al.*, 1989). Moreover, binding of p105RB to SV40 large T antigen requires amino acids 105–114 (DeCaprio *et al.*, 1988), a region that encompasses the E1A region 2 and is necessary for transforming activity. Since sequence homologies between HPV16-E7, AdE1A, and SV40 large T antigen include this region 2 (Dyson *et al.*, 1989), it has been suggested that these three nuclear oncoproteins participate in DNA transformation through at least one common mechanism—binding to p105RB.

This latter assumption has been further supported by the fact that the E7 proteins of the nononcogenic HPV6b and HPV11 bind p105RB with lower affinity than the oncogenic HPV16 and HPV18 (Münger *et al.*, 1989b; Gage *et al.*, 1990). By analogy with large T antigen, this binding to p105RB might be regulated by phosphorylation; only the hypophosphorylated p105RB is able to bind to SV40 large T antigen (Ludlow *et al.*, 1989; DeCaprio *et al.*, 1989). This hypophosphorylated form of p105RB is mainly present in the G0 and G1 phases of the cell cycle and thus probably represents the p105RB form that suppresses cell proliferation (Mihara *et al.*, 1989). Since the purified HPV16-E7 protein was later demonstrated to preferentially bind to the hypophosphorylated p105RB protein (Imai *et al.*, 1991), all of the three transforming proteins E7, AdE1A, and large T antigen may therefore act at least in part by this common mechanism. By analogy with AdE1A, this binding may be sufficient to induce DNA synthesis (Howe *et al.*, 1990) and dissociate the complex formed by the hypophosphorylated p105RB protein and the transcription factor E2F (Bandara and La Thanghe,

1991; Chellappan *et al.*, 1991). Both effects may be linked, since the interaction of Rb with E2F (Chellappan *et al.*, 1991; Chittenden *et al.*, 1991) inhibits DNA binding activity of E2F (Bagchi *et al.*, 1991) and consequently E2F-dependent expression of genes involved in the control of cell proliferation. Moreover, p105RB also binds c-myc and N-myc proteins, which are necessary for cell cycle progression from G1 to S phase (Rustgi *et al.*, 1991). This progression might be favored by an increased concentration of free myc proteins in the presence of HPV16-E7 protein, since the interaction of myc proteins with p105RB is competitively inhibited by binding of HPV16-E7 protein to p105RB (Rustgi *et al.*, 1991).

The possibility of other mechanisms mediating transforming activity of E7 protein comes from the fact that E7 protein possesses at least two functional domains (Rawls *et al.*, 1990). The first domain induces DNA synthesis in quiescent rodent cells, and maps to the N-terminal portion of the molecule, which contains sequences related to the AdE1A-conserved domains 1 and 2 required for cell transformation and RB binding. The second domain binds zinc, has transactivating activities possibly involving interaction with the E2F transcription factor (Rawls *et al.*, 1990), and maps to the C-terminal portion of the molecule, which contains the Cys-X-X-Cys motifs and two serine residues (31 and 32) that can be phosphorylated by casein kinase II (Firzlaff *et al.*, 1989; Barbosa *et al.*, 1990). Apparently, RB binding (involving the N-terminus of E7) and phosphorylation (involving the C terminus) are separate and independent biological activities, both of which contribute to E7-mediated transformation (Barbosa *et al.*, 1990). Furthermore, the E7 protein of oncogenic HPV16 and HPV18 binds p105RB protein with a higher affinity than the nononcogenic HPV6 and HPV11 (Münger *et al.*, 1989b; Gage *et al.*, 1990), and the rate of phosphorylation of HPV18-E7 is twofold faster than HPV16-E7 and fourfold faster than HPV6-E7 (Barbosa *et al.*, 1990).

To summarize, one can speculate that the E7 protein of oncogenic HPVs exhibits transforming activity through at least two complementary mechanisms. First, through the N terminus, it releases E2F transcription factor from its complex with p105RB. Second, through its C terminus (Rawls *et al.*, 1990), it interacts with the E2F factor to modulate the expression of specific genes involved in the control of cell proliferation. This interaction is dependent upon phosphorylation levels of E7 protein.

Regulation of E6/E7 Expression

Structure of the Regulatory Region

In view of the major role of E6/E7 gene products in the transforming activity of some genital HPVs, an understanding of the regulatory circuits controlling this expression appears critical. Considerable effort was devoted to this goal and the initiation site of E6/E7 transcription was localized at nucleotide

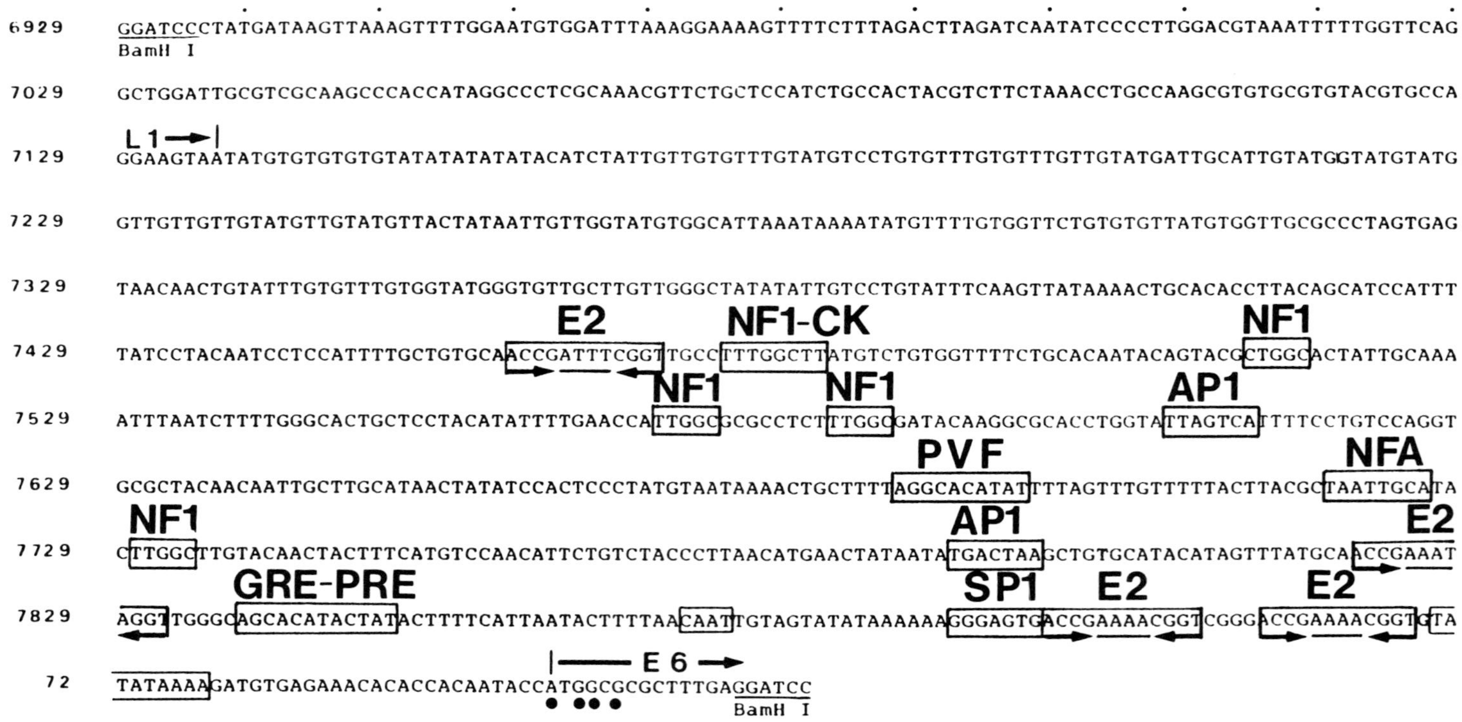

```
6929   GGATCCCTATGATAAGTTAAAGTTTTGGAATGTGGATTTAAAGGAAAAGTTTTCTTTAGACTTAGATCAATATCCCCTTGGACGTAAATTTTTGGTTCAG
       BamH I

7029   GCTGGATTGCGTCGCAAGCCCACCATAGGCCCTCGCAAACGTTCTGCTCCATCTGCCACTACGTCTTCTAAACCTGCCAAGCGTGTGCGTGTACGTGCCA

       L1 →|
7129   GGAAGTAATATGTGTGTGTGTATATATATATACATCTATTGTTGTGTTTGTATGTCCTGTGTTTGTGTTTGTTGTATGATTGCATTGTATGGTATGTATG

7229   GTTGTTGTTGTATGTTGTATGTTACTATAATTGTTGGTATGTGGCATTAAATAAAATATGTTTTGTGGTTCTGTGTGTTATGTGGTTGCGCCCTAGTGAG

7329   TAACAACTGTATTTGTGTTTGTGGTATGGGTGTTGCTTGTTGGGCTATATATTGTCCTGTATTTCAAGTTATAAAACTGCACACCTTACAGCATCCATTT

                                     E2         NF1-CK                                  NF1
7429   TATCCTACAATCCTCCATTTTGCTGTGCAACCGATTTCGGTTGCCTTTGGCTTATGTCTGTGGTTTTCTGCACAATACAGTACGCTGGCACTATTGCAAA
                                            NF1        NF1                          AP1
7529   ATTTAATCTTTTGGGCACTGCTCCTACATATTTTGAACCATTGGCGCGCCTCTTTGGCGATACAAGGCGCACCTGGTATTAGTCATTTTCCTGTCCAGGT
                                                                 PVF                              NFA
7629   GCGCTACAACAATTGCTTGCATAACTATATCCACTCCCTATGTAATAAAACTGCTTTTAGGCACATATTTTAGTTTGTTTTTACTTACGCTAATTGCATA
       NF1                                                       AP1                                  E2
7729   CTTGGCTTGTACAACTACTTTCATGTCCAACATTCTGTCTACCCTTAACATGAACTATAATATGACTAAGCTGTGCATACATAGTTTATGCAACCGAAAT
             GRE-PRE                                             SP1     E2            E2
7829   AGGTTGGGCAGCACATACTATACTTTTCATTAATACTTTTAACAATTGTAGTATATAAAAAAGGGAGTGACCGAAAACGGTCGGGACCGAAAACGGTGTA

                                         |—— E6 →
72     TATAAAAGATGTGAGAAACACACCACAATACCATGGCGCGCTTTGAGGATCC
                                                        BamH I
```

FIGURE 2 Nucleotide sequence of the HPV18 long control region (HPV18-LCR). Note the presence of binding sites for ubiquitous transcription factors (AP1, SP1, NF1), for the specific factor PVF, for hormone receptors (GRE-PRE), and for the virally encoded E2 protein.

97 for HPV16 (Smotkin and Wettstein, 1987) and 105 for HPV18 (Thierry *et al.,* 1987). Subsequently, the notion emerged that all genital HPVs—namely HPVs 6, 11, 16, 18, 31, and 33—have assembled in the LCR (Fig. 2), a cluster of similar transcription factor binding sites, possibly reflecting particular functional requirements and evolutionary conservation in spite of overall genomic divergence (Chong *et al.,* 1990) (Fig. 3). Binding sites have been identified for NF1 (Chin *et al.,* 1989; Gloss *et al.,* 1989a; Cripe *et al.,* 1990; Sibbet and Campo, 1990), Sp1 (Gloss and Bernard, 1990), and AP1 transcription factors (Garcia-Carranca *et al.,* 1988; Chan *et al.,* 1990; Cripe *et al.,* 1990; Offord and Beard, 1990; Sibbet and Campo, 1990). These latter AP1 binding sites mediate the stimulation of E6/E7 gene transcription by tumor promoters (Gius and Laimins, 1989; Chan *et al.,* 1990). The ubiquitous CCAAT motif was observed, as well as the consensus sequence AARCCAAA also present in the involucrin and cytokeratin promoters (Cripe *et al.,* 1987; Sibbet and Campo, 1990). Furthermore, the sequence AGGCACATAT, found in the enhancer of all six genital HPVs, appears necessary for AP1- and NF1-driven enhancement and binds a novel transcription factor called papilloma virus factor (PVF) (Chong *et al.,* 1990). Royer *et al.* (1991) also identified Oct-1 and AP2 interactions with the enhancer of both HPV16 and HPV18. Finally, glucocorticoid responsive elements (GRE) and binding sites for the virally encoded E2 protein were localized.

			NFA		NF1	
HPV-6	:	-T	TAAAAGCA	TTT	TTGGC	TT
HPV-11	:	-T	TAAAAGCA	TTT	TTGGC	TT
HPV-16	:	-C	TAATTGCA	TAT	TTGGC	AT
HPV-18	:	-C	TAATTGCA	TAC	TTGGC	TT
HPV-31	:	-T	TGATTGCA	GTG	CTGGC	TT
HPV-33	:	-T	TAAGTGCA	GTT	TTGGC	TT

			PVF	
HPV-6	:	-G	CAGCACATTT	TT
HPV-11	:	-G	CTGAACATTT	TT
HPV-16	:	-T	AGGCACATAT	TT
HPV-18	:	-T	AGGCACATAT	TT
HPV-31	:	-C	ATGCACATAT	AT
HPV-33	:	-T	AGGCACATAT	TT

			SP1	E2		E2		
HPV-6	:	-AAT	AGGAGGG	ACCGAAAACGGT	TCA	ACCGAAAACGGT	TGTA	TATAAA
HPV-11	:	-AAG	AGGAGGG	ACCGAAAACGGT	TCA	ACCGAAAACGGT	TATA	TATAAA
HPV-16	:	-TAA	GGGCGTA	ACCGAAATCGGT	TGA	ACCGAAACCGGT	TAG	TATAAAA
HPV 18	:	-AAA	GGGAGTA	ACCGAAAACGGT	CGGG	ACCGAAAACGGT	GTA	TATAAAA
HPV-31	:	-GTA	GGGAGTG	ACCGAAAGTGGT	GA	ACCGAAAACGGT	TGG	TATATAA
HPV-33	:	-GTA	GGGTGTA	ACCGAAAGCGGT	TCA	ACCGAAAACCGT	GCA	TATATAA

FIGURE 3 Sequence comparison of NFA (for NF1-associated factor), NF1, PVF, SP1 and E2 binding sites present in the LCR of HPV type 6, 11, 16, 18, 31, and 33. Note that the sequences of NFA and PVF binding sites highlight some differences between HPV type 6 and 11 on the one hand, and HPV type 16, 18, 31, and 33 on the other hand.

The HPV16 enhancer shows preference for epithelial cells (Cripe *et al.*, 1987; Gloss *et al.*, 1989b), and the LCR is functionally able to drive, in an epithelial cell-specific fashion, the transcription from the natural promoter of HPV18 (Bernard *et al.*, 1989) and HPV16 (Cripe *et al.*, 1990). Moreover, the LCR has been identified as the major determinant of the differential immortalization activities of HPV16 and HPV18 (Romanczuk *et al.*, 1991). But, unexpectedly, nuclear extracts from human keratinocytes and cervical carcinoma cells protect the same viral *cis* sequences in DNase footprinting experiments as extracts from other cells in which the enhancer functions poorly (Gloss *et al.*, 1989b; Cripe *et al.*, 1990). The tissue preference of the HPV enhancer, even among epithelial cells from different tissues (Steinberg *et al.*, 1989), could thus be determined by (i) the delicate balance of ubiquitous transcription factors as suggested by the functional dependence of the HPV16 enhancer on the cooperation of NF1, AP1, and PVF (Chong *et al.*, 1990), (ii) the existence of specific members—for example, AP1 (Offord and Beard, 1990)—of the ubiquitous families, or (iii) by either positive or negative second-order factors binding by protein–protein interactions to ubiquitous factors.

Role of E2 Protein

If many host cell transcription factors interact with the HPV-LCR, the protein encoded by the E2 ORF also interacts with the LCR. It is an essential viral enhancer factor and studies performed on the prototype BPV1-E2 protein show that it binds as a dimer to the sequence ACCGNNNNCGGT (Dostatni *et al.*, 1988; Moskaluk and Bastia, 1989); this sequence is present fourfold in the LCR of genital HPVs (Chong *et al.*, 1990). Analysis of the alignment of the amino acid sequences of ten E2 proteins reveals three distinct regions: two partially conserved domains at the N and C termini of the proteins and a variable region in the middle (Giri and Yaniv, 1988). The N-terminal part is predicted to be an acidic amphipathic alpha helix and is important for transcriptional activation (Giri and Yaniv, 1988; Haugen *et al.*, 1988; McBride *et al.*, 1989), while the DNA-binding and dimerization domain has been localized to the C-terminal part (Dostatni *et al.*, 1988; McBride *et al.*, 1989).

Although the E2 protein is generally recognized as a transactivator (Spalholz *et al.*, 1985; Phelps and Howley, 1987; Haugen *et al.*, 1987; Hawley-Nelson *et al.*, 1988; Hirochika *et al.*, 1988; Thierry and Yaniv, 1987; Thierry *et al.*, 1990; Spalholz *et al.*, 1991), Thierry and Yaniv (1987) showed that the prototype BPV1-E2 protein behaves as a transrepressor of transcription from the HPV18-LCR. The short E2-TR transcriptional repressor encoded by the 3′ portion of BPV1-E2 ORF (Lambert *et al.*, 1987; McBride *et al.*, 1988) is apparently not involved in this effect. In fact, in human keratinocytes, the HPV18-E2 protein also behaves as a transcrip-

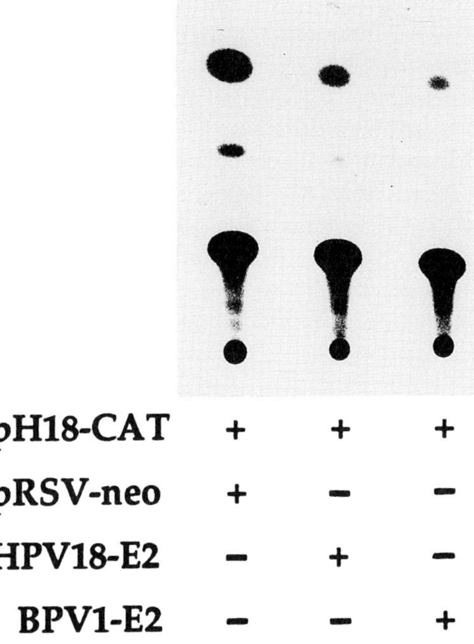

FIGURE 4 HPV18-E2 protein is a transcriptional repressor of the p105 HPV18-E6 promoter. The plasmid pH18-CAT (10 mg) containing the *cat* gene under the control of the HPV18-LCR and the natural p105 E6 promoter was cotransfected into human keratinocytes with 5 mg of either neomycin resistance (pRSV-neo) BPV1-E2, or HPV18-E2 gene expression vectors. Note that the *cat* activity, as evidenced by acetylation of ^{14}C-labeled chloramphenicol substrate, is repressed in the presence of either BPV1- or HPV18-E2 protein.

tional repressor (Fig. 4) of the homologous HPV18-E6/E7 p105 promoter (Bernard *et al.*, 1989; Romanczuk *et al.*, 1990). This inhibitory activity of E2 protein on homologous promoters has been confirmed for HPV11-E2 (Chin *et al.*, 1989) and HPV16-E2 (Romanczuk *et al.*, 1990). This effect is probably of importance since, as previously mentioned, oncogenic HPV DNA always integrates in the host genome with a disruption of E2-ORF. This would result in the loss of E2 transcriptional regulation and, consequently, a derepressed regulation of the HPV16-p97 and HPV18-p105 promoters and deregulated expression of E6 and E7 oncoproteins.

How does E2 protein inhibit the activity of p97 and p105 promoters? Two E2-binding sites are located between the CCAAT and the TATA boxes, immediately upstream of the transcription start site, in the LCR of the all genital HPVs (Chan *et al.*, 1990; Gloss *et al.*, 1989a; Romanczuk *et al.*, 1990). Mutational analysis has shown that E2 repression on p97 and p105 promoters is mediated through the E2-binding site immediately proximal to the TATA box (Romanczuk *et al.*, 1990). The E2 protein could thus exert its repressing effect by interfering with the binding of a TATA motif-binding factor such as the TFIID ubiquitous transcription factor. It could also potentially exert its repressing effect by modulating the binding of sp1 transcription factor to the sequence NGGNGN, which results, in the absence of E2 protein, in the activation of E6/E7 promoter (Gloss and Bernard, 1990). This sequence is located immediately upstream of the second E2-binding site between the CCAAT and the TATA boxes of genital HPV-LCR. In line of this possible interference, direct interaction between the BPV1-E2 protein and sp1 has been reported (Li *et al.*, 1991) and a competitive binding of BPV1-E2 and sp1 has been described, which results in the negative regulation of a BPV1 constitutive enhancer (Vande Pol and Howley, 1990).

Role of Chromosome 11

The role of E6 and E7 proteins from oncogenic HPVs in immortalization and transformation was described previously. In fact, one could state that transformation is a multistep process, involving not only viral gene products but also cellular events (e.g., chromosome rearrangement or loss) resulting in activation of protooncogenes and failed control of suppressor genes. Chromosomal abnormalities are often present in primary cervical carcinoma cells, most frequently affecting chromosomes 11, 1, 7, and 3 (Atkin and Baker, 1982, 1984). Deletion in the short arm of chromosome 11 (namely in the 11p13 region) has been associated with the development of Wilm's tumor (Riccardi *et al.*, 1980) suggesting the localization of a tumor suppressor gene on 11p13. This hypothesis has been confirmed by the suppression of the tumorigenic phenotype of Wilm's tumor cells (Geiser and Stanbridge, 1987) and HeLa cells (Saxon *et al.*, 1986; Srivatsan *et al.*, 1986) upon microcell transfer of a normal chromosome 11 into these cells. Conversely, human embryonic fibroblasts with a deletion (11p11.11p15) in the short arm of chromosome 11 acquire anchorage-independent growth upon transfection of the early region of HPV16-DNA (Smits *et al.*, 1988). This effect requires the presence of E6/E7 ORF but is abolished when the E2 ORF is left intact in the transfected DNA. This supports (see related section on the role of E2 protein) the concept that, *in vivo,* E2 might exert a trans-repressor function with respect to E6/E7, unless the E2 ORF is disrupted. Nevertheless, it appears that the short arm of chromosome 11 harbors a gene that can normally control transformation by HPV16-E6/E7 proteins.

An insight into the mechanism involved in this process comes from studies using HeLa/fibroblast (Bosch *et al.*, 1990) and HPV18-CAT-SiHa/human keratinocyte hybrid cells (Rösl *et al.*, 1991). In both cases, the extinction of the regulatory region of HPV18 is observed, with suppression of E6/E7 (Bosch *et al.*, 1990) and CAT expression (Rösl *et al.*, 1991) under the control of HPV18-LCR. This strongly suggests that *trans*-acting negative regulatory factors (encoded by a gene present on chromosome 11?) derive from the nontumorigenic cell partner of the hybrids and that an intracellular surveillance mechanism for HPV gene expression exists in nontumorigenic cells, which must be overcome for the onset of the transformed state.

Hormonal Control

Glucocorticoids and Estrogens

Sequence analysis has identified in the HPV16-LCR an element (TGTACAttgTGTCAT) with a large degree of homology to the partially palindromic sequence GGTACANNNTGTTCT (Gloss *et al.,* 1987), the consensus sequence of the glucocorticoid responsive elements (GRE) of genes known to be regulated by this hormone (Strähle *et al.,* 1987). Since then, GREs have been identified in the LCR of several HPVs involved in anogenital neoplasia, namely HPV6, 11, 16, 18, and 33 (Chan *et al.,* 1989). The fragment of HPV16-LCR harboring the GRE confers strong inducibility by dexamethasone to the promoter p97, and direct binding of the rat glucocorticoid receptor to this GRE has been observed (Gloss *et al.,* 1987). This points to a possible role of steroid hormones in the regulation of expression of E6/E7 proteins and in HPV-induced cell transformation. This hypothesis is strengthened by the evidence for oncogenic transformation of primary cells with a combination of HPV16 DNA and the activated form of *Ha-ras* oncogene only in the presence of dexamethasone (Pater *et al.,* 1988). Moreover, estrogen differentially stimulates transcription of HPV16 in SiHa cells, which contain a single copy insertion of HPV16 genome (Mitrani-Rosenbaum *et al.,* 1989). In fact, in SiHa cells, the HPV16-GRE mediates an increase in transcripts encoding E6 and E7 transforming proteins in response to dexamethasone and progesterone treatment (Chan *et al.,* 1989). The antiprogesterone and antiglucocorticoid RU486 interferes with both hormonal responses. This hormonal regulation is not restricted to HPV16 since similar effects are observed with HPV18 and HPV11 (Chan *et al.,* 1989). In concert, these findings suggest that the GRE/PREs are an integral part of the regulatory circuit of gene expression in genital HPVs. Since these HPVs infect epithelial cells that bear progesterone receptors (Sanborn *et al.,* 1976), GRE/PREs could confer some advantage to the virus and hormonal induction of viral gene expression might help the virus to overcome the intracellular surveillance system (see previous section) and constitute one of numerous events cooperating in the etiology of cervical carcinogenesis.

Retinoids

A very scarce literature suggests that retinoids may decrease the copy number of papillomaviruses in infected cells. These studies, however, have been restricted to bovine papillomavirus (Gassenmaier *et al.,* 1985; Li *et al.,* 1988; Tsang *et al.,* 1988; Stich *et al.,* 1990) and it is difficult to extend results on BPV-infected cells to human cells carrying nonintegrated HPV.

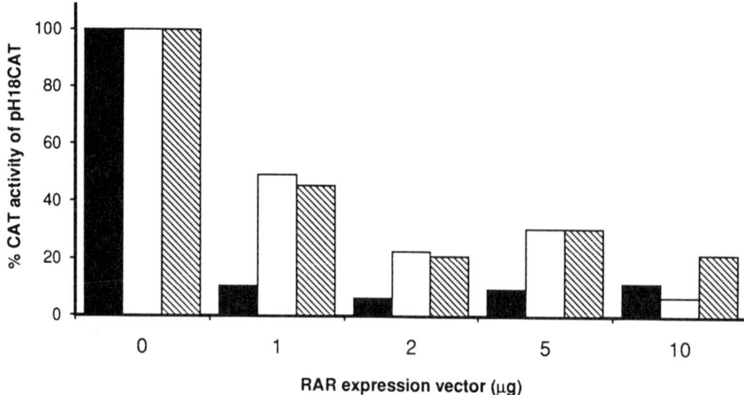

FIGURE 5 Repression of p105 HPV18-E6 promoter in the presence of retinoic acid receptors (RAR) and $10^{-7}M$ retinoic acid. 1 μg of the pH18-CAT plasmid (see Fig. 4) was cotransfected into HeLa cells with increasing amounts of either RARα(■), RARβ(□), or RARγ(▨) expression vectors. Note the dose-dependent inhibition of p105 promoter activity in the presence of RAR.

However, retinoic acid seems to be able to specifically modulate transcription from the HPV18-LCR in human keratinocytes (Bernard *et al.*, 1990b) Furthermore, HPV16-immortalized human keratinocytes are more sensitive than normal human keratinocytes to growth inhibition by retinoic acid. The expression of HPV16-E6 and E7 ORFs is two to three times lower in the presence of retinoic acid (L. Pirisi and co-workers, unpublished results). In HeLa cells, a repressing effect is observed, which is dependent upon the presence of specific retinoic acid receptors (Fig. 5). This suggests that retinoic acid may exert some control at the transcriptional level. Whether or not this effect is mediated through the formation of a nonproductive complex with Jun/AP1 (Schüle *et al.*, 1991) or specific responsive elements present in the HPV18-LCR remains to be established.

Discussion and Perspective

The goal of this chapter has been to update the knowledge about the role of human papillomaviruses in malignant transformation. Because of the high frequency of occurrence of genital lesions and cervical cancer, I deliberately focused on HPV16 and HPV18, the most common HPVs associated with genital carcinomas. A concensus points to the possibility that these two viruses might be considered to be causative agents of these cancers and not only etiological agents. The molecular mechanisms involved in this process have begun to be understood, and among them, the direct interaction of E6 and E7 proteins with the p53 and p105RB antioncogenes, respectively, has been established (Fig. 6). Furthermore, one of the most important features found in genital carcinoma cells is the deregulated synthesis of these E6/E7 proteins, as a possible consequence of the loss of E2 protein synthesis and/or an altered response to glucocorticoids, estrogens, and perhaps retinoids. The understanding of such mechanisms gives possible indications for further design of therapeutic strategies and agents, as exemplified by the synthesis of

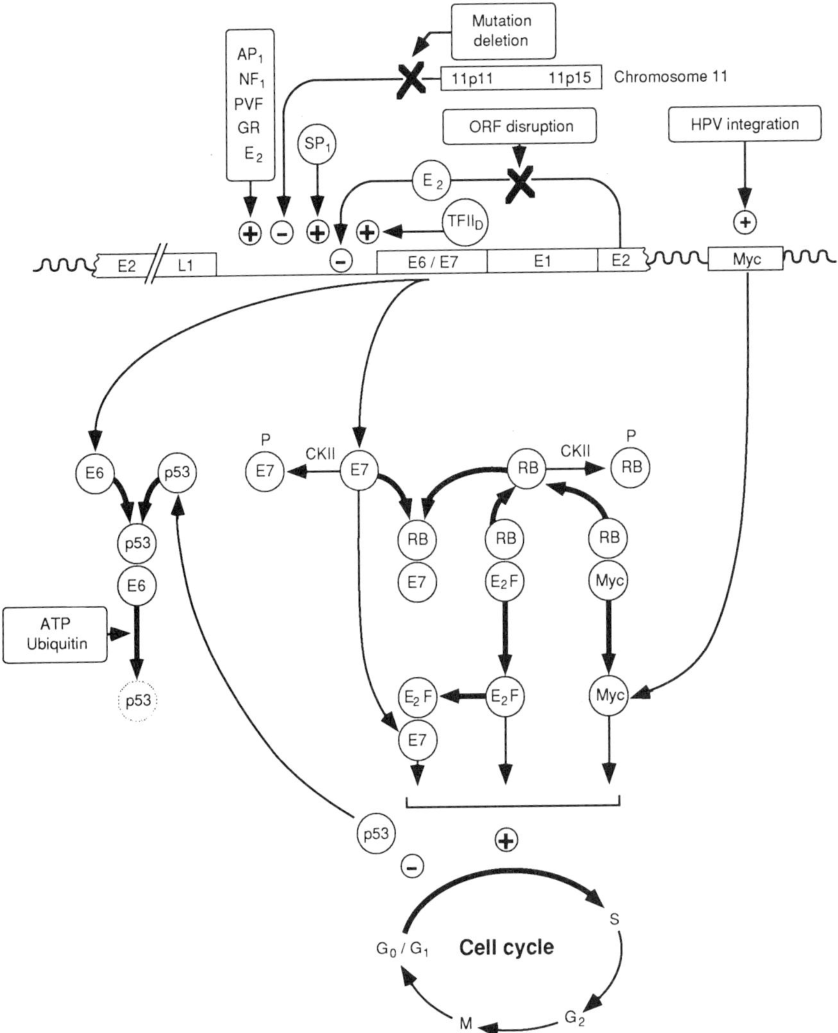

FIGURE 6 Schematic working hypothesis: synthetic representation of some pathways leading to the E6/E7 induced deregulation of the cell cycle in HPV-transformed keratinocytes. The dark arrows highlight the molecular events (protein–protein association or dissociation) favored by the unregulated expression of E6 and E7 proteins. This scheme stresses (i) the mechanisms involved in the deregulation of E6/E7 protein synthesis and (ii) the association of E6 and E7 proteins with p53 and RB^{105} antioncogenes, respectively, and the possible effects on E2F factor and *myc* levels.

peptides that are potent antagonists of E7 binding to the p105RB protein (Jones *et al.*, 1990). However, the progression to malignant transformation is a multistep process and it remains difficult to evaluate the weight of individual events (e.g., chromosomal rearrangements) directly linked to the virus or linked to the cell host. (A possible such rearrangement could be deletion of chromosome 11 and activation of *c-myc* synthesis.)

In the absence of adequate therapeutics, and despite the tremendous progress in the understanding of the biology of these viruses, I would like to stress the necessity of regular medical checkups and the importance of preventive behavior in combatting the spread of these sexually transmitted viruses.

Addendum

The most exciting developments achieved since year 1991 deal with the effects of retinoids on HPV-immortalized keratinocytes and E6/E7 expression. Agarwal *et al.* (1991) showed that natural and synthetic retinoids suppressed the differentiation of HPV-transformed cervical cells and may thus be useful in controlling the incidence and/or progression of HPV-associated cervical tumors. In the same line, Pirisi *et al.* (1992) reported HPV16-immortalized keratinocytes exhibited an increased sensitivity to growth control by retinoic acid (RA), when compared to normal human foreskin keratinocytes. Since the expression of E6/E7 gene products was also found inhibited by RA (Pirisi *et al.*, 1992), it was suggested that this increased sensitivity to growth control by RA may be mediated by an inhibition of the expression of gene products (E6/E7) required for the maintenance of continuous growth. It was then recently confirmed that retinoic acid receptors (RARs) exerted a direct repressing effect on the promoter activity of HPV18-LCR (Barstsch *et al.*, 1992; Bernard *et al.*, 1993). A putative responsive element was localized in the HPV18-LCR from nt 7768 to nt 7784, with a sequence 5′AGTTCAtgttaAGTTCA (opposite strand) very similar to the RA-responsive element 5′GGTTCAccgaaAGTTCA found in the RARβ gene (Bernard *et al.*, 1993). Although the repression mechanism remains to be established, it could involve the nearby AP1 binding site (nt 7791 to nt 7798), recently shown to be essential for HPV18 transcription in keratinocytes (Thierry *et al.*, 1992), or the 5′AATTGCAT Oct-1 sequence (nt 7720 to nt 7728) present in the central enhancer of HPV18-LCR (Hoppe-Seyler *et al.*, 1991) to which regulatory elements involved in RAR-mediated repression have been recently ascribed (Bartsch *et al.*, 1992). Interestingly, RA-induced down regulation of the IL-2 and IgH promoter was recently shown to be mediated by RARs, via *cis*-regulatory sequences containing octamer motifs (Felli *et al.*, 1991) similar to that present in the

HPV18-LCR, and which maps to the NFA sequence found in the central enhancer of type 6, 11, 16, and 18 genital papillomaviruses (Chong *et al.*, 1990). These data represent a decisive progress in the understanding of the control of E6/E7 transforming gene expression and may generate possible clues for future therapeutic approaches.

References

Agarwal, C., Rorke, E. A., Irwin, J. C., and Eckert, R. C. (1991). Immortalization by human papillomavirus type 16 alters retinoid regulation of human ectocervical epithelial cell differentiation. *Cancer Res.* **51**, 3982–3989.

Androphy, E. J., Schiller, J. T., and Lowy, D. R. (1985). Identification of the protein encoded by the E6 transforming gene of bovine papillomavirus. *Science* **230**, 442–445.

Androphy, E. J., Hubbert, N. L., Schiller, J. T., and Lowy, D. R. (1987). Identification of the HPV16-E6 protein from transformed mouse cells and human cervical carcinoma cell lines. *EMBO J.* **6**, 989–992.

Asselineau, D., Bernard, B. A., Bailly, C., and Darmon, M. (1985). Epidermal morphogenesis and induction of the 67-kDa keratin polypeptide by culture at the liquid–air interface. *Exp. Cell. Res.*, **159**, 536–539.

Atkin, N. B. and Baker, M. C. (1982). Nonrandom chromosome changes in carcinoma of the cervix uteri. I. Nine near-diploid tumours. *Cancer Genet. Cytogenet.* **7**, 209–222.

Atkin, N. B. and Baker, M. C. (1984). Nonrandom chromosome changes in carcinoma of the cervix uteri. II. Ten tumours in the triploid–tetraploid range. *Cancer Genet. Cytogenet.* **13**, 189–207.

Bagchi, S., Weinmann, R., and Raychaudhuri, P. (1991). The retinoblastoma protein copurified with E2F-I, and E1A-regulated inhibitor of the transcription factor E2F. *Cell* **65**, 1063–1072.

Baker, C. C., Phelps, W. C., Lindgren, V., Braun, M. J., Gonda, M. A., and Howley, P. M. (1987). Structural and transcriptional analysis of human papillomavirus type 16 sequences in cervical carcinoma cell lines. *J. Virol.* **61**, 962–971.

Bandara, L. R. and La Thangue, N. B. (1991). Adenovirus E1a prevents the retinoblastoma gene product from complexing with a cellular transcription factor. *Nature* **351**, 494–497.

Banks, L., Spence, P., Androphy, E. J., Hubbert, N. L., Matlashewski, G., Murray, A., and Crawford, L. (1987). Identification of human papillomavirus type 18 E6 polypeptide in cells derived from human cervical carcinomas. *J. Gen. Virol.* **68**, 1351–1359.

Banks-Schlegel, S. P. and Howley, P. M. (1983). Differentiation of human epidermal cells transformed by SV40. *J. Cell Biol.* **96**, 330–337.

Barbosa, M. S., Lowy, D. R., and Schiller, J. T. (1989). Papillomavirus polypeptides E6 and E7 are zinc-binding proteins. *J. Virol.* **63**, 1404–1407.

Barbosa, M. S., Edmonds, C., Fisher, C., Schiller, J. T., Lowy, D. R., and Vousden, K. H. (1990). The region of the HPV E7 oncoprotein homologous to adenovirus E1a and SV40 large T antigen contains separate domains for Rb binding and casein kinase II phosphorylation. *EMBO J.* **9**, 153–160.

Barbosa, M. S., Vass, W. C., Lowy, D. R., and Schiller, J. T. (1991). *In vitro* biological activities of the E6 and E7 genes vary among human papillomaviruses of different oncogenic potential. *J. Virol.* **65**, 292–298.

Bargonetti, J., Friedman, P. N., Kern, S. E., Vogelstein, B., and Prives, C. (1991). Wildtype but not mutant p53 immunopurified proteins bind to sequences adjacent to the SV40 origin of replication. *Cell* **65**, 1083–1091.

Bartsch, D., Boye, B., Baust, C., zur Hausen, H., and Schwarz, E. (1992). Retinoic acid-mediated repression of human papillomavirus 18 transcription and different ligand regulation of the retinoic acid receptor β gene in non-tumorigenic and tumorigenic HeLa hybrid cells. *EMBO J.* **11**, 2283–2291.

Bedell, M. A., Jones, K. H., and Laimins, L. A. (1987). The E6–E7 region of human papillomavirus type 18 is sufficient for transformation of NIH 3T3 and Rat-1 cells. *J. Virol.* **61**, 3635–3640.

Bedell, M.A., Jones, K.H., Grossman, S. R., and Laimins, L. A. (1989). Identification of human papillovirus type 18 transforming genes in immortalized and primary cells. *J. Virol.* **63**, 1247–1255.

Bernard, B. A., Robinson, S., Sémat, A., and Darmon, Y. M. (1985). Reexpression of fetal characters in SV40 transformed human keratinocytes. *Cancer Res.* **45**, 1707–1716.

Bernard, H.-U., Oltersdorf, T., and Seedorf, K. (1987). Expression of the human papillomavirus type 18 E7 gene by a cassette-vector system for the transcription and translation of open reading frames in eukaryotic cells. *EMBO J.* **6**, 133–138.

Bernard, B. A., Bailly, C., Lenoir, M. C., Darmon, M., Thierry, F., and Yaniv, M. (1989). The human papillomavirus type 18 (HPV 18) gene product is a repressor of the HPV18 regulatory region in human keratinocytes. *J. Virol.* **63**, 4317–4324.

Bernard, B. A., Bailly, C., Lenoir, M. C., and Darmon, M. Y. (1990a). Modulation of HPV-18 and BPV-1 transcription in human keratinocytes by simian virus 40 large T antigen and adenovirus type 5 E1a antigen. *J. Cell. Biochem.*, **42**, 1–10.

Bernard, B. A., Magnaldo, T., Lenoir, M. C., Bailly, C., and Darmon, M. (1990b). Modulation of papillomavirus transcription by viral transeffectors and retinoids. *In* "Papillomaviruses" (P. Howley and T. Broker, eds.), pp. 481–490. Wiley-Liss, Inc., New York.

Bernard, B. A., Gerst, C., Lenoir, M. C., Delescluse, C., and Darmon, M. (1993). Retinoic acid responsive element in the long control region of human papillomavirus type 18. In "Pharmacology and the skin, from Molecular Biology to Therapeutics. Karger AG, Basel." (B. A. Bernard and B. Shroot, eds.), in press.

Bosch, F. X., Schwarz, E., Boukamp, P., Fusenig, N. E., Bartsch, D., and Zur Hausen, H. (1990). Suppression *in vivo* of human papillomavirus type 18 E6–E7 gene expression in nontumorigenic HeLa χ fibroblast hybrid cells. *J. Virol.* **64**, 4743–4754.

Boshart, M., Gissmann, L., Ikenberg, H., Kleinheinz, A., Scheurlen, W., and zur Hausen, H. (1984). A new type of papillomavirus DNA, its presence in genital cancer biopsies and in cell lines derived from cervical cancer. *EMBO J.* **3**, 1151–1157.

Braun, L., Dürst, M., Mikumo, R., and Grupposo, P. (1990). Differential response of non-tumorigenic and tumorigenic human papillomavirus type 16-positive epithelial cells to transforming growth factor β1. *Cancer Res.* **50**, 7324–7332.

Chan, W.-K., Klock, G., and Bernard, H.-U. (1989). Progesterone and glucocorticoid response elements occur in the long control regions of several human papillomaviruses involved in anogenital neoplasia. *J. Virol.* **63**, 3261–3269.

Chan, W.-K., Chong, T., Bernard, H.-U., and Klock, G. (1990). Transcription of the transforming genes of the oncogenic human papillomavirus-16 is stimulated by tumor promoters through AP1 binding sites. *Nucleic Acids Res.* **18**, 763–769.

Chellappan, S. P., Hiebert, S., Mudryj, M., Horowitz, J. M., and Nevins, J. R. (1991). The E2F transcription factor is a cellular target for the RB protein. *Cell* **65**, 1053–1061.

Chin, M. T., Broker, T. R., and Chow, L. T. (1989). Identification of a novel constitutive enhancer element and an associated binding protein: implications for human papillomavirus type 11 enhancer regulation. *J. Virol.* **63**, 2967–2976.

Chittenden, T., Livingston, D. M., and Kaelin, W. J. Jr. (1991). The T/E1A-binding domain of the retinoblastoma product can interact selectively with a sequence-specific DNA-binding domain. *Cell* **65**, 1073–1082.

Chong, T., Chan, W.-K., and Bernard, H.-U. (1990). Transcriptional activation of human papillomavirus 16 by nuclear factor I, AP1, steroid receptors and a possibly novel transcription factor, PVF: a model for the composition of genital papillomavirus enhancers. *Nucleic. Acids Res.* **18,** 465–470.

Choo, K. B., Pan, C. C., and Han, S. M. (1987). Integration of human papillomavirus type 16 into cellular DNA of cervical carcinoma: preferential deletion of the E2 gene and invariable retention of the long control region and the E6/E7 open reading frames. *Virology* **161,** 259–261.

Choo, K.-B., Lee, H.-H., Pan, C.-C., Wu, S.-M., Liew, L.-N., Cheung, W.-F., and Han, S.-H. (1988). Sequence duplication and internal deletion in the integrated human papillomavirus type 16 genome cloned from a cervical carcinoma. *J. Virol.* **62,** 1659–1666.

Ciechanover, A. and Schwartz, A. L. (1989). How are substrates recognized by the ubiquitin-mediated proteolytic system? *Trends Biochem. Sci.* **14,** 483–488.

Cole, S. T. and Danos, O. (1987). Nucleotide sequence and comparative analysis of the human papillomavirus type 18 genome. Phylogeny of papillomaviruses and repeated structure of the E6 and E7 gene products. *J. Mol. Biol.* **193,** 599–608.

Couturier, J., Sastre-Garau, X., Schneider-Maunoury, S., Labib, A., and Orth, G. (1991). Integration of papillomavirus DNA near *myc* genes in genital carcinomas and its consequences for protooncogene expression. *J. Virol.* **65,** 4534–5438.

Cripe, T. P., Haugen, T. H., Turk, J. P., Tabatabai, F., Schmid, III, P. G., Dürst, M., Gissmann, L., Roman, A., and Turek, L. P. (1987). Transcriptional regulation of the human papillomavirus-16 E6–E7 promoter by a keratinocyte-dependent enhancer, and by viral E2 *trans*-activator and repressor gene products: implications for cervical carcinogenesis. *EMBO J.* **6,** 3745–3753.

Cripe, T. P., Alderborn, A., Anderson, R. D., Parkkinen, S., Bergman, P., Haugen, T. H., Petersson, U., and Turek, L. P. (1990). Transcriptional activation of the human papillomavirus-16 p97 promoter by an 88-nucleotide enhancer containing distinct cell-dependent and AP-1-responsive modules. *The New Biologist* **2,** 450–463.

Crook, T., Almond, N., Murray, A., Stanley, M., and Crawford, L. (1989a). Constitutive expression of *c-myc* oncogene confers hormone independence and enhanced growth-factor responsiveness on cells transformed by human papillomavirus type 16. *Proc. Natl. Acad. Sci. U.S.A.* **86,** 5713–5717.

Crook, T., Morgenstern, J. P., Crawford, L., and Banks, L. (1989b). Continued expression of HPV16 E7 protein is required for maintenance of the transformed phenotype of cells cotransformed by HPV-16 plus *EJ-ras*. *EMBO J.* **8,** 513–519.

Danos, O. and Yaniv, M. (1987). E6 and E7 gene products evolved by amplification of a 33-amino-acid peptide with a potential nuclear-acid-binding structure. *Cancer Cells* **5,** 145–149.

DeCaprio, J. A., Ludlow, J. W., Figge, J., Shew, J.-Y., Huang, C.-M., Lee, W.-H., Marsilio, E., Paucha, E., and Livingston, D. M. (1988). SV40 large tumor antigen forms a specific complex with the product of the retinoblastoma susceptibility gene. *Cell* **54,** 275–283.

DeCaprio, J. A., Ludlow, J. W., Lynch, D., Furukawa, Y., Griffin, J., Piwnica-Worms, H., Huang, C.-M., and Livingston, D. (1989). The product of the retinoblastoma susceptibility gene has properties of a cell cycle regulatory element. *Cell* **58,** 1085–1095.

Dostatni, N., Thierry, F., and Yaniv, M. (1988). A dimer of BPV-1 E2 containing a protease resistant core interacts with its DNA target. *EMBO J.* **7,** 3807–3816.

Dürst, M., Gissmann, L., Ikenberg, H., and zur Hausen, H. (1983). A papillomavirus DNA from cervical carcinoma and its prevalence in cancer biopsy samples from different geographic regions. *Proc. Natl. Acad. Sci. U.S.A.* **80,** 3812–3815.

Dürst, M., Kleinheinz, A., Hotz, M., and Gissmann, L. (1985). The physical state of human papillomavirus type 16 DNA in benign and malignant genital tumors. *J. Gen. Virol.* **66,** 1515–1522.

Dürst, M., Croce, C. M., Gissmann, L., Schwarz, E., and Huebner, K. (1987a). Papillomavirus sequence integrate near cellular oncogenes in some cervical carcinomas. *Proc. Natl. Acad. Sci. U.S.A.* **84,** 1070–1974.

Dürst, M., Dzarlieva-Petrusevska, R. T., Boukamp, P., Fusenig, N. E., and Gissmann, L. (1987b). Molecular and cytogenetic analysis of immortalized human primary keratinocytes obtained after transfection with human papillomavirus type 16 DNA. *Oncogene* **1,** 251–256.

Dürst, M., Bosch, F. X., Glitz, D., Schneider, A., zur Hausen, H. (1991). Inverse relationship between human papillomavirus (HPV) type 16 early gene expression and cell differentiation in nude mouse epithelial cysts and tumors induced by HPV-positive human cell lines. *J. Virol.* **65,** 796–804.

Dyson, N., Howley, P. M., Münger, K., and Harlow, E. (1989). The human papillomavirus virus-16 E7 oncoprotein is able to bind to the retinoblastoma gene product. *Science* **243,** 934–936.

Dyson, N., Bernards, R., Friend, S. H., Gooding, L. R., Hassell, J. A., Major, E. O., Pipas, J. M., Vandyke, T., and Harlow, E. (1990). Large T antigens of many polyomaviruses are able to form complexes with the retinoblastoma protein. *J. Virol.* **64,** 1353–1356.

El Awady, M. K., Kaplan, J. B., O'Brien, S. J., and Burk, R. D. (1987). Molecular analysis of integrated human papillomavirus 16 sequences in the cervical cancer cell line SiHA. *Virology* **159,** 389–398.

Felli, M. P., Vacca, A., Meco, D., Screpanti, I., Farina, A. R., Maroder, M., Martinotti, S., Petrangeli, E., Frati, L., and Gulino, A. (1991). Retinoic acid-induced down-regulation of the interleukin-2 promoter via cis-regulatory sequences containing an octamer motif. *Mol. Cell. Biol.* **11,** 4771–4778.

Ferguson, B., Kripple, B., Andrisani, O., Jones, N., Westphal, H., and Rosenberg, M. (1985). E1a 13S and 12S mRNA products made in *Escherichia coli* both function as nucleus localized transcription activators but do not directly bind to DNA. *Mol. Cell. Biol.* **5,** 2653–2661.

Fields, S. and Jang, S. K. (1990). Presence of a potent transcription activating sequence in the p53 protein. *Science* **249,** 1046–1049.

Finlay, C. A., Hinds, P. W., and Levine, A. J. (1989). The p53 proto-oncogene can act as a suppressor of transformation. *Cell* **57,** 1083–1093.

Firzlaff, J. M., Galloway, D. A., Eisenman, R. N., and Lüscher, B. (1989). The E7 protein of human papillomavirus type 16 is phosphorylated by casein kinase II. *The New Biologist* **1,** 44–54.

Gage, J. R., Meyers, C., and Wettstein, F. O. (1990). The E7 proteins of the nononcogenic human papillomavirus type 6b (HPV-6b) and of the oncogenic HPV-16 differ in retinoblastoma protein binding and other properties. *J. Virol.* **64,** 723–730.

Garcia-Carranca, A., Thierry, F., and Yaniv, M. (1988). Interplay of viral and cellular proteins along the long control region of human papillomavirus type 18. *J. Virol.* **62,** 4321–4330.

Gassenmaier, A., Lammel, M., Kleiner, E., and Pfister, H. (1985). Treatment of bovine-papillomavirus type-1 (BPV)-transformed mouse cells with aromatic retinoids and retinoic acid. *Arch. Dermatol. Res.* **278,** 79–81.

Geiser, A. G. and Stanbridge, E. J. (1987). Introduction of a normal human chromosome 11 into a Wilm's tumor cell line controls its tumorigenic expression *Science* **236,** 175–180.

Giri, I. and Danos, O. (1986). Papillomavirus genomes: from sequence data to biological properties. *Trends Genet.* **9,** 227–232.

Giri, I. and Yaniv, M. (1988). Structural and mutational analysis of E2 *trans*-activating proteins of papillomaviruses reveals three distinct functional domains. *EMBO J.* **7,** 2823–2829.

Gissmann, L. (1984). Papillomaviruses and their association with cancer in animals and in man. *Cancer Surv.* **3,** 161–181.

Gius, D. and Laimins, L. A. (1989). Activation of human papillomavirus type 18 gene expression by herpes simplex virus type 1 viral transactivators and a phorbol ester. *J. Virol.* **63**, 555–563.

Gloss, B. and Bernard, H.-U. (1990). The E6/E7 promoter of human papillomavirus type 16 is activated in the absence of E2 proteins by a sequence-aberrant sp1 distal element. *J. Virol.* **64**, 5577–5584.

Gloss, B., Bernard, H. U., Seedorf, K., and Klock, G. (1987). The upstream regulatory region of the human papillomavirus virus-16 contains an E2 protein-independent enhancer which is specific for cervical carcinoma cells and regulated by glucocorticoid hormones. *EMBO J.* **6**, 3735–3743.

Gloss, B., Yeo-Gloss, M., Meisterernst, M., Rogge, L., Winnacker, R. L., and Bernard, H.-U. (1989a). Clusters of nuclear factor I binding sites identify enhancers of several papillomaviruses, but alone are not sufficient for enhancer function. *Nucleic Acids Res.* **17**, 3519–3533.

Gloss, B., Chong, T., and Bernard, H.-U. (1989b). Numerous nuclear proteins bind the long control region of human papillomavirus type 16: a subset of 6 of 23 DNase I-protected segments coincides with the location of the cell-type-specific enhancer. *J. Virol.* **63**, 1142–1152.

Grossman, S. R., Mora, R., and Laimins, L. A. (1989). Intracellular localization and DNA-binding properties of human papillomavirus type 18 E6 protein expressed with a baculovirus vector. *J. Virol.* **63**, 366–374.

Halbert, C. L., Demers, G. W., and Galloway, D. A. (1991). The E7 gene of human papillomavirus type 16 is sufficient for immortalization of human epithelial cells. *J. Virol.* **65**, 473–478.

Haugen, T. H., Cripe, T. P., Ginder, G. D., Karin, M., and Turek, L. P. (1987). *trans*-Activation of an upstream early gene promoter of bovine papilloma virus-1 by a product of the viral E2 gene. *EMBO J.* **6**, 145–152.

Haugen, T. H., Lurek, L. P., Mercurio, F. M., Cripe, T. P., Olson, B. J., Anderson, R. D., Seidl, D., Karin, M., and Schiller, J. (1988). Sequence-specific and general transcriptional activation by the bovine papillomavirus-1 E2 *trans*-activator require an N-terminal amphipathic helix-containing E2 domain. *EMBO J.* **7**, 4245–4253.

Hawley-Nelson, P., Androphy, E. J., Lowy, D. R., and Schiller, J. T. (1988). The specific DNA recognition sequence of the bovine papillomavirus E2 protein is an E2-dependent enhancer. *EMBO J.* **7**, 525–531.

Hawley-Nelson, P., Vousden, K. H., Hubbert, N. L., Lowy, D. R., and Schiller, J. T. (1989). HPV16 E6 and E7 proteins cooperate to immortalize human foreskin keratinocytes. *EMBO J.* **8**, 3905–3910.

Hirochika, H., Hirochika, R., Broker, T. R., and Chow, L. T. (1988). Functional mapping of the human papillomavirus type 11 transcriptional enhancer and its interaction with the *trans*-acting E2 proteins. *Genes & Dev.* **2**, 54–67.

Hoppe-Seyler, F., Butz, K., and zur Hausen, H. (1991). Repression of the human papillomavirus type 18 enhancer by the cellular transcription factor Oct-1. *J. Virol.* **65**, 5613–5618.

Howe, J. A., Mymryk, J. S., Egan, C., Branton, P. E., and Bayley, S. T. (1990). Retinoblastoma growth suppressor and a 300-kDa protein appear to regulate cellular DNA synthesis. *Proc. Natl. Acad. Sci. U.S.A.* **87**, 5883–5887.

Hudson, J. B., Bedell, M. A., McCance, D. J., and Laimins, L. A. (1990). Immortalization and altered differentiation of human keratinocytes *in vitro* by the E6 and E7 open reading frames of human papillomavirus type 18. *J. Virol.* **64**, 519–526.

Hurlin, P. J., Kaur, P., Smith, P. P., Perez-Reyes, N., Blnton, R. A., and McDougall, J. K. (1991). Progression of human papillomavirus type 18-immortalized human keratinocytes to a malignant phenotype. *Proc. Natl. Acad. Sci. U.S.A.* **88**, 570–574.

Imai, Y., Tsunokawa, Y., Sugimura, T., and Terada, M. (1989). Purification and DNA-binding properties of human papillomavirus type 16 E6 protein expression in *Escherichia coli*. *Biochem. Biophys. Res. Commun.* **164,** 1402–1410.

Imai, Y., Matsushima, Y., Sugimura, T., and Terada, M. (1991). Purification and characterization of human papillomavirus type 16 E7 protein with preferential binding capacity to the underphosporylated form of retinoblastoma gene product. *J. Virol.* **65,** 4966–4972.

Inagaki, Y., Tsunokawa, Y., Takeke, N., Nawa, H., Nakanishi, S., Terada, M., and Sugimura, T. (1988). Nucleotide sequences of cDNAs for human papillomavirus type 18 transcripts in HeLa cells. *J. Virol.* **62,** 1640–1646.

Jones, R. E., Wegrzyn, R. J., Patrick, D. R., Balishin, N. L., Vuocolo, G. A., Riemen, M. W., Defeo-Jones, D., Garsky, V. M., Heimbrook, D. C., Oliff, A. (1990). Identification of HPV-16 E7 peptides that are potent antagonists of E7 binding to the retinoblastoma suppressor protein. *J. Biol. Chem.* **265,** 12782–12785.

Kaur, P. and McDougall, J. K. (1988). Characterization of primary human keratinocytes transformed by human papillomavirus type 18. *J. Virol.* **62,** 1917–1924.

Kaur, P. and McDougall, J. K. (1989). HPV-18 immortalization of human keratinocytes. *Virology* **173,** 302–310.

Kern, S. E., Kinzler, K. W., Bruskin, A., Friedman, P. N., Pries, C., and Vogelstein, B. (1991). Sequence-specific binding of p53 to DNA. *Science* **252,** 1708–1711.

Kirchner, H. (1986). Immunobiology of human papillomavirus infection. *Prog. Med. Virol.* **33,** 1–41.

Klug, A. and Rhodes, D. (1987). 'Zinc fingers': a novel protein motif for nucleic acid recognition. *Trends Biochem. Sci.* **12,** 464–469.

Kreider, J. W. (1980). Neoplastic progression of the Shope rabbit papilloma. *In* "Viruses in Naturally Occurring Cancers" (M. Essex, G. Todaro, and H. Zur Hausen, eds.). Cold Spring Harbor Conf. on Cell Proliferation, Vol. 7, pp. 283–300. Cold Spring Harbor.

Lambert, P. F., Spalholz, B. A., and Howley, P. M. (1987). A transcriptional repressor encoded by BPV-1 shares a common carboxy-terminal domain with the E2 transactivator. *Cell* **50,** 69–78.

Lane, D. P. and Crawford, L. V. (1979). T antigen is bound to a host protein in SV40-transformed cells. *Nature* **278,** 261–263.

Lazo, P. A., DiPaolo, J. A., and Popescu, N. C. (1989). Amplification of the integrated viral transforming genes of human papillomavirus 18 and its 5′-flanking cellular sequence located near the *myc* protooncogene in HeLa cells. *Cancer Res.* **49,** 4305–4310.

Le, J.-Y. and Defendi, V. (1988). A viral–cellular junction fragment from a human papillomavirus type 16-positive tumor is competent in transformation of NIH 3T3 cells. *J. Virol.* **62,** 4420–4426.

Lee, E. Y.-H., To, H., Shew, J.-Y., Bookstein, R., Scully, P., and Lee, W.-H. (1988). Inactivation of the retinoblastoma susceptibility gene in human breast cancers. *Science* **241,** 218–221.

Levine, A. J., Momand, J., and Finlay, C. A. (1991). The p53 tumour suppressor gene. *Nature* **351,** 453–456.

Li, G., Tsang, S. S., and Stich, H. F. (1988). Changes in DNA copy numbers of bovine papillomavirus type 1 after termination of retinoic acid treatment. *J. Natl. Cancer Inst.* **80,** 567–570.

Li, R., Knight, J. D., Jackson, S. P., Tjian, R., and Botchan, M. R. (1991). Direct interaction between sp1 and the BPV enhancer E2 protein mediates synergistic activation of transcription. *Cell* **65,** 493–505.

Linzer, D. I. H. and Levine, A. J. (1979). Characterization of a 54kDa cellular SV40 tumor antigen present in SV40-transformed cells and uninfected embryonal carcinoma cells. *Cell* **17,** 43–52.

Ludlow, J. W., DeCaprio, J. A., Huang, C.-M., Lee, W.-H., Paucha, E., and Livingston, D. M.

(1989). SV40 large T antigen binds preferentially to an underphosphorylated member of the retinoblastoma susceptibility gene product family. *Cell* **56**, 57–65.

Matlashewski, G., Schneider, J., Banks, L., Jones, N., Murray, A., and Crowford, L. (1987). Human papillomavirus type 16 DNA cooperates with activated *ras* in transforming primary cells. *EMBO J.* **6** 1741–1746.

McBride, A. A., Schlegel, R., and Howley, P. M. (1988). The carboxy-terminal domain shared by the bovine papillomavirus E2 transvactivator and repressor proteins contains a specific DNA binding domain. *EMBO J.* **7**, 533–539.

McBride, A. A., Byrne, J. C., and Howley, P. M. (1989). E2 polypeptides encoded by bovine papillomavirus type 1 form dimers through the common carboxyl-terminal domain: transcactivation is mediated by the conserved amino-terminal domain. *Proc. Natl. Acad. Sci. U.S.A.* **86**, 510–514.

McCance, D. J. (1986). Human papillomaviruses and cancer. *Biochim. Biophys. Acta* **823**, 195–205.

McCance, D. J., Kopan, R., Fuchs, E., and Laimins, L. A. (1988). Human papillomavirus type 16 alters human epithelial cell differentiation *in vitro*. *Proc. Natl. Acad. Sci. U.S.A.* **85**, 7169–7173.

Mihara, K., Cao, X.-R., Yen, A., Chandler, S., Driscoll, B., Murphree, A. L., T'Ang, A., and Fung, Y.-K. T. (1989). Cell cycle-dependent regulation of phosphorylation of the human retinoblastoma gene product. *Science* **246**, 1300–1303.

Mitrani-Rosenbaum, S., Tsvieli, R., and Tur-Kaspa, R. (1989). Oestrogen stimulates differential transcription of human papillomavirus type 16 in SiHa cervical carcinoma cells. *J. Gen. Virol.* **70**, 2227–2332.

Moskaluk, C. A. and Bastia, D. (1989). The bovine papillomavirus type 1 transcriptional activator E2 protein binds to its DNA recognition sequence as a dimer. *Virology* **169**, 236–238.

Münger, K., Phelps, W. C., Bubb, V., Howley, P. M., and Schlegel, R. (1989a). The E6 and E7 genes of the human papillomavirus type 16 together are necessary and sufficient for transformation of primary human keratinocytes. *J. Virol.* **63**, 4417–4421.

Münger, K., Werness, B. A., Dyson, N., Phelps, W., Harlow, E., and Howley, P. M. (1989b). Complex formation of human papillomavirus E7 proteins with the retinoblastoma tumor suppressor gene product. *EMBO J.* **8**, 4099–4105.

Offord, E. A. and Beard, P. (1990). A member of the activator protein 1 family found in keratinocytes but not in fibroblasts required for transcription fro a human papillomavirus type 18 promoter. *J. Virol.* **64**, 4792–4798.

Oltersdorf, T., Seedorf, K., Röwekamp, W., and Gissmann, L. (1987). Identification of human papillomavirus type 16 E7 protein by monoclonal antibodies. *J. Gen. Virol.* **68**, 2933–2938.

Orth, G. (1986). Epidermodysplasia verruciformis: a model for understanding the oncogenicity of human papillomaviruses. *In* "Papillomaviruses" (D. Evered and S. Clark, eds.) CIBA Foundation Symp. Vol. 120. pp. 157–174. John Wiley & Sons, New York.

Parkin, D. M., Stjernsward, J., and Muir, C. S. (1984). Estimates of the worldwide frequency of twelve major cancers. *Bull. WHO* **62**, 163–182.

Pater, M. M. and Pater, A. (1985). Human papillomavirus types 16 and 18 sequences in carcinoma cell lines of the cervix. *Virology* **145**, 313–318.

Pater, M. M., Hughes, G. A., Hyslop, D. A., Nakshatri, H., and Pater, A. (1988). Glucocorticoid-dependent oncogenic transformation by type 16 but not type 11 human papillomavirus DNA. *Nature* **335**, 832–835.

Pecoraro, G., Morgan, D., and Defendi, V. (1989). Differential effects of human papillomavirus type 6, 16, and 18 DNAs on immortalization and transformation of human cervical epithelial cells. *Proc. Natl. Acad. Sci. U.S.A.* **86**, 563–567.

Pfister, H. (1984). Biology and biochemistry of papillomaviruses. *Rev. Physiol. Biochem. Pharmacol.* **99**, 111–182.

Phelps, W. C. and Howley, P. M. (1987). Transcriptional *trans*-activation by the human papillomavirus type 16 E2 gene product. *J. Virol.* **61**, 1630–1638.

Phelps, W. C., Yee, C. L., Münger, K., and Howley, P. M. (1988). The human papillomavirus type 16 E7 gene encodes transactivation and transformation functions similar to those of adenovirus E1A. *Cell* **53**, 539–547.

Pirisi, L., Creek, K. E., Doniger, J., and DiPaolo, J. A. (1988). Continuous cell lines with altered growth and differentiation properties originate after transfection of human keratinocytes with human papillomavirus type 16 DNA. *Carcinogenesis* **9**, 1563–1579.

Pirisi, L., Yasumoto, S., Feller, M., Doniger, J., and DiPaolo, J. A. (1987). Transformation of human fibroblasts and keratinocytes with human papillomavirus type 16 DNA. *J. Virol.* **61**, 1061–1066.

Pirisi, A., Batova, A., Jenkins, G. R., Hodam, J. R., and Creek, K. E. (1992). Increased sensitivity of human keratinocytes immortalized by human papillomavirus type 16 DNA to growth control by retinoids. *Cancer Res.* **52**, 187–193.

Popescu, N. C., Amsbauch, S. C., and DiPaolo, J. A. (1987). Human papillomavirus type 18 DNA is integrated at a single chromosome site in cervical carcinoma cell line SW756. *J. Virol.* **51**, 1682–1685.

Rawls, J. A., Pusztai, R., and Green, M. (1990). Chemical synthesis of human papillomavirus type 16 E7 oncoprotein: autonomous protein domains for induction of cellular DNA synthesis and for trans activation. *J. Virol.* **64**, 6121–6129.

Raycroft, L., Wu, H., and Lozano, G. (1990). Transcriptional activation by wildtype but not transforming mutants of the p53 antioncogene. *Science* **249**, 1049–1051.

Reich, N. C., Oren, M., and Levine, A. J. (1983). Two distinct mechanisms regulate the levels of a cellular tumor antigen. *Mol. Cell. Biol.* **3**, 2143–2150.

Riccardi, V. M., Hittner, H. M., Francke, U., Yunis, J. J., Ledbetter, D., and Borges, W. (1980). The aniridia-Wilms tumour association: the critical role of chromosome band 11p13. *Cancer Genet. Cytogenet.* **2**, 131–137.

Romanczuk, H., Thierry, F., and Howley, P. (1990). Mutational analysis of *cis* elements involved in E2 modulatin of human papillomavirus type 16 p97 and type 18 p105 promoters. *J. Virol.* **64**, 2849–2859.

Romanczuk, H., Villa, L. L., Schlegel, R., and Howly, P. M. (1991). The viral transcriptional regulatory region upstream of the E6 and E7 genes is a major determinant of the differential immortalization activities of human papillomavirus types 16 and 18. *J. Virol.* **65**, 2739–2744.

Royer, H.-D., Freyaldenhoven, M. P., Napierski, I., Spitkovsky, D. D., Bauknecht, T., and Dathan, N. (1991). Delineation of human papillomavirus type 18 enhancer binding proteins: the intracellular distribution of a novel octamer binding protein p92 is cell cycle regulated. *Nucleic Acids Res.* **19**, 2363–2371.

Rustgi, A., Dyson, N., and Bernards, R. (1991). Amino-terminal domains of *c-myc* and *N-myc* proteins mediate binding to the retinoblastoma gene product. *Nature* **352**, 542–544.

Rösl, F., Achtstätter, T., Bauknecht, T., Hutter, K.-J., Futterman, G., and Zur Hausen, H. (1991). Extinction of the HPV18 upstream regulatory region in cervical carcinoma cells after fusion with nontumorigenic human keratinocytes under nonselective conditions. *EMBO J.* **10**, 1337–1345.

Sanborn, B. M., Held, B., and Kuo, H. S. (1976). Hormonal action in human cervix. *J. Steroid Biochem.* **7**, 665–674.

Sato, H., Watanabe, S., Furuno, A., and Yoshiike, K. (1989). Human papillomavirus type 16 E7 protein expressed in *Escherichia coli* and monkey COS-1 cells: immunofluorescence detection of the nuclear E7 protein. *Virology* **170**, 311–315.

Saxon, P. J., Srivatsan, E. S., and Stanbridge, E. J. (1986). Introduction of human chromosome

11 via microcell transfer controls tumorigenic expression of HeLa cells. *EMBO J.* **5**, 3461–3466.

Scheffner, M., Werness, B. A., Huibregtse, J. M., Levine, A. J., and Howley, P. M. (1990). The E6 oncoprotein encoded by human papillomavirus types 16 and 18 promotes the degradation of p53. *Cell* **63**, 1129–1136.

Schlegel, R., Phelps, W. C., Zhang, Y.-L., and Barbosa, M. (1988). Quantitative keratinocyte assay detects two biological activities of human papillomavirus DNA and identifies viral types associated with carcinoma. *EMBO J.* **7**, 3181–3187.

Schneider-Gädicke, A. and Schwarz, E. (1986). Different human cervical carcinoma cell lines show similar transcription patterns of human papillomavirus type 18 early genes. *EMBO J.* **5**, 2285–2292.

Schneider-Gädicke, A., Kaul, S., Schwarz, E., Gausephol, H., Frank, R., and Bastert, G. (1988). Identification of the human papillomavirus type 18 E6* and E6 proteins in nuclear protein fractions from human cervical carcinoma cells grown in the nude mouse or *in vitro*. *Cancer Res.* **48**, 2969–2974.

Schneider-Maunoury, S., Croissant, O., and Orth, G. (1987). Integration of human papillomavirus type 16 DNA sequences: a possible early event in the progression of genital tumors. *J. Virol.* **61**, 3295–3298.

Schüle, R., Rangarajan, P., Yang, N., Kliewer, S., Ransone, L. J., Bolado, J., Verma, I. M., and Evans, R. M. (1991). Retinoic acid is a negative regulator of AP-1-responsive genes. *Proc. Natl. Acad. Sci. U.S.A.* **88**, 6092–6096.

Schwarz, E., Freese, U. K., Gissmann, L., Mayer, W., Roggenbuck, B., Stremlau, A., and zur Hausen, H. (1985). Structure and transcription of human papillomavirus sequences in cervical carcinoma cells. *Nature* **314**, 111–114.

Seedorf, K., Krammer, G., Dürst, M., Suhai, S., and Rowekamp, W. G. (1985). Human papillomavirus type 16 DNA sequence. *Virology* **145**, 181–185.

Seedorf, K., Oltersdorf, T., Krämmer, G., and Röwekamp, W. (1987). Identification of early proteins of the human papilloma viruses type 16 (HPV 16) and type 18 (HPV 18) in cervical carcinoma cells. *EMBO J.* **6**, 139–144.

Shirasawa, H., Tomita, Y., Sekiya, S., Takamizawa, H., and Simizu, B. (1987). Integration and transcription of human papillomavirus type 16 and 18 sequences in cell lines derived from cervical carcinomas. *J. Gen. Virol.* **68**, 583–591.

Sibbet, G. J. and Campo, M. S. (1990). Multiple interactions between cellular factors and the non-coding region of human papillomavirus type 17. *J. Gen. Virol.* **71**, 2699–1707.

Smits, H. L., Raadsheer, E., Rood, I., Mehendale, S., Slater, R. M., Van der Nooraa, J., and Ter Schegget, J. (1988). Induction of anchorage-independent growth of human embryonic fibroblasts with a deletion in the short arm of chromosome 11 by human papillomavirus type 16 DNA. *J. Virol.* **62**, 4538–4543.

Smotkin, D. and Wettstein, F. O. (1987). The major human papillomavirus protein in cervical cancers is a cytoplasmic phosphoprotein. *J. Virol.* **61**, 1686–1689.

Spalholz, B. A., Yang, Y. -C., and Howley, P. M. (1985). Transactivation of a bovine papillomavirus transcriptional regulatory element by the E2 gene product. *Cell* **42**, 183–191.

Spalholz, B. A., Vande Pol, S. B., and Howley, P. M. (1991). Characterization of the *cis* elements involved in basal and E2-transactivated expression of the bovine papillomavirus p2443 promoter. *J. Virol.* **65**, 743–753.

Srivatsan, E. S., Benedict, W. F., and Stanbridge, E. J. (1986). Implication of chromosome 11 in the suppression of neoplastic expression in human cell hybrids. *Cancer Res.* **46**, 6174–6179.

Steinberg, B. M., Auborn, K. J., Brandsma, J. L., and Taichman, L. B. (1989). Tissue site-specific enhancer function of the upstream regulatory region of human papillomavirus type 11 in cultured keratinocytes. *J. Virol.* **63**, 957–960.

Stich, H. F., Tsang, S. S., and Palcic, B. (1990). The effect of retinoids, carotenoids, and phenolics on chromosomal instability of bovine papillomavirus DNA-carrying cells. *Mutat. Res.* **241,** 387–393.

Storey, A., Pim, D., Murray, A., Osborn, K., Banks, L., and Crawford, L. (1988). Comparison of the *in vitro* transforming activities of human papillomavirus types. *EMBO J.* **7,** 1815–1820.

Strähle, U., Klock, G., and Schütz, G. (1987). A DNA sequence of 15 base pairs is sufficient to mediate both glucocorticoid and progesterone induction of gene expression. *Proc. Natl. Acad. Sci. U.S.A.* **84,** 7871–7875.

Taichman, L. B., Breitburd, F., Croissant, O., and Orth, G. (1984). The search of a culture system for papillomavirus. *J. Invest. Dermatol.* **83,** 2A.

Takebe, N., Tsunokawa, Y., Nozawa, S., Terada, M., and Takashi, S. (1987). Conservation of E6 and E7 regions of human papillomavirus types 16 and 18 present in cervical cancers. *Biochem. Biophys. Res. Commun.* **143,** 837–844.

Tanaka, A., Noda, T., Yajima, H., Hatanaka, M., and Ito, Y. (1989). Identification of a transforming gene of human papillomavirus type 16. *J. Virol.* **63,** 1465–1469.

Thierry, F. and Yaniv, M. (1987). The BPV1-E2 *trans*-activating protein can be either an activator or a repressor of the HPV18 regulatory region. *EMBO J.* **6,** 3391–3397.

Thierry, F., Heard, J. M., Dartmann, K., and Yaniv., M. (1987). Characterization of a transcriptional promoter of human papillomavirus 18 and modulation of its expression by simian virus 40 and adenovirus early antigens. *J. Virol.* **61,** 134–142.

Thierry, F., Dostatni, N., Arnos, F., and Yaniv, M. (1990). Cooperative activation of transcription by bovine papillomavirus type 1 E2 can occur over a large distance. *Mol. Cell. Biol.* **10,** 4431–4437.

Thierry, F., Spyron, G., Yaniv, M., and Howley, P. (1992). Two AP1 sites binding *junB* are essential for human papillomavirus type 18 transcrition in keratinocytes. *J. Virol.* **66,** 3740–3748.

Tsang, S. S., Li, G., and Stich, H. F. (1988). Effect of retinoic acid on bovine papillomavirus (BPV) DNA-induced transformation and nuber of BPV DNA copies. *Int. J. Cancer* **42,** 94–98.

Tsunokawa, Y., Takebe, N., Kasamatsu, T., Terada, M., and Sugimura, T. (1986). Transforming activity of human papillomavirus type 16 DNA sequences in a cervical cancer. *Proc. Natl. Acad. Sci. U.S.A.* **83,** 2200–2203.

Vande Pol, S. C. and Howley, P. M. (1990). A bovine papillomavirus constitutive enhancer is negatively regulated by the E2 repressor through competitive binding for a cellular factor. *J. Virol.* **64,** 5420–5429.

Ward, P., Coleman, D. V., and Malcolm, A. D. B. (1989). Regulatory mechanisms of the papillomaviruses. *Trends Genet.* **5,** 97–99.

Watanabe, S., Kanda, T., and Yoshiike, K. (1989). Human papillomavirus type 16 transformation of primary human embryonic fibroblasts requires expression of open reading frames E6 and E7. *J. Virol.* **63,** 965–969.

Watanabe, S., Kanda, T., Sato, H., Furuno, A., and Yoshiike, K. (1990). Mutational analysis of human papillomavirus type 16 E7 functions. *J. Virol.* **64,** 207–214.

Werness, B. A., Levine, A. J., and Howley, P. M. (1990). Association of human papillomavirus types 16 and 18 E6 proteins with p53. *Science* **248,** 76–79.

Whyte, P., Buchkovich, K. J., Horowith, J. M., Friend, S. H., Raybuck, M., Weinberg, R. A., and Harlow, E. (1988a). Association between an oncogene and an antioncogene: the adenovirus E1A proteins bind to the retinoblastoma gene product. *Nature* **334,** 124–129.

Whyte, P., Ruley, H. E., and Harlow, E. (1988b). Two regions of the adenovirus early region 1A proteins are required for transformation. *J. Virol.* **62,** 257–265.

Whyte, P., Williamson, N. M., and Harlow, E. (1989). Cellular targets for transformation by the adenovirus E1A proteins. *Cell* **56,** 67–75.

Wilczynski, S. P., Pearlman, L., and Walker, J. (1988). Identification of HPV16 early genes retained in cervical carcinomas. *Virology* **166,** 624–627.

Woodworth, C. D., Bowden, P. E., Doniger, J., Pirisi, L., Barnes, W., Lancaster, W. D., and DiPaolo, J. A. (1988). Characterization of normal human exocervical epithelial cells immortalized in vitro by papillomavirus types 16 and 18 DNA. *Cancer Res.* **48,** 4620–4628.

Woodworth, C. D., Doniger, J., and DiPaolo, J. A. (1989). Immortalization of human foreskin keratinocytes by various human papillomavirus DNAs corresponds to their association with cervical carcinoma. *J. Virol.* **63,** 159–164.

Woodworth, C. D., Notario, V., and DiPaolo, J. A. (1990a). Transforming growth factors beta1 and 2 transcriptionally regulate human papillomavirus (HPV) type 16 early gene expression in HPV-immortalized human genital epithelial cells. *J. Virol.* **64,** 4767–4775.

Woodworth, C. D., Waggoner, S., Barnes, W., Stoler, M. H., and DiPaolo, J. A. (1990b). Human cervical and foreskin epithelial cells immortalized by human papillomavirus DNAs exhibit dysplastic differentiation *in vitro*. *Cancer Res.* **50,** 3709–3715.

Yasumoto, S., Burkhardt, A. L., Doniger, J., and DiPaolo, J. (1986). Human papillomavirus type 16 DNA-induced malignant transformation of NIH 3T3 cells. *J. Virol.* **57,** 572–577.

Yee, C., Krishnan-Hewlett, I., Baker, C. C., Schlegel, R., and Howley, P. M. (1985). Presence and expression of human papillomavirus sequences in human cervical carcinoma cell lines. *Am. J. Pathol.* **119,** 361–366.

Yutsudo, M., Okamoto, Y., and Hakura, A. (1988). Functional dissociation of transforming genes of human papillomavirus type 16. *Virology* **166,** 594–597.

zur Hausen, H. (1989). Papillomaviruses in anogenital cancer as a model to understand the role of viruses in human cancers. *Cancer Res.* **49,** 4677–4681.

8

Transgenic Mouse Models for the Study of the Skin

Catherine Cavard, Alain Zider, and Pascale Briand

Introduction

The epidermis is a stratified squamous epithelium with appendages—hair follicles, sweat, and sebaceous glands—that forms the protective covering of the skin. Its major cellular component is the epidermal keratinocyte, which provides an environment for the other cell types such as melanocytes, Langerhans cells, Merkel cells, and possibly lymphocytes. The epidermis is thus not a uniform tissue but is an ectodermal derivative harboring a number of

Molecular Biology of the Skin: The Keratinocyte

other cell populations of neuroectodermal or mesenchymal origin. In addition to its mechanical protective function, this tissue also forms the barrier of the skin, preventing the entrance of toxic substances and microorganisms and the loss of water and electrolytes. The multiple functions assured by the epidermis involve many different cell types on the basis of their nature and state of differentiation. The permanent remodeling of the skin leads to a continuous migration and differentiation of basal proliferating cells that renders difficult the establishment of valuable *in vitro* systems.

The use of transgenic mouse systems circumvents some of the problems that face investigators. These models allow the study of the regulatory sequences of epidermis-specific genes, as well as the effect of transgene expression in different cell types at a defined moment during the life of the mouse. In addition, cell lineage analysis by toxigenetic strategies and inactivation of specific genes by homologous recombination in ES cells are powerful methods for the dissection of complex processes that underly epidermis differentiation and for the creation of animal models of skin diseases.

Transgenic Mouse Technology

Retroviral Gene Transfer

Infection of mouse embryos with recombinant retroviruses has been used to transfer genetic information into mice (Fig. 1). After some cleavages (up to the morula stage), embryos become infectable by retroviruses. The DNA provirus integrates as a single copy in the host genome in a manner that prevents rearrangement in the flanking DNA at the site of integration. This is of considerable advantage when attempting to identify the host gene disrupted by insertion of the proviral DNA. The primary disadvantages of the use of retroviruses for gene transfer are the size limitation of the transduced DNA and the interference between the viral and the transgene sequences resulting in abnormal levels of expression.

Microinjection in One-Cell Embryos

Microinjection in one-cell embryos (Fig. 2) has been the most widely and successfully used method for generating transgenic mice. Typically, multiple DNA molecules organized in a tandem array integrate stably into the host genome. The principal advantage of direct microinjection of DNA is the efficiency of generating transgenic lines that express the transgene in a predictable manner. Transgenic systems have allowed the dissection of gene enhancer elements using chimeric transgenes in which the *cis*-acting sequences drive the expression of reporter genes such as *cat* (chloramphenicol acetyltransferase) or *lacZ*. Once defined, these regulatory sequences may be

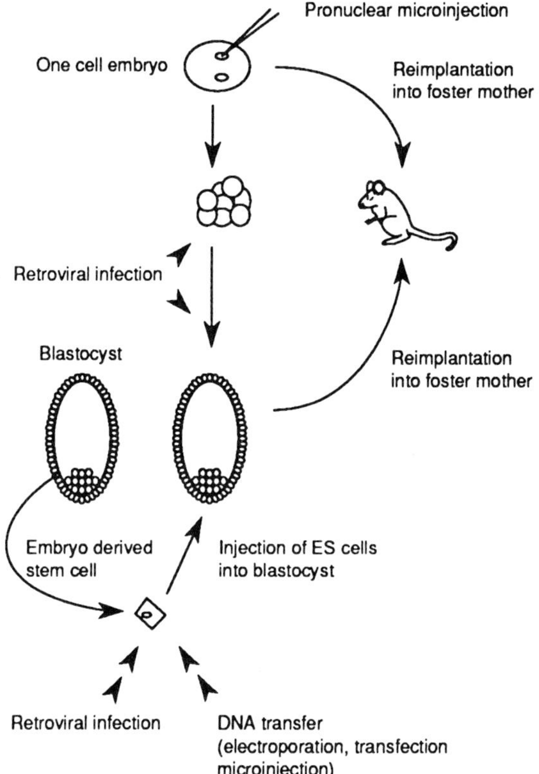

FIGURE 1 Methods of producing transgenic mice.

used to target the expression of desired genes (such as oncogenes) to obtain models of cancers, genes coding for toxins to eliminate a particular cell type, or for various proteins to investigate their cytopathological effect.

Homologous Recombination in Embryonic Stem Cells

In contrast to the previously described techniques that consist of random addition of genetic information, homologous recombination allows precise targeting of the integration (and thus disruption) of a given gene. At the present time, the homologous recombination frequency is too low to permit attempts to inject constructs into fertilized eggs and then to screen transgenic mice for site-specific changes. Fortunately, another approach is possible. Pluripotent embryonic stem (ES) cell lines established from the inner cell mass of the blastocyst can be manipulated *in vitro* and will resume normal embryonic development following reintroduction into a host embryo and subsequent development in a foster mother. The ES cells can contribute to

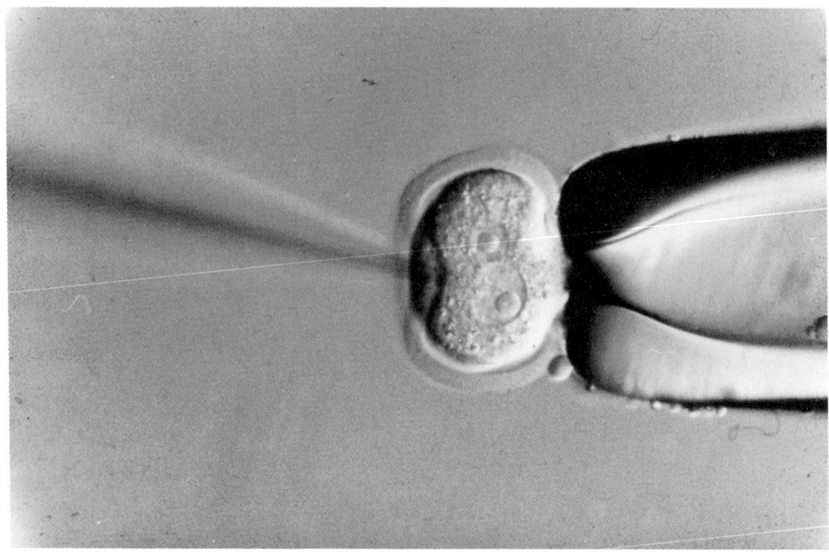

FIGURE 2 A fertilized one-cell mouse embryo, showing the two pronuclei: maternal and paternal. The embryo is being restrained by a holding pipet, using a slight vacuum on the zona pellucida. A glass capillary needle contains the solution of DNA.

the germ cell lineage as well as somatic tissues in chimeric animals thus formed, allowing them to serve as vehicles for creating heritable changes in mice. The strategy is based on the introduction of a gene into ES cells, selection of the rare cells in which homologous recombination has occurred *in vitro,* and the use of these cells to generate mouse strains heterozygous for the altered genes. Such mice can be bred to obtain homozygous animals for the targeted genetic change and the functional consequences of the alteration assessed.

The ES Cells: An *in vitro* Model of Embryonic Development

In addition to their use for functional studies *in vivo,* ES cells provide a powerful *in vitro* model of embryonic development. In suspension culture, ES cells differentiate to organized structures termed "embryoid bodies" containing an outer layer of endoderm and an inner layer of ectoderm separated by a basal lamina. These embryoid bodies expand into large cystic structures reminiscent of the visceral yolk sac. During mouse embryogenesis, keratin 8 (K8) and keratin 18 (K18) are the first intermediate filaments proteins to be expressed. They are detected just before the morula stage and are restricted to the simple epithelium of the early embryo, including the trophectoderm, the parietal, and the visceral endoderm. The mouse K8 and its type I keratin partners, K18 and K19, are induced in extraembryonic endoderm *in vivo,* and after *in vitro* differentiation of ES cells or embryonal carcinoma cells into embryoid bodies. Inactivation of both K8 alleles by homologous recombination in ES cells has lead to surprising results (Baribault and Oshima, 1991). The mutated ES cells differentiate to embryoid bodies with normal epithelia albeit in the absence of filament formation. Indeed, no polymerization is observed for the K18 and K19 type I keratins. The fact that K8 is not essential to the formation of extra embryonic endodermal epithelium, and that apparently no other type II keratin can replace it in filament formation with K18 and K19, raises several questions. Specifically, as in many instances already reported, it has become evident that the pattern of expression of the targeted gene is not necessarily a good indicator of the real requirement of the gene product. In addition, the function of the K8 *in vivo* may be more important and more subtle than during the *in vitro* differentiation of ES cells. And finally, it cannot be excluded that a partial redundancy of information might exist, leading to minimal structures not detectable during analysis but sufficient to assure a normal epithelium formation. Some of these questions might be resolved by the generation of transgenic animals carrying the targeted alleles of the K8 gene. Preliminary results concerning the embryos carrying the targeted alleles of the K8 genes indicate that they develop through gastrulation and organogenesis without apparent defect. However, no homozygous mouse line has been established suggesting a perinatal lethality (*Baribault et al.,* 1992).

Transgenic Mice in Identification of Sequences Playing a Role during Epidermal Differentiation

Two major problems confront investigators attempting to identify sequences regulating gene expression during epidermal differentiation. First, transfection efficiencies of primary cultures of epidermal cells are extremely low. Second, the inability to induce high level expression of endogenous genes such as differentiation-specific keratins *ex vivo* persists. To circumvent these problems, transgenic mice carrying either cloned genomic fragments or regulatory sequences fused to reporter genes present a valuable alternative that has been used to study keratin gene expression (Table 1).

Human Keratin Gene Characterization

Keratin 14

K14 (type I keratin; 50 kDa) is one of a pair of keratins expressed in the basal layer of all stratified squamous epithelia. As a basal cell undergoes a commitment to differentiate, it down-regulates the expression of the keratins specific of the basal layer and induces a new set of keratins that is dependent upon the particular stratified squamous epithelium (K1, K2, K10, and K11

TABLE 1 **Epidermis-Expressed Transgenes Using Cellular Promoters**

Transgene promoter coding sequences[a]	Localization of expression	Phenotype	References
hK14/P-tagged hK14 cDNA	basals cells of epidermis, tongue, esophagus		Vassar *et al.*, 1989
hK14/*cat*	*idem*		Leask *et al.*, 1990
hK14/TGFα	*idem*	epidermal thickening, psoriatic lesions and skin papillomas	Vassar and Fuchs, 1991
hK14/K14 cDNA CΔ135	*idem*	neonatal mortality, basal cell cytolysis	Vassar *et al.*, 1991
hK14/K14-truncated cDNAs	*idem*	basal cell cytolysis	Coulombe *et al.*, 1991b
hK1	basal and suprabasal cells of epidermis		Rosenthal *et al.*, 1991
bK10/Ha-*ras*	suprabasal layer of epidermis, forestomach, and tongue	hyperkeratosis papillomas	Bailleul *et al.*, 1990
mUHS K/*cat*	hair follicle		McNab *et al.*, 1990
sKIF type II	*idem*	cyclic hair loss and regrowth	Powell and Rogers, 1990
m-tyrosinase minigene	melanocytes	pigmented	Beermann *et al.*, 1990, 1991
truncated α-A-crystallin (insertional mutagenesis)		allelic to *downless*	Shawlot *et al.*, 1989
m-tyrosinase/T	pigment cells	melanomas	Bradl *et al.*, 1991
m-metallothionein/*ret*	pigment cells	*idem*	Iwamoto *et al.*, 1991

[a]b, Bovine; h, human; m, mouse; s, sheep; *cat,* chloramphenicol acetyltransferase; K, keratin; KIF, keratin intermediate filament; TGF, transforming growth factor; UHS, ultra-high-sulfur.

in the epidermis; K3 and K12 in the corneal cell; K4 and K13 in the esophageal cell). The regulation of expression of the K14 gene has been investigated by both *in vitro* and *in vivo* experiments (Vassar *et al.*, 1989). An intronless construct comprising the cDNA of human K14 fused to the antigenic portion of substance P driven by 2500 bp of K14 5′ genomic sequence was transfected into different culture lines positive or negative for endogenous K14 expression. In this case, expression of the construct, determined by immunological methods using an antibody against substance P, was detected in all cell types used. When studied in transgenic mice, this construct is properly expressed in a tissue-specific manner in the basal cells of the tongue, epidermis, and esophagus, following the endogenous K14 expression pattern. This expression seems to be appropriately down-regulated since it declines markedly as cells move to the suprabasal layer of the epidermis as is observed for endogenous K14. Taken in concert, these observations suggest that factors allowing expression of this construct are present in many cell types. Nevertheless, the mechanisms of the down-regulation are functional only in the *in vivo* system. In addition, the transgenic mouse experiments show that the 5′ sequences contained in the construct are able to regulate the expression in a tissue-specific fashion.

Determination of the sequences necessary for the tissue-specific regulation of this gene has been further investigated by both *in vitro* and *in vivo* experiments using constructs carrying modified versions of the sequence 5′ to the K14 gene coupled to a chloramphenicol acetyltransferase (*cat*) reporter gene or to the intronless construct previously described. These experiments lead to the characterization of a keratinocyte-specific transcriptional activation domain, which is found in several other genes expressed in keratinocytes and recognized by a nuclear factor most commonly found in this cell type (Leask *et al.*, 1990). This extensive work on the K14 gene has demonstrated the need to set up both *in vitro* and *in vivo* systems, the latter to provide the final demonstrative step of the experiments.

Keratin 1

An identical approach was chosen to study keratin 1 (type II, M_r 67,000), which with keratin 10 (type I, M_r 59,000) is expressed as a pair in all suprabasal layers of the epidermis. Generation of transgenic mice for the human K1 gene has demonstrated that the 12-kb genomic fragment containing the K1 coding region along with 1.2 kb of 5′- and 4.3 kb of 3′-flanking DNA is sufficient to direct epidermis-specific expression (Rosenthal *et al.*, 1991). However, analysis of expression in transgenic neonates has shown that the transgene is inappropriately expressed compared with the mouse K1 gene. A significant part of the basal cells prematurely expressed the human K1 before the commitment to differentiate. In addition, establishment of primary cultures from transgenic keratinocytes demonstrated that normal negative regulation [by either retinoic acid, TPA (12-O-tetradecanoylphorbol-13-acetate), or Harvey viral *ras* oncogene] of the endogenous murine K1 gene did not occur using the human genomic fragment. However, the expression of both human and murine genes was induced by culture in elevated levels of Ca^{2+}, although the optimum concentration was different for both genes. These results suggest that some of the elements necessary for proper regulation of the human K1 gene are missing in the injected construct. Production of transgenic mice with additional 5′-flanking regions indicate that regulatory elements responsive to retinoic acid repression are located 3000 bp upstream relative to the transcription start site (Rothnagel *et al.*, 1991).

Murine Keratin Gene Characterization

Ultra-High-Sulfur Keratin

Hair development and growth are also tightly regulated phenomena for which *in vitro* studies are highly limited. To gain information in the regulation of genes that might trigger the hair growth cycle, preliminary injections were performed using the promoter region of a murine ultra-high-sulfur keratin gene fused to the *cat* gene (McNab *et al.*, 1990). Expression of this gene is tightly regulated during the hair cycle and it may also provide an indicator of hair differentiation control. A chimeric construct including 671 bp of the ultra-high-sulfur keratin gene 5′ region was appropriately expressed in a tissue-specific manner during the anagen phase of hair growth in transgenic mice. Since the 5′ region used contains the primary control elements necessary for appropriate expression, the next phase of study will be the targeted expressions of specific genes that may modify the development of the hair follicle.

Transgenic Mice in Defining Mechanisms Regulating Epidermal Differentiation

In vivo analysis of biological mechanisms can be performed by inducing perturbation. This can be achieved by introducing genes coding for normal or abnormal products—supposedly involved in differentiation—linked to specific promoters (Table 1).

Induced Perturbation by the Surexpression of a Keratin Gene

Among the keratin genes, the keratin intermediate filament (IF) genes are regulated so as to ensure an equivalent abundance of both type I and type II IF proteins for coordinate assembly into filaments. The sheep wool keratin IF type II gene has been introduced into the germline of mice and the consequences of its expression upon filament formation in the transgenic mice were investigated (Powell and Rogers, 1990). In mice, follicle activity is synchronized and waves of new hair growth occur in regular cycles. Two high copy number transgenic lines exhibit a cyclic pattern of hair loss and regrowth, mimicking some natural hair-loss mutants. Analysis of the protein composition demonstrates that there is a decrease in high glycine/tyrosine proteins and high sulfur proteins, the major components of the hair keratin matrix. This suggests that overexpression of the transgene in cortical cells disrupts the normal ratio of filament to filament-associated protein. This imbalance between filament keratin protein and filament-associated protein leads to weakened fibers and premature hair loss.

Induced Perturbation by the Expression of a Growth Factor

In vivo demonstration of the tissue-specificity of enhancer elements may allow their use to study the effect of targeted expression of a given gene. The previously described K14 5′ genomic sequences were used to drive the expression of a rat TGF-α (transforming growth factor α) cDNA in transgenic animals (Vassar and Fuchs, 1991a). This potent mitogen belonging to the epidermal growth factor (EGF) family is synthesized by epidermal cells and stimulates keratinocyte growth and cell migration. With the regulatory sequences of K14 used, an overexpression of the autocrine factor was obtained in the basal cells of stratified squamous epithelia and led to regional thickening of the epidermis. One unanticipated result was the developmental and regional feature of the perturbation. Gross phenotypic abnormalities of the skin were observed in neonatal TGF-α transgenic mice whose severity correlated with the level of transgene expression. A very high expression of the growth factor in the skin lead to death of the animals. Interestingly, the proliferation rate, determined by BrdU incorporation, was only modified in the basal layer while TGF-α mRNA was also detected in the first few suprabasal layers. Furthermore this normalized 5 weeks after birth. This localization and the time course of the TGF-α effect on proliferation correlated perfectly with the presence of EGF receptors at the surface of the basal cells. Surprisingly, changes also took place in upper layers where cells had lost most of their EGF receptors. This suggests that either the effects of TGF-α may be long lasting (i.e., able to persist after commitment of cells) or that TGF-α induces secondary changes in basal cells that may in turn affect suprabasal cells. In adults, the histological abnormalities persist only in restricted areas and include epidermal thickening of ear and footpad, genital epithelial aberrations, psoriatic lesions, and skin papillomas. This latter lesion is almost exclusively found in areas prone to self-induced mechanical abrasion or wounds. The same authors have shown that these papillomas did not present mutations in the Ha-*ras* gene, suggesting a central role for TGF-α overexpression in skin tumorigenesis (Vassar *et al.*, 1992). One important point is the minor effect of TGF-α on the overall program of terminal differentiation. The different cell populations are increased, keratinocytes are enlarged and the keratin patterns exhibited are very similar to those seen for various hyperproliferative epidermal diseases. Thus, the changes in keratin expression do not seem to perturb the morphology of terminal differentiation. This result had been predicted previously by several authors who proposed that altered keratin expression might be due to environmental changes rather than to hyperproliferation *per se*. In addition to

TGF-β and retinoids, TGF-α may now be included in this list of environmental factors.

Previous studies have identified increased levels of TGF-α and other cytokines in psoriatic skin. However, these do not explain at which step of development of disease this phenomenon is involved. Interestingly, transgenic rats expressing HLA-B27 and human β2-microglobulin transgenes have been shown to develop psoriatiform skin and nail lesions resembling those found in humans with HLA-B27-associated reactive arthritis (Hammer *et al.*, 1990). The epidermal alterations are accompanied by a leukocytic infiltration which is a hallmark of psoriasis. This typical feature of psoriasis is not observed in TGF-α transgenic mice since they display elevated levels of TGF-α and, consequently, epidermal hyperplasia. Exceptionally, localized regions of the outer ear skin, where mechanical irritation and secondary bacterial infection are frequent, exhibit characteristics of psoriatic plaques. Taken together, these observations suggest that overexpression of TGF-α alone is not sufficient for the development of the disease but may play a role in its progression in conjunction with other earlier events such as HLA expression.

Microinjection As a Tool to Attribute Gene Function

Transfer of genes in mutant mice can also be used to demonstrate that a gene is directly responsible for one function. Injection of tyrosinase sequences in albino mice provides a good example of such a strategy.

Among epidermal cells other than keratinocytes, the melanocyte (or more particularly melanogenesis and its control) has been extensively studied in recent years by workers of various disciplines. More than 130 mutations affecting coat color and pigmentation at more than 50 loci have been described in the house mouse. The albino or *c*-series of alleles are characterized by a deficiency or an alteration in the structure of tyrosinase. It catalyzes the hydroxylation of tyrosine to 3,4-dihydroxyphenylalanine (dopa) and the oxydation of dopa to dopaquinone; dopaquinone is further metabolized to phaeomelanin or eumelanin. Until recently, it was not clear whether the *c*-locus encoded tyrosinase or a *trans*-regulatory protein that modulates tyrosinase activity. Isolation of genomic or cDNA sequences for the tyrosinase was the first step toward the characterization of the mutations at this locus. However, the ultimate proof that a functional tyrosinase gene is encoded at the *c*-locus involved the rescue of the albino phenotype in the transgenic mice. A minigene construct was injected into fertilized eggs of the albino mouse strain NMRI, containing 5′ upstream sequences and the first intron of the mouse (chinchilla) tyrosinase gene, the first exon and 60 bp of the second exon of the tyrosinase coding sequence, part of the second exon and exons 3–5 of the tyrosinase cDNA, SV40 splice and polyadenylation sites, and the small *t* gene intron. Five transgenic mice displayed pigmentation in skin and eyes (Beermann *et al.*, 1990). This demonstrated that the *c*-locus encodes the structural gene for tyrosinase and is not a regulator for tyrosinase expression. *In situ* hybridizations showed that the transgene was expressed in a cell type-specific manner that overlaps with the known expression of the endogenous gene. Further analysis will be required to define the regulatory elements for the melanocyte-specific expression and for expression during development.

Because the expression of the tyrosinase gene is easily detectable by the pigmented phenotype it confers, this gene is a useful tool for transgenesis investigation. One illustration of this was the use of an analogous construct carrying a tyrosinase minigene as a coinjected marker for production of transgenic mice in the NMRI strain. Coinjection of two constructs led to transgenic mice carrying both transgenes at a single chromosomal site, and thus the detection of the modified phenotype allowed a rapid identification of the transgenics, circumventing classical DNA analysis (Beermann *et al.*, 1991).

Insertional Mutagenesis

The integration of transgenic DNA into the genome of mouse embryos can cause insertional inactivation of the flanking genetic locus and may provide, in some rare cases, a molecular tag for identifying new genes involved in development and differentiation. One line of mice carrying a truncated version of a transgene (the regulatory sequences of the murine α-A-crystallin gene fused to the SV40 early region) showed in its homozygous offspring a reproducible mutant phenotype unrelated to transgene expression (Shawlot *et al.*, 1989). The affected mice had thin greasy fur, lacked guard hairs, and had a bald patch behind the ears. Hair follicles were absent in the tail and in the skin behind the ears, suggesting a defect in hair follicle induction. In addition, the absence of Meibomian (an exocrine gland of the eyelid) glands was observed in these mice. This recessive mutation was shown to be allelic to a previously described mutation termed downless (*dl*) that maps to chromosome 10. This locus seems to be essential in the morphogenesis of a subset of epidermally derived structures. Cloning of the integration site will thus be useful in the characterization of the *dl* gene and its transcript. However, as it had been previously described by other workers, the rearrangement at the transgene insertion site could hamper the cloning and the further identification of the gene. Mutated genes have been identified more easily in transgenic mice generated by retroviral infection. Nevertheless, only 5% of transgenic strains generated by retroviral infection of embryos exhibit an overt phenotype. To avoid extensive breeding of transgenic lines to homozygotes, new methods allowing screening and selection for mutations *in vitro* have been developed. These are based on the use of new vectors termed "entrapment vectors" and of pluripotent ES cells (Skarnes, 1990). The constructs used comprise a reporter gene, the *lacZ* gene, whose expression is dependent on the *cis*-acting regulatory sequences of an endogenous cellular gene. After integration of the vector nearby or within a gene, the expression of the reporter gene should reflect the normal expression of the endogenous gene. There are two groups of entrapment vectors, "gene trap" and "enhancer trap" vectors. Gene trap vectors lack any promoter element, and the activation of the reporter gene requires the integration of the construct inside a transcription unit to create a fusion transcript. Enhancer trap vectors consist of a minimal promoter element linked to the reporter gene whose expression is dependent on endogenous enhancer elements. Introduction of these vectors in ES cells allows preliminary screening of the targeting events and selection of the expressing cells. ES cells can be reintroduced into host blastocysts to generate transgenic animals in which the reporter activity may be further studied before the molecular cloning of the mutated locus.

Transgenic Mouse Models of Disease

Targeted Expression in Keratinocytes

Initial attempts to derive transgenic mice expressing the human Ha-*ras* gene in the basal layer of the epidermis using the keratin 5 (K5) promoter were unsuccessful, indicating the determining effect of the tissue microenvironment on the expression of a particular gene. Expression of this oncogene in the suprabasal layer of the epidermis was later achieved using the promoter region of the bovine K10 gene (Bailleul *et al.*, 1990). The K10 gene is expressed suprabasally in the epidermis and in the forestomach, and to a lesser extent in the tongue. Expression of Ha-*ras* in a differentiating cell compartment within the epidermis led to the development of hyperkeratosis in the skin and in the forestomach and to the appearance of highly differentiated papillomas at specific sites (apparently as a consequence of wound healing following irritation). This latter observation suggests that cells within the epidermis that have already progressed to this stage of differentiation may constitute a target for the initiating event of the tumoral process. The hyperkeratosis observed in the keratinized part of the forestomach may be associated with growth retardation of the transgenic mice. Two hypotheses may account for this phenomenon. First, hyperkeratosis may lead to an inefficient processing of food intake. A more attractive explanation would be an induction of factors involved in the digestive process, and more particularly, of TGF-α, which is known to be secreted by cells expressing activated oncogenes. A comparison between the studies involving *in vivo* expression of the Ha-*ras* oncongene in transgenic mice clearly shows that the response of particular cell types to the *ras* oncogene is variable resulting in either repression of differentiation or hyperplasia (Quaife *et al.*, 1987; Andres *et al.*, 1987). In the case of the K10-*ras* mice, the epidermis undergoes morphological changes apparently without a significant increase in the proliferative cell layers. Expression of the *ras* oncogene causes a dramatic change in the differentiation pattern of the epidermis, increasing the apparent degree of differentiation, and sensitizes epidermal cells similarly to the effect of a wounding stimulus resulting in a preneoplastic state.

Targeted Expression in Other Cell Types

Spontaneous malignant melanomas are very rare in laboratory animals. Production of animal models that develop melanomas was achieved by injecting inbred C57BL/6 fertilized eggs with a construct containing the SV40 early region encoding the large (T) and the small (t) tumor antigens fused to the 5′ sequences of the mouse tyrosinase gene (Bradl *et al.*, 1991). In all the transgenic mice obtained, the coat was lighter than the black color normally observed in the C57BL/6 strain, indicating a reduced melanization. Ocular melanomas arose in these mice chiefly in the retinal pigment epithelium and also in the choroid. The eye tumors were invasive and metastasized to local and distant sites. Primary skin melanomas were much less frequent and occurred later than do ocular melanomas (12 weeks versus 2–4 weeks, respectively). In humans, by contrast, skin melanomas are generally more frequent and occur earlier than ocular melanomas. This may be due to earlier activation of the transgene in pigment cells of the eye than in the skin and/or to the greater exposure of the human skin to ultraviolet radiation. The mouse tumors are histopathologically similar to those observed in humans, with a preponderance of epithelioid cells, spindle cells, and with premelanosomes and melanosomes. Albeit that differences exist with human pathology, these mice provide a useful model for studying the progression, metastasis, and preliminary treatments of melanomas and for the generation of cell lines that maintain a particular differentiated state. Indeed, melanocyte cell lines derived from these transgenic animals have been described recently. These transgenic cells, after one exposure to UVB (1.75 mJ/cm^2), formed numerous foci that led to malignant melanomas in graft hosts (Larue, *et al.*, 1992).

Aberrant melanogenesis and melanocytic tumors have also been observed in a published model as a consequence of the injection of a fragment containing *ret* oncogene cDNA under the control of the promoter/enhancer of the mouse metallothionein I gene (MT), which is known to function in almost all tissues (Iwamoto *et al.*, 1991). The *ret* proto-oncogene encodes a receptor-type tyrosine kinase and is often expressed in human cell lines or tumors of neuroectodermal origin. The *ret* oncogene was activated by DNA rearrangement of the *ret* proto-oncogene during the process of an NIH3T3 transfection assay and its oncogenicity was demonstrated in transgenic mice using the mouse mammary tumor virus promoter. Among the 17 transgenic founders obtained after injection of MT-*ret* cDNA construct, four showed hyperpigmented skin, and melanocytic tumors developed in three of the four lines.

It seems difficult to attribute the development of aberrant melanogenesis and melanocytic tumors to the function of either the metallothionein regulatory sequences or to the *ret* product. It is more conceivable that the unexpected expression of the transgene is due to the sole combination of both metallothionein and *ret* sequences, as was frequently observed in the transgenic studies. By contrast with the preceding model (SV40 early region controlled by the mouse tyrosinase promoter), in which ocular tumors originated from the retinal pigment epithelium and coat color became hypopigmented, these transgenic mice (MT/*ret*) showed ocular tumors in the choroidal pigment cells, melanocytic tumors in the dermis, and a hyperpigmented skin. Both these models (by virtue of differences in the cell types involved—due to the oncogenes and to the regulatory sequences used) provide two distinct pathways for disecting the mechanisms of the multistep transformation of pigment cells, a phenomenon which is rather frequent in humans but for which mammalian models of study are rare. In addition, these mice constitute experimental tools to design more suitable somatic gene therapy approaches to treatment of human melanomas (Rosenberg *et al.*, 1990).

Targeted Expression of Mutated Proteins

If the concept that an extensive network of intermediate filaments composed of keratin is required for the architecture and the cellular organization of differentiated tissues is commonly accepted, direct *in vivo* demonstration of this notion has nonetheless been hampered by the lack of known genetic diseases involving keratin mutations. However, *in vitro* studies involving transfection of epithelial carcinoma cell lines with truncated keratin genes (Coulombe *et al.*, 1990) and retroviral-mediated keratin cDNA transfer in NIH3T3 cells (Lu and Lane, 1990) have shown that protein and filament stabilization and filament network organization constitutes much of the regulation of the keratins. For example, this work demonstrates that the production of a small amount of mutant keratin (e.g., keratin 14 missing 135 amino acid residues at the carboxyl end and thus missing segments of the central α-helical rod domain termed "CΔ135") causes a dominant alteration in filament formation.

The cDNA encoding this truncated K14 keratin (CΔ135) was placed under the control of the 5′ regulatory sequences of the K14 gene whose functionality had been previously tested in transgenic mice (Vassar *et al.*, 1989), and the resulting chimeric gene was injected into fertilized eggs (Vassar *et al.*, 1991). Among the 11 resulting transgenic mice obtained, eight had a high incidence of neonatal mortality and showed gross skin abnormalities. Blistering and basal cell cytolysis were observed. The linkage of the epidermal abnormalities developed by these mice to either the lack of an extensive keratin filament network *per se*, or to the large aggregates of mutant keratin detected in the basal cells, is unclear. Elucidation of the underlying mechanism may require the generation of null animal mutants for the K14 gene or for its coexpressed partner in the basal cell compartment, the K5 gene.

The striking resemblance between the marked abnormalities in the transgenic mice and patients exhibiting skin disorders known as epidermolysis bullosa simplex (EBS) led to a search for a defect in the basal K14- and K5-expressed genes in affected individuals. The keratin filament networks and keratin proteins from three patients having Dowling-Meara EBS (this severe EBS form is characterized by large cytoplasmic clumps of tonofilaments) and from three patients having Koebner EBS (milder EBS form with rare disorganization of tonofilaments) were analyzed. In addition, basal epidermal keratin mRNAs and the corresponding genes were studied for two Dowling-Meara EBS patients (Coulombe *et al.*, 1991a). Keratin networks from the Dowling-Meara EBS patients showed pronounced abnormalities and *in vitro*-assembled 10-nm filaments from purified keratins were significantly shorter than normal, resembling those formed with bacterially expressed normal human K5 and deletion or point mutant K14 proteins. For two of these affected individuals, sequence analysis of the K14 and K5 cDNAs corresponding to the entire central α-helical rod segments revealed point mutations (encoding arg→cys and arg→his substitutions) affecting the same amino acid 125—which is extremely conserved among species—on the K14 cDNA. The functional demonstration—that mutations affecting arg 125 can perturb keratin network formation and filament assembly—was achieved by engineering the transversion leading to the arg→cys substitution into a substance P-tagged K14 cDNA sequence which was cloned downstream from the SV40 early promoter and enhancer sequences. The resulting plasmid, when transfected into cultured keratinocytes, resulted in the expression of a mutant keratin that exhibited a disrupted endogenous keratin network similar to that in the cultured keratinocytes from the two Dowling-Meara EBS patients.

More recently, additional transgenic mouse studies were conducted involving COOH-terminal (missing 50 amino acid residues from the carboxy terminus "CΔ50") and/or NH_2-terminal (NΔ117–CΔ42) truncated K14 mutants known to affect keratin network formation to varying degrees (Coulombe *et al.*, 1991b). Transgenic mice expressing high levels of NΔ117–CΔ42 and low levels of CΔ50 exhibit biochemical and morphological features of the milder forms of EBS. This strongly suggests that both mild and severe forms of EBS are due to mutations occuring in a single gene, namely the K14 gene. In addition, these results demonstrate that the filament network perturbation is the essential component of this disease. Indeed, the mutants having the less severe effects do not display the large aggregates of keratins in the basal cells—previously supposed to be involved in cytolysis—but still present the tonofilament disorganization. This thus eliminates the previously proposed hypothesis concerning the first CΔ135 transgenic mice that predicted that large clumps of keratin proteins would be at the origin of basal cell cytolysis. Finally, these analyses may allow the assessment of the structural role of the keratin network. However, it should be emphasized that the linkage between the mutations and the phenotypes observed in the mouse may not be a strict representation of human disease and, in particular, the more subtle mutations may arise in the K14 gene but also in its K5 partner.

It is remarkable that this study demonstrated—by the characterization of the K14 regulatory sequences described above and by identification of human mutations—the essential role of transgenic models that may play an essential role in the clarification of each step of an *in vitro* molecular study and, in addition, may also constitute the basis of a novel concept for unsolved etiologies of known syndromes.

Transgenic Models of Viral Diseases

As a mean of circumventing the virus–receptor barrier, transgenic methodology has allowed the production of animals expressing viral gene products in tissues which may not be normally infected by the virus.

Transgenic Mice Expressing Bovine Papillomavirus 1 Gene Products

Bovine papillomavirus 1 (BPV-1) infects cutaneous tissues of cattle and induces benign fibropapillomas with proliferative squamous epithelial and dermal fibroblast components. The virus infects both dermal fibroblasts and epidermal epithelial cells where its genome replicates as an autonomous plasmid. Virus particles are produced only in terminally differentiated epidermal keratinocytes where the viral genome replicates very efficiently. In contrast, replication is lowered in proliferating dermal fibroblasts and in squamous epidermal cells. The entire BPV-1 genome has been injected into the germ line in a topology designed to allow excision and extrachromosomal replication in somatic tissues (Lacey *et al.*, 1986). Transgenic mice harboring this insertion are phenotypically normal during development and early adult life. Skin tumors—initially benign and finally malignant and locally invasive—appear at 8 months of age and arise in multiple locations on the body. In addition, thickening and hardening of the skin and hair loss are observed. Analysis of protuberant tumors and abnormal skin show a proliferation of dermal fibroblasts and a disorganization of the dermal layer similar to that observed in dermatofibromas and fibromatoses in humans. The epidermis presents two different histopathologies. First, in abnormal skin associated with the thickened dermal layer, an atrophic epidermis with significant loss of hair follicles and glands is observed suggesting that the normal nutritive and structural function of the dermis is impaired. In association with tumors, the epidermis showed hyperplasia; thickening and disorganization of the layers are observed. DNA analysis from affected mice reveals that the BPV-1 genomes are integrated in the normal tissues but are present as extrachromosomal plasmids in the abnormal skin tissues and in the dermal fibroblastic tumors. Analysis of derived cultures from normal skin, from areas of abnormal skin, and from protuberant tumors indicate cell-heritable characteristics (Sippola-Thiele *et al.*, 1989). During latency, the transgenome is transcriptionally inactive. Activation of viral gene expression correlates with the presence of extrachromosomal DNA in skin and in particular in dermal fibroblasts with a predilection for tumor formation at sites of wounding and irritation. The tumor tissues contain a high copy number of viral DNA and the viral genome is transcriptionally active. Expression of the BPV-1 oncoproteins is necessary as an initial step in tumor formation (proliferation stage) but apparently insufficient for progression to malignancy since the levels of viral transcription in the fibrosarcomas cells are similar to those observed for the aggressive fibromatosis cells *in vivo* and *in vitro*. This suggests that additional nonviral events must be involved in this multistep tumor formation. Since tumor progression associated with the papillomaviruses in nature also appears to involve additional cellular genetic events, this phenomenon in the BPV transgenic mice may illustrate some of the stages observed naturally. The ability to identify distinct stages and to establish cell lines representative of each stage may allow a better analysis of the progression at the molecular level.

Transgenic Mice As a Model for the Study of the HIV-1 LTR

One of the characteristics of HIV-1 (human immunodeficiency virus type 1) is the extended time frame from initial infection to disease manifestation. In addition to HIV-1 regulatory proteins, physical, chemical, and other viral agents may act *in vitro* to activate the HIV-1 long terminal repeat (LTR). Since activation signals are required for the establishment of productive HIV infection *in vitro,* it is likely that activators contribute to the conversion of a latent or chronic infection to a productive one *in vivo*. The identification of factors that induce *in vivo* viral expression could be of use in controlling the evolution of the disease. To study the cellular and exogenous factors that may activate HIV-1 LTR *in vivo*, we have generated transgenic mice in which the reporter gene *lacZ* encoding β-galactosidase is placed under the control of the HIV-1 U3R region of the LTR (Cavard *et al.*, 1990). Two lines of transgenic mice spontaneously express the transgene in the lens, tongue, epidermis, and (more weakly) forestomach and esophagus. In the epidermis, expression is associated with both hair follicles and cells interspersed throughout the stratum corneum (Fig. 3A).

The same pattern of expression observed in adults is found in embryos. Activity in the lens is detectable from gestational age 11 (G11) and persists until birth. Besides lens staining, an intense external staining of the embryo is apparent at G16. Examination of the fetuses reveals that the *lacZ* gene is strongly expressed over almost the entire skin surface. In addition to keratinized cells, a few cells surrounding the hair follicle are also stained (Fig. 3B).

Previous *in vitro* experiments led to the hypothesis that UV irradiation could affect HIV-1 expression *in vivo*. To test this hypothesis, irradiation at 280–320 nm (UV-B rays, 18 k*J*/m^2) and 254 nm (UV-C rays, 600 k*J*/m^2)

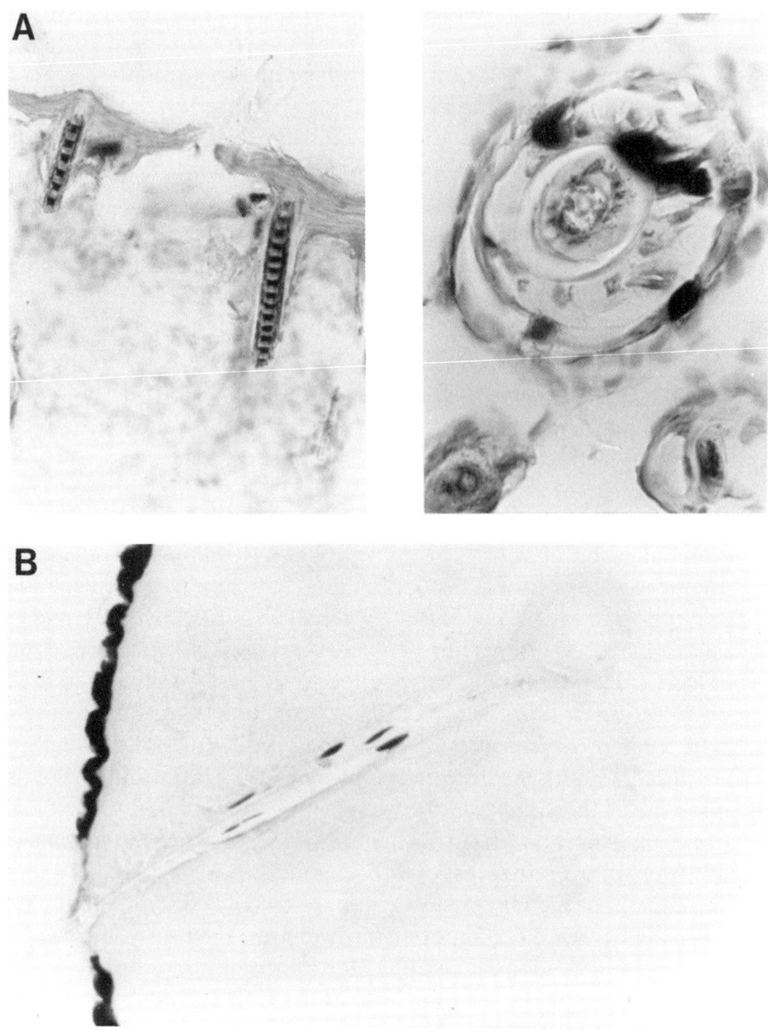

FIGURE 3 *In situ* assay of bacterial β-galactosidase activity in sections prepared from a transgenic mouse line carrying the HIV-1 U3R region linked to the *lacZ* gene (Cavard *et al.*, 1990). A, (*left*) transverse section of dorsal skin of a one-month-old transgenic mouse, magnification ×62.5; (*right*) cross-section of dorsal skin of a one-month-old transgenic mouse, magnification ×125. B, transverse section of skin of a 16-day-old transgenic embryo, magnification ×25.

were performed on 10-day-old mice. In both cases, a significant increase in β-galactosidase activity compared to control animals was observed on skin biopsy sections. This effect was detected within 24 hr after irradiation and persisted 24 hr at the same level.

Other models involving transgenic mice carrying the HIV-1 LTR linked to different reporter genes have been devised. They include the use of the chloramphenicol acetyltransferase gene (Leonard *et al.*, 1989; Frucht *et al.*, 1991) or the luciferase gene (Morrey *et al.*, 1991) and also the *lacZ* gene (Morrey *et al.*, 1991). In this latter case, the expression in the epidermis was analyzed qualitatively (i.e., by *in situ* β-galactosidase assay) and quantitatively (i.e., by luciferase assay on whole cellular extracts). HIV-1 directed expression in the skin was activated a maximum of 1,000- to 2,000-fold above background level when exposed to either UV-B irradiation or UV-A irradiation in combination with psoralen (8-methoxypsoralen applied at 10 mg/ml). In addition, exposure of transgenic mice pretreated with psoralen to sunlight for 2 hr lead to a 20-fold increase above background levels. Taken together, these results clearly demonstrate that UV rays are exogenous activators of the HIV-1 promoter *in vivo*. The cascade of events triggered by irradiation has not yet been elucidated at the molecular level. Its analysis may be complicated by the multiplicity of the cell types involved in the response (i.e., keratinocytes and/or skin dendritic cells). In addition, the nature of cells in the epidermis expressing the transgene remains controversial depending on the reporter genes and consequently on the methods used (determination of the reporter activity *in situ* or on tissue homogenates). However, in our hands, we can assume that β-galactosidase activity is predominantly located in the keratinocytes whether or not activated by UV rays. We cannot exclude a preferential revelation of β-galactosidase activity in these cell types (or an instability of this enzyme in cell types in the epidermis other than keratinocytes) masking another expression site in the epidermis. In any case, expression of LTR-directed genes in mouse epidermis is not a particularly surprising phenomenon since, in a previous model of transgenic mice carrying the entire HIV-1 genome, expression of HIV was detected by *in situ* hybridization in the outer two-thirds of this tissue and was associated with both hair follicles and with cells interspersed throughout the stratum spinosum (Leonard *et al.*, 1988).

The predominance of transgene expression, induced or not, in epidermis could reflect a particular distribution of transcription factors in this tissue. Among the binding sites for transcription factors present in the HIV-1 LTR, the enhancer core sequence has been shown to play a crucial role in viral expression and to interact with at least two classes of proteins that bind to the κB motif. One class is represented by NF-κB and related factors such as κBF1. The other class of proteins recognizes binding sites by a zinc finger motif and includes transcription factors such as AGIE-BP1 or MBP-1 which have been found to regulate the expression of the angiotensinogen gene.

To determine whether some of these factors could be involved in the pattern of expression, we have designed electrophoretic mobility shift experiments using nuclear extracts from murine epidermal cells: the murine epidermal cell line PAM 212 and the human epidermal cell line TR146. The oligonucleotide used covers the two κB sites of the HIV-1 LTR. As positive and negative controls of NF-κB binding activity, we used nuclear extracts from HeLa cells activated (or not) by UV-C rays, since it had been previously described that exposure of HeLa cells to UV-C rays induces NF-κB activation.

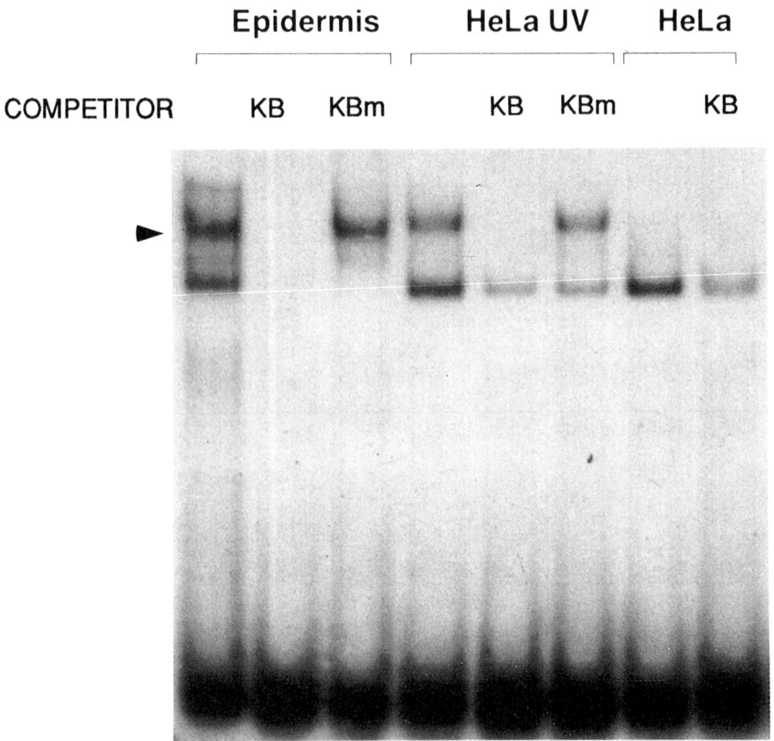

FIGURE 4 Identification of NF-κB-like binding activity in murine epidermal cells. 8-day-old mice were sacrificed and the epidermis was separated from dermis by trypsin treatment. Epidermal sheet was removed and placed in a complete media containing 10% fetal calf serum to produce a suspension of single cells. Briefly, 5 μg of nuclear extracts prepared as described by Stein *et al.* (1989) was incubated at room temperature with the P^{32}-labeled NF-κB oligonucleotide (κB) with or without unlabeled competitor oligonucleotides added at 100-fold molar excess (wildtype or mutated oligonucleotides, κBm). Delayed fragments were separated from free DNA by electrophoresis through a 5% nondenaturing polyacrylamide gel. As negative and positive controls of NF-κB binding activity, we have used nuclear extracts from HeLa cells induced or not by UV-C rays (20 J/m^2).

NF-κB oligonucleotide $_{-105}$AAGGGACTTTCCGCTGGGGACTTTCCAG$_{-79}$
κBm $_{-105}$AA<u>CTC</u>ACTTTCCGCTG<u>CTC</u>ACTTTCCAG$_{-79}$

(Relative to the HIV-1 transcription start site.)

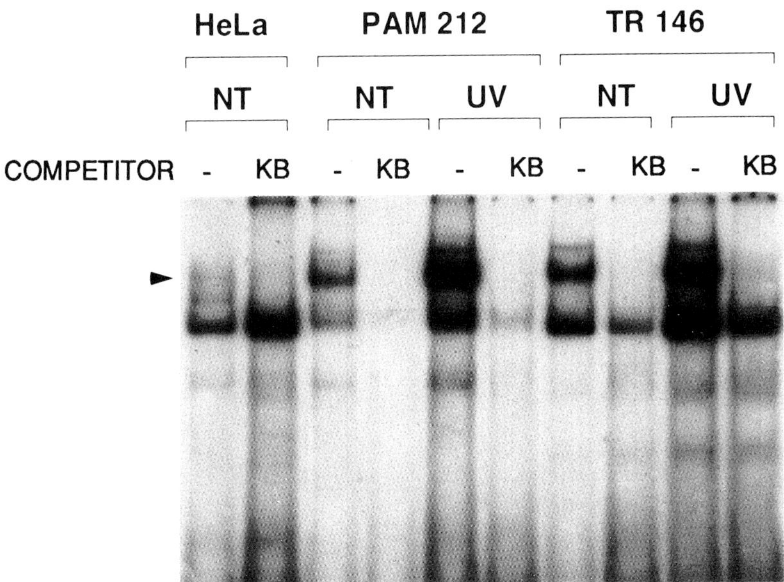

FIGURE 5 Effect of UV-C rays on an NF-κB-like binding activity in epidermal cell lines. Nuclear extracts prepared as described by Stein *et al.* (1989) were incubated at room temperature with the P^{32}-labeled NF-κB oligonucleotide with or without unlabeled competitor oligonucleotides added at 100-fold molar excess (κB). Delayed fragments were separated from free DNA by electrophoresis through a 5% nondenaturing polyacrylamide gel. PAM 212 and TR 146 are, respectively, murine and human epidermal cell lines (kindly provided by Michel Darmon, CIRD, Sophia-Antipolis). Where indicated, cells were irradiated with 20 J/m^2 of UV-C and nuclear extracts were prepared 6 hr after irradiation. NT, not treated.

The experiments showed that a factor recognizing the NF-κB binding sites of the HIV-1 LTR was present in keratinocytes prepared from mouse epidermis (Fig. 4). The presence of this activity in murine and human epidermal cell lines (Fig. 5) supports the hypothesis that keratinocytes themselves contain one or more factors able to promote and to regulate LTR expression in skin and that additional cells, like dendritic cells, might not be involved in this phenomenon in mice.

UV irradiation of epidermal cell lines notably increased NF-κB-like binding activity as previously described by Stein *et al.* (1989) in HeLa cells (Fig. 5). An increase of lower amplitude was observed in nuclear extracts prepared from the skin of mice treated with UV-B (data not shown). The transient transfection assays of modified LTR in HeLa cells also demonstrated the involvement of the NF-κB sites in the UV induction and were consistent with previously published data (Stein *et al.*, 1989). Secondly, to study the regulation acting on an integrated gene, we generated transgenic mice carrying the *lacZ* gene under the control of the partially deleted LTR. All the transgenic lines and unexpectedly those carrying the LTR deleted for the κB sites displayed a UV-inducible epidermal expression. This suggests that, in mice, the UV induction might be mediated through other sites than the κB sites and might depend on another level of regulation involving changes of the chromatin state (Zider *et al.*, 1993).

References

Andres, A. C., Schonenberger, C. A., Groner, B., Henninghausen, L., Le Meur, M., and Gerlinger, P. (1987). Ha-*ras* oncogene expression directed by a milk protein gene promoter: tissue specificity, hormonal regulation, and tumor induction in transgenic mice. *Proc. Natl. Acad. Sci. U.S.A.* **84,** 1299–1303.

Bailleul, B., Surani, A., White, S., Barton, S. C., Brown, K., Blessing, M., Jorcano, J., and Balmain, A. (1990). Skin hyperkeratosis and papilloma formation in transgenic mice expressing a *ras* oncogene from a suprabasal keratin promoter. *Cell.* **62,** 697–708.

Baribault, H., and Oshima, R. G. (1991). Polarized and functional epithelia can form after the targeted inactivation of both mouse keratin 8 alleles. *J. Cell. Biol.* **115,** 1675–1684.

Baribault, H., Price, J., and Oshima, R. G. (1992). Keratin 8 targeted gene maltivation leads to perinatal lethality and do not prevent the formation of functional extraembryonic epithelia, *in vitro* and *in vivo*. Abstracts of papers presented at the 1992 meeting on mouse molecular genetics. August 26–August 30, 1992. Cold Spring Harbor, New York.

Beermann, F., Ruppert, S., Hummler, E., Bosch, F. X., Müller, G., Rüthner, U., and Schütz, G. (1990). Rescue of the albino phenotype by introduction of a functional tyrosinase gene into mice. *EMBO J.* **9,** 2819–2826.

Beermann, F., Ruppert, S., Hummler, E., and Schütz, G. (1991). Tyrosinase as a marker for transgenic mice. *Nucleic Acids Res.* **19,** 958.

Bradl, M., Klein-Szanto, A., Porter, S., and Mintz, B. (1991). Malignant melanoma in transgenic mice. *Proc. Natl. Acad. Sci. U.S.A.* **88,** 164–168.

Cavard, C., Zider, A., Vernet, M., Bennoun, M., Saragosti, S., Grimber, G., and Briand, P. (1990). *In vivo* activation by ultraviolet rays of the human immunodeficiency virus type 1 long terminal repeat. *J. Clin. Invest.* **86,** 1369–1374.

Coulombe, P. A., Chan, Y. M., Albers, K., and Fuchs, E. (1990). Deletion in epidermal keratins that lead to alterations in filament organization and assembly: *in vivo* and *in vitro* studies. *J. Cell. Biol.* **111,** 3049–3064.

Coulombe, P. A., Hutton, M. E., Letai, A., Hebert, A., Paller, A. S., and Fuchs, E. (1991a). Point mutations in human keratin 14 genes of *epidermolysis bullosa simplex* patients: genetic and functional analysis. *Cell.* **66,** 1301–1311.

Coulombe, P. A., Hutton, M. E., Vassar, R., and Fuchs, E. (1991b). A function for keratins and a common thread among different types of *epidermolysis bullosa simplex* diseases. *J. Cell. Biol.* **115,** 1661–1674.

Frucht, D. M., Lamperth, L., Vicenzi, E., Belcher, J. H., and Martin, M. A. (1991). Ultraviolet radiation increases HIV-long terminal repeat-directed expression in transgenic mice. *AIDS Res. Hum. Retroviruses.* **7,** 729–733.

Hammer, R., Maika, S., Richardson, J., Tang, J., and Taurog, J. (1990). Spontaneous inflammatory disease in transgenic rats expressing HLA-B27 and human β2m: an animal model of HLA-B27-associated human disorders. *Cell.* **63,** 1099–1112.

Iwamoto, T., Takahashi, M., Ito, M., Hamatani, K., Ohbayashi, M., Wajjwalku, W., Isobe, K., and Nakashima, I. (1991). Aberrant melanogenesis and melanocytic tumor development in transgenic mice that carry a metallothionein/*ret* fusion gene. *EMBO. J.* **10,** 3167–3175.

Lacey, M., Alpert, S., and Hanahan, D. (1986). Bovine papillomavirus genome elicits skin tumours in transgenic mice. *Nature (London).* **322,** 609–612.

Larue, L., Dougherty, N., and Mintz, B. (1992). Genetic predisposition of transgenic mouse melanocytes to melanoma results in malignant melanoma after exposure to a low ultraviolet B intensity nontumorigenic for normal melanocytes. *Proc. Natl. Acad. Sci. U.S.A.* **89**, 9534–9538.

Leask, A., Rosenberg, M., Vassar, R., and Fuchs, E. (1990). Regulation of a human epidermal keratin gene: sequences and nuclear factors involved in keratinocyte-specific transcription. *Genes & Dev.* **4**, 1985–1998.

Leonard, J. M., Abramczuk, J. W., Pezen, D. S., Rutledge, R., Belcher, H. J., Hakim, F., Shearer, G., Lamperth, L., Travis, W., Fredrickson, T., Notkins, A. L., and Martin, M. A. (1988). Development of disease and virus recovery in transgenic mice containing HIV proviral DNA. *Science.* **242**, 1665–1670.

Leonard, J., Khillan, J. S., Gendelman, H. E., Adachi, A., Lorenzo, S., Westphal, H., Martin, M. A., and Meltzer, M. S. (1989). The human immunodeficiency virus long terminal repeat is preferentially expressed in Langerhans cells in transgenic mice. *AIDS Res. Hum. Retroviruses.* **5**, 421–430.

Lu, X., and Lane, E. B. (1990). Retrovirus-mediated transgenic keratin expression in cultured fibroblasts: specific domain functions in keratin stabilization and filament formation. *Cell.* **62**, 681–696.

McNab, A. R., Andrus, P., Wagner, T. E., Buhl, A. E., Waldon, D. J., Kawabe, T. T., Rea, T. J., Groppi, V., and Vogeli, G. (1990). Hair expression of chloramphenicol acetyl transferase in transgenic mice under the control of an ultra-high-sulfur keratin promoter. *Proc. Natl. Acad. Sci. U.S.A.* **87**, 6848–6852.

Morrey, J. D., Bourn, S. M., Bunch, T. D., M. K. Jackson, Sidwell, R. W., Barrows, L. R., Daynes, R. A., and Rosen, C. A. (1991). *In vivo* activation of human immunodeficiency virus type 1 long terminal repeat by UV type A (UV-A) light plus psoralen and UV-B light in the skin of transgenic mice. *J. Virol.* **65**, 5045–5051.

Powell, B. C., and Rogers, G. E. (1990). Cyclic hair loss and regrowth in transgenic mice overexpressing an intermediate filament gene. *EMBO. J.* **9**, 1485–1493.

Quaife, C. J., Pinkert, C. A., Ornitz, D. M., Palmiter, R. D., and Brinster, R. L. (1987). Pancreatic neoplasia induced by *ras* expression in acinar cells in transgenic mice. *Cell.* **48**, 1023–1034.

Rosenberg, S. A., Aebersold, P., Cornetta, K., Kasid, A., Morgan, R. A., Moen, R., Karson, E. M., Lotze, M. T., Yang, J. C., Topalian, S. L., Merino, M. J., Culver, K., Miller, A. D., Blaese, R. M., and Anderson, W. F. (1990). Gene transfer into humans: immunotherapy of patients with advanced melanoma using tumor infiltrating lymphocytes modified by retroviral gene transduction. *N. Engl. J. Med.* **323**, 570–578.

Rosenthal, D. S., Steinert, P. M., Chung, S., Huff, C. A., Johnson, J., Yuspa, S. H., and Roop, D. R. (1991). A human epidermal differentiation-specific keratin gene is regulated by calcium but not negative modulators of differentiation in transgenic mouse keratinocytes. *Cell Growth and Differentiation* **2**, 107–113.

Rothnagel, J. A., Greenhalgh, D. A., Bowden, P. E., Longley, M. A., Bundman, D. S., and Roop, D. R. (1991). Use of transgenic mice to identify regulatory sequences 5′ to the human K1 keratin gene. Abstracts for the 1991 annual meeting of the society for investigative dermatology. Seattle, Washington. *J. Invest. Dermatol.* **96**, 541.

Shawlot, W., Siciliano, M. J., Stallings, R. L., and Overbeek, P. A. (1989). Insertional inactivation of the downless gene in a family of transgenic mice. *Mol. Biol. Med.* **6**, 299–307.

Sippola-Thiele, M., Hanahan, D., and Howley, P. M. (1989). Cell-heritable stages of tumor progression in transgenic mice harboring the bovine papillomavirus type 1 genome. *Mol. Cell. Biol.* **9**, 925–934.

Skarnes, W. C. (1990). Entrapment vectors : a new tool for mammalian genetics. *Biotechnology.* **8**, 827–831.

Stein, B., Rahmsdorf, H. J., Steffen, A., Litfin, M., and Herrlich, P. (1989). UV-induced DNA damage is an intermediate step in UV-induced expression of human immunodeficiency virus type 1, collagenase, *c-fos* and metallothionein. *Mol. Cell. Biol.* **9**, 5169–5181.

Vassar, R., and Fuchs, E. (1991). Transgenic mice provide new insights into the role of TGF-α during epidermal development and differentiation. *Genes & Dev.* **5**, 714–727.

Vassar, R., Rosenberg, M., Ross, S., Tyner, A., and Fuchs, E. (1989). Tissue-specific and differentiation-specific expression of a human K14 keratin gene in transgenic mice. *Proc. Natl. Acad. Sci. U.S.A.* **86**, 1563–1567.

Vassar, R., Coulombe, P. A., Degenstein, L., Albers, K., and Fuchs, E. (1991). Mutant keratin expression in transgenic mice causes marked abnormalities resembling a human genetic skin disease. *Cell.* **64**, 365–380.

Vassar, R., Hutton, M. E., and Fuchs, E. (1992). Transgenic overexpression of transforming growth factor α bypasses the need for C-Ha-*ras* mutations in mouse skin tumorigenesis. *Mol. Cell. Biol.* **12**, 4643–4653.

Zider, A., Mashhour, B., Fergelot, P., Grimber, G., Vernet, M., Hazan, U., Couton, D., Briand, P., and Cavard, C. (1993). Dispensable role of the NF-κB sites in the UV-induction of the HIV-1 LTR in transgenic mice. *Nucleic Acids Res.* **21**(1), 79–86.

9

Keratinocytes as a Target for Gene Therapy

Joseph M. Carroll, Elizabeth S. Fenjves, Jonathan A. Garlick, and Lorne B. Taichman

Introduction

The age of gene therapy is upon us and the field of keratinocyte biology may be profoundly changed as a result. The impetus for gene therapy stems from an improved understanding of the correlation between human disease states and defective or missing gene products coupled with new knowledge and capabilities in cell cultivation, tissue grafting, and gene transfer. The goal of gene therapy is to permanently introduce new genetic material into autologous somatic cells for the purpose of expressing a therapeutic gene product.

Gene therapy has moved beyond the conceptual stage to application in humans. Several clinical trials using gene therapy have been underway since 1989 (Anderson, 1992). These trials have focused primarily on hematopoietic-derived cells. One example is the attempt to treat metastatic melanoma by infusion of autologous tumor infiltrating lymphocytes (TIL) engineered to secrete high levels of tumor necrosis factor (TNF) (Rosenberg, 1990). TIL cells cultured from a biopsy of melanoma tissue are altered by the introduction of the TNF gene under the control of a strong promoter. When infected back into the donor, it is hoped that the altered TIL will traffic to the sites of metastatic disease and secrete high levels of TNF leading to tumor destruction. It is too early at this junction to make definitive conclusions from these trials. These experiments will help determine the effectiveness as well as the safety of introducing genetically altered cells back into a donor.

In this chapter, we discuss gene therapy and how it may be applied to the keratinocyte. Many of the capabilities required for this endeavor are already well established, such as the growth of keratinocytes in culture and the ability to use these cultures as stable autologous grafts. Other areas need to be developed for keratinocyte gene therapy to become a practical reality. For example, gene expression in this cell type is currently a subject of intense investigation, but relatively little is known about foreign gene transfer into keratinocytes. Data is also relatively scant on the secretory activity of keratinocytes. We will review some of the salient features of these aspects of keratinocyte biology and how they relate to possible applications in epithelial gene therapy.

Keratinocyte Gene Therapy

Several features of keratinocytes make them particularly suitable target cells for gene therapy:

1. Conditions under which keratinocytes are rapidly grown and repeatedly subcultured *in vitro* have been delineated (Rheinwald and Green, 1975) making manipulation in culture possible.
2. Cultured keratinocytes have been used as autologous grafts in the treatment of burns with long-term graft survival (Gallico *et al.*, 1984) and normalization of epidermal and connective tissue morphology (Compton *et al.*, 1989).
3. The putative epidermal stem cell is present in culture (Barrandon and Green, 1987) and may be more easily targeted for gene transfer than stem cells of other tissues.

4. The epidermis is a stratified tissue composed of differentiated and undifferentiated cells, and it may be possible to direct expression of a therapeutic gene to a specific cell strata.
5. In the event that a deleterious reaction occurs, a graft of genetically-modified keratinocytes can be removed.

There are several general applications for keratinocyte gene therapy as illustrated in Figure 1. First, genetic or acquired epidermal diseases that would be alleviated by expression of a single gene product in diseased keratinocytes might be treated by keratinocyte gene therapy. For example, the disease xeroderma pigmentosum is caused by an inherited defect in a DNA repair enzyme and is associated with a high risk of epidermal cancer. Repair may be restored within grafts of autologous keratinocytes modified to express the normal repair enzyme. This form of therapy would not be practical over a wide surface area but could be applied to selective sites at high risk for neoplastic transformation. Second, genetically altered keratinocytes could be used to secrete a gene product that may have either local or systemic distribution. For example, keratinocytes modified to secrete high levels of a cytokine such as interferon gamma might be useful in treating certain disorders of the underlying connective tissue. Third, gene therapy could be used to modify the catabolic phenotype of keratinocytes. In this way, they could metabolize and/or eliminate toxic substrates present at high levels in the

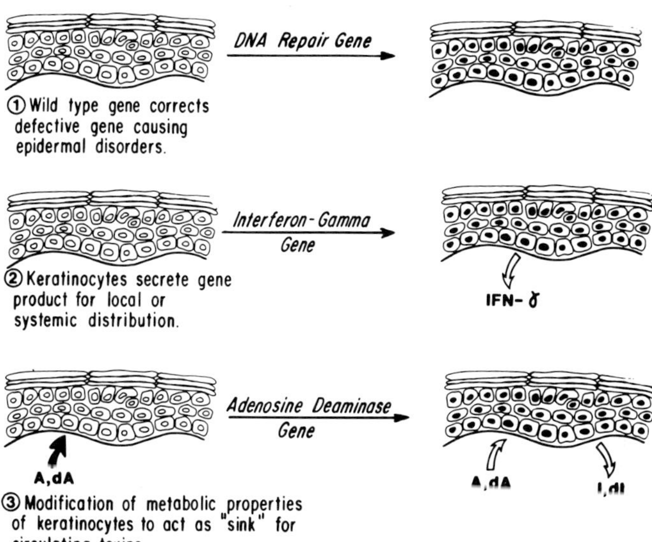

FIGURE 1 Potential applications of keratinocytes for gene therapy.

circulation. An example of this approach might be to use genetically modified keratinocytes to catabolize the high amounts of circulating deoxyadenosine in adenosine deaminase deficiency.

Keratinocyte Stem Cells as Gene Therapy Targets

Gene transfer to defined populations of keratinocytes is a prerequisite for epidermal gene therapy. The stem cell, which is responsible for cell replacement in epidermis, is the target of such therapy. An intermediate between the keratinocyte stem cell and terminally differentiating cells is the amplifying cell which undergoes limited cycles of replication (Potten, 1975). Genetic alteration of amplifying cells would result in a transient change in phenotype. By targeting the stem cell, gene expression in this cell and all of its descendants would be permanently altered.

Although there is no specific assay for keratinocyte stem cells, the fact that cultured keratinocytes can form a long-term autologous graft indicates that these cultures include stem cells (Gallico *et al.*, 1984). Barrandon and Green (1987) have shown that holoclones, which have the greatest proliferative potential, are likely to be the stem cells in culture. If one could enrich for holoclones prior to gene transfer, one presumably would have a better chance of targeting the stem cell population.

Gene Transfer

Considerations for Gene Therapy

Gene transfer is a major factor in keratinocyte gene therapy and within this context several criteria need to be considered. First, the gene must be introduced into keratinocytes at a very high efficiency. This is a requirement because stem cells are likely to be present at a low frequency and may not be targeted in sufficient numbers if gene transfer rates are low. Second, it is imperative that integration of the newly introduced gene occur in a high number of cells that receive the gene. Failure to integrate would result in loss of the gene during repeated cycles of cell replication. Although dominant selectable genes such as the gene for neomycin resistance (Davies and Jimenez, 1980) allow recovery of rare transformants that integrate and stably express the transferred gene, it is likely that the growth required to select such cells prior to grafting would measurably reduce the life expectancy of such cells after grafting. Third, it is imperative that there be no rearrangement or deletion of the integrated gene that would reduce or eliminate activity of the encoded protein. And fourth, it is important that the actual number of copies of integrated genes be predictable. Variability in copy number from cell to cell could lead to deleterious effects in cells that acquire a high copy number and would introduce uncertainty in the behavior of the grafts. A number of techniques is available and these have already been used to effect gene transfer in both *in vitro* and *in vivo*, and these will be discussed in detail.

Methods of Gene Transfection in Culture

Several methods to affect gene transfer with plasmid DNA into cultured keratinocytes are available. These include complex formation with calcium phosphate, polybrene, or lipids and electroporation to facilitate uptake of DNA into the cell (reviewed and compared in Jiang *et al.*, 1991). In the study of Jiang *et al.* (1991), complexing the DNA with polybrene or lipids resulted in the highest levels of transient gene expression. However, in these methods for gene transfer, expression diminishes within several days. Stable, long-term expression occurs only if the newly introduced DNA integrates into the host chromosome. However, only a small number of cells actually integrate the DNA, and in these cells, copy number is highly variable with deletions and rearrangements common. Cotransfection with a dominant drug resistance gene, such the gene for neomycin B resistance (Davies and Jimenez, 1980) or hygromycin resistance (Sugden *et al.*, 1985), allows recovery of rare transformants. However, senescence in selected cells remains a problem as discussed above. Transfection, as a method for gene transfer with or without selection, is more applicable to basic studies of gene expression than to gene therapy.

Retroviral Transduction

The limitations in gene transfer using plasmid DNA are largely overcome with the use of retroviral vectors. Retroviruses are RNA viruses which, upon infecting a cell, reverse transcribe their genome to a DNA copy and efficiently integrate a single copy as an intact, uninterrupted molecule. Transcription of the integrated viral genes, RNA processing, and encapsidation are all regulated by *cis*-acting viral sequences as well as cell factors (Varmus, 1988).

To prepare infectious retroviral vector, retroviral DNA lacking the *cis*-acting sequences required for encapsidation is first introduced into a line of 3T3 cells (Miller, 1990). The resulting packaging cells constitutively express viral gene functions required for production of infectious virus, but the RNA transcribed from this provirus cannot be encapsidated. Retrovector DNA is prepared by replacing viral genes essential for replication with DNA sequences encoding the gene(s) of interest while leaving the *cis*-acting sequences required for transcription, processing, and encapsidation intact (Cepko, 1989). When the retrovector DNA is introduced into the packaging

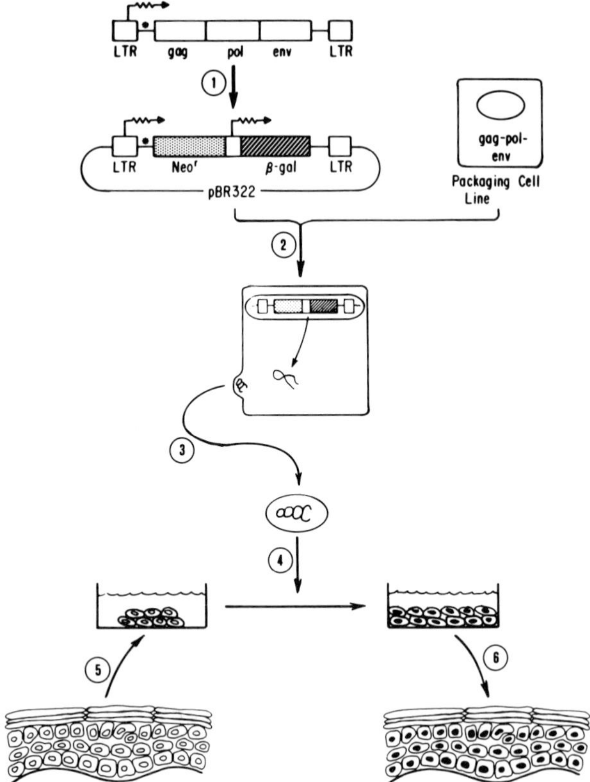

FIGURE 2 Steps for generation of retrovirus vectors. Packaging lines are generated by transfecting a fibroblast cell line (NIH3T3) with retrovirus DNA that lacks the packaging sequence required for its encapsidation into virions, yet encodes genes for retroviral enzymatic and structural protein (*gag, pol, env*). The retrovirus vector is constructed by replacing genes for these viral proteins with genes intended for transfer (*step 1*), thereby rendering it replication-deficient. To generate infectious retrovirus, this plasmid is introduced into the producer cell line (*step 2*). This retroviral genome is encapsidated, since it contains an intact packaging sequence, and released as an infectious retroviral virion (*step 3*). The infectious retroviral vector thus generated can infect replicating keratinocytes *in vitro* (*step 4*) that were grown in culture from an excisional biopsy (*step 5*). Upon reaching confluence, cells are grafted back to the donor and express the transferred genes (*step 6*).

cells, sequences containing the gene of interest are transcribed into RNA, encapsidated, and exported to yield an infectious viral particle (Gilboa *et al.*, 1986). The steps for generating infectious retrovirus vectors are summarized in Figure 2.

Retroviral vector transduction of keratinocytes in culture can be very efficient. At viral titers greater than 5×10^6 CFU/ml, all clonogenic cells are successfully infected (Garlick *et al.*, 1991). Clonal derivatives of the trans-

TABLE 1 **Selected Examples of the Use of Retroviral-Mediated Transduction in Keratinocytes**

Transferred gene	Target cell	Conclusions	Reference
Human growth hormone	Neonatal keratinocytes	Secretion of active hormone	Morgan *et al.* (1987)
Adenovirus E1A protein	Adult skin keratinocytes	Paraclones transformed	Barrandon *et al.* (1989)
TGF-α	Mouse keratinocytes	TGF-α overexpression did not transform cells	Finzi *et al.* (1990)
β-Galactosidase	Neonatal keratinocytes	Stem cells transduced	Garlick *et al.* (1991)
Neo phosphotransferase	Canine keratinocytes	Neo^r maintained in grafted cells for 130 days	Flowers *et al.* (1990)

duced cells can be expanded to 10^9 to 10^{10} cells before senescence, and each colony has been shown to harbor a single copy of retrovector DNA per cell. The high yield of cells from these experiments is likely to indicate that stem cells are transduced. Selected examples of retroviral-mediated transduction are presented in Table 1.

There are several variables to consider in using this means of gene transfer. First, the importance of high titers has been discussed above and improvements in vector design and infectious titers are steadily being made (Miller and Rosman, 1989). However, it is important to keep in mind that the inserted sequences must be engineered so that they do not lower viral titers. For example, the presence of a strong polyadenylation signal in the gene being inserted is likely to cause premature termination of the RNA in the packaging cell line and thus reduce viral titers of the full-length viral clones. (Miller *et al.*, 1988). In addition, the size of the gene of interest may be a limiting factor in the design of retrovirus vectors. DNA fragments of less than 5 kb in size are generally used (Gilboa *et al.*, 1986). Since retroviruses will only transduce replicating cells (Miller *et al.*, 1990), keratinocytes must be grown under conditions that will maximize replication.

DNA Virus-Mediated Transduction

Two DNA viruses may be useful as transducing agents and may provide certain advantages for gene transfer to keratinocytes. Adenovirus has a tropism for respiratory and oral epithelial cells (Aneskievich and Taichman, 1988) and replication-defective vectors have been engineered to introduce the human gene for the cystic fibrosis transmembrane conductance regulator into rat airway epithelium (Rosenfeld *et al.*, 1992). These vectors do not require cell replication in order to transduce and integrate into the cell genome. For gene therapy applications in keratinocytes, this feature may allow targeting of stem cells when they are quiescent. However, these vectors may also result in infection of nonreplicating, terminally differentiated cells limiting expression to the transit times of these cells. A recent adaptation of classical transduction has been developed in which the new gene is physically linked to the external surface of the adenovirus particle (Wagner *et al.*, 1992). Human papillomaviruses (HPVs) are epitheliotropic and are distinguished by the fact that, in benign papillomatous lesions, multiple copies of viral DNA replicate as stable episomes in basal cells (Broker and Botchan, 1986). Genes transferred with a bovine papillomavirus-based vector remain episomal (DiMaio *et al.*, 1982), thus avoiding insertional mutagenesis at the site of integration and providing a multicopy pool of vector genomes for expression. An HPV vector may now be possible with the development of methods for achieving episomal replication of HPV DNA in cultured keratinocytes (Mungal *et al.*, 1992).

In Vivo Gene Transfer

Techniques to affect the introduction of genes directly into animals (*in vivo*) without having to resort to manipulations in culture have been explored. *In vivo* gene transfer has been achieved using DNA–$CaPO_4$ precipitates (Benvenisty and Reshef, 1986), DNA encapsulated within liposomes (Nicolau *et al.*, 1983), or by direct injection of DNA and RNA (Wolff, 1990). More recently, DNA was introduced into the liver and skin of mice using microprojectiles to "shoot" the DNA directly into the animal (Williams *et al.*, 1991). In all of these experiments, high initial levels of expression are observed, although expression is often short-lived; stable expression (beyond one month) is not usually achieved. Additionally, all of these methods inherently have problems of low levels of stable integration and rearrangement/deletion events. Because these types of protocols are used directly on the animal, drug selection and characterization of individual clones are not possible. Culver *et al.* (1982) injected retroviral-producing cells directly into tissue of the rat and showed that neighboring cells were successfully transduced.

Expression of Transferred Genes in Keratinocytes

Viral Promoters

An important aspect of stable gene expression is the choice of promoter to regulate transcription. High levels of transcription are achieved through the use of viral promoters such as the LTRs of retroviruses (Wong *et al.*, 1987;

Miller *et al.*, 1988). These promoters yield constitutively high levels of gene expression in keratinocytes (Morgan *et al.*, 1987). However, viral promoters do not usually sustain high levels of expression *in vivo* for prolonged periods of time. The ability of cells to down-regulate viral promoters has been observed in genetically altered keratinocytes (Teumer *et al.*, 1990). Transcriptional silencing of an endogenous promoter located downstream of a viral promoter in an expression cassette has also been observed (Emerman and Temin, 1986; Bowtell *et al.*, 1988).

Endogenous Promoters

The problems encountered with long-term expression using viral promoters may be solved by the use of endogenous cellular promoters. While it is possible to utilize a strong endogenous gene promoter such as the β-actin promoter, which is active in a variety of tissues, it may be desirable to use tissue-specific promoters to restrict expression to certain cell types to prevent promiscuous expression in non-target tissue. The use of a particular endogenous promoter to drive foreign gene expression appears to have no deleterious effect on expression of the endogenous gene from which the promoter was derived (Vassar *et al.*, 1989).

Keratin promoters have successfully been used to direct tissue- and strata-specific expression of foreign genes in transgenic mice (Vassar *et al.*, 1989; Bailleul *et al.*, 1990; Rosenthal *et al.*, 1991). Depending on the specific gene therapy protocol, it may be desirable to direct gene expression to a specific strata of the epidermis. Expression in the basal layers can be achieved through the use of a basal cell-specific promoter (e.g., K14 promoter). For higher levels of production of a therapuetic gene product, it may be preferable to localize expression to the more transcriptionally active suprabasal layers (e.g., K1/K10 promoters). Both of these promoters would secure high levels of expression in stratified epithelia for a variety of purposes, while limiting expression to these cell types. It may also be desirable to modulate these promoters by including inducible genetic elements such as a those that would respond to steroids or retinoids (Miller *et al.*, 1984; Evans, 1988).

Stability of Gene Expression in Tissue

Stable gene expression will be a major hurdle for any gene therapy endeavor. Expression of transduced genes in culture is relatively stable, although instances of instability are known. For example, when cells are transduced with a retrovirus containing two selectable genes, each regulated by a different promoter, selection in a medium that utilizes one of the genes usually results in suppression of the other nonselected gene (Emerman and Temin, 1984). This transcriptional interference is thought to be epigenetic in nature.

Additionally, the newly acquired DNA can be physically lost from the genome (Miller *et al.*, 1988).

Loss of expression often occurs when genetically transformed cells are transplanted back to the host. An established line of epidermal keratinocytes transfected with DNA encoding human growth hormone regulated by the herpes virus thymidine kinase promoter initially showed high levels of hormone production when grafted onto mice, but serum levels decreased to baseline over several weeks (Teumer *et al.*, 1990). Activity was prolonged somewhat when the mouse metallothionein promoter was employed, but hormone levels were noted to decline throughout the observation period. In another report (Flowers *et al.*, 1990), canine keratinocytes were transduced in culture with the Neo^r gene and implanted in subcutaneous sites in the donor animal. Although Neo^r keratinocytes could be cultured from the graft for the duration of the experiment (120 days), their numbers decreased throughout that time. A detailed analysis of this phenomenon was carried out by Palmer *et al.* (1991) who transduced rat fibroblasts with the genes for Neo^r and human adenosine deaminase (hADA) and transplanted the cells into isogeneic rats as dermal equivalents. hADA activity was present in the graft at 2 weeks posttransplantation, but by one month was undetectable. Semiquantitative PCR analysis of DNA extracted 8.5 months posttransplantation showed that hADA DNA had indeed persisted at undiminished levels. Furthermore, vector positive fibroblasts could be cultured from the tissue at 8.5 months but these cells failed to express the Neo^r phenotype. These studies indicate that while the transduced cells survive, viral encoded genes are gradually inactivated. The issue of stability of foreign gene expression in tissue is not fully resolved.

Effect of Expressing a Foreign Gene in Keratinocytes

Little is known about how expression of a foreign protein will affect keratinocyte function in an epithelium. Transforming growth factor-α (TGF-α) has been introduced by retrovirus into mouse keratinocytes, and despite overexpression of this potent mitogen, no toxic effect or transformation was observed in the transduced cells (Finzi *et al.*, 1990). In transgenic mice with TGF-α introduced into their germ line, pathological responses were seen in the epidermis ranging from hyperplasia and papillomatosis to leukocyte infiltration and granular layer loss (Vassar and Fuchs, 1991). Keratinocytes transduced with the hGH gene and grafted beneath the skin in athymic mice form a normal differentiating epithelium, indistinguishable from the epithelium generated by nontransduced cells that are similarly grafted (Morgan and Eden, 1991).

Marker genes introduced into keratinocytes may help elucidate possible effects of foreign gene expression on keratinocytes. Canine keratinocytes retrovirally transduced with the Neo^r gene showed no detectable phenotypic changes even after transplantation (Flowers *et al.*, 1990). Human keratinocytes transduced with the marker gene β-galactosidase undergo no alterations in growth potential nor differentiation (Garlick *et al.*, 1991). Furthermore, transgenic mice expressing β-galactosidase in keratinocytes do not undergo changes or transformations when compared with normal mouse keratinocytes (Sanes *et al.*, 1986). Information on the consequences of overexpression of a foreign gene in keratinocytes is an area of great importance and must continue to be explored.

Secretory and Catabolic Activity of Keratinocytes for Gene Therapy

Keratinocytes Induced to Secrete New Gene Products

Keratinocytes are not traditionally considered to be secretory cells. Recent evidence suggests that keratinocytes produce and secrete a complex network of cytokines that have local effects on the epidermis and dermis as well as more distal effects (Luger and Schwarz, 1990). For example, interleukin I (IL-1) synthesized by keratinocytes is known to have an effect on keratinocyte proliferation, fibroblast proliferation, synthesis of collagenase, thymocyte proliferation, stimulation of other interleukins, stimulation of B-cell function, and fever induction (Kupper, 1989).

It is not clear whether all proteins secreted by keratinocytes have the capacity to reach the systemic circulation. In one instance, human apolipoprotein E (apoE), which is naturally secreted by keratinocytes in culture (Gordon *et al.*, 1989), was detected in the serum of athymic mice bearing grafts of human keratinocytes (Fenjves *et al.*, 1989). Systemic uptake of proteins secreted by keratinocytes may not occur in all instances. In two separate transgenic mouse lines harboring IL-6 (Turksen *et al.*, 1992) and TGF-α transgenes (Vassar and Fuchs, 1991), little if any secreted protein was detected in the circulation.

The capacity of genetically modified keratinocytes to secrete a gene product into the circulation has been examined in two instances. A squamous cell carcinoma line of keratinocytes, SCC9, was transfected with the gene for apoE. When these cells were injected subepidermally into nude mice, they formed tumors that secreted the protein. Apo E was detected in the serum of these animals (Fenjves *et al.*, 1990). Another squamous carcinoma cell line, SCC13, was induced to secrete hGH following transfection with the hGH gene (Teumer *et al.*, 1990). When sheets of these cells were transplanted beneath the skin of mice, hGH was present in serum in the initial posttransplant period. These experiments were performed with established lines of human keratinocytes. Although primary keratinocytes have been transduced and can be induced to secrete hGH in culture (Morgan *et al.*, 1987), their capacity to deliver hGH to the circulation has not been experimentally demonstrated in animals.

That keratinocytes readily secrete soluble products and can be induced to secrete others underlines the fact that these are active, secretory cells with biochemical and immune functions.

Genetic Modification of the Catabolic Potential of Keratinocytes

Many inborn errors of metabolism include, as part of their pathogenesis, high levels of toxic precursors in the circulation. Autologous grafts of keratinocytes transformed by the gene for the normal allele might function as a "metabolic sink" and metabolize circulating toxic precursors to innocuous products. This type of application was proposed by Flowers *et al.* (1990) for genetically altered fibroblasts in reducing (deoxy)adenosine levels in the circulation of patients with adenosine deaminase deficiency.

Safety Considerations of Keratinocyte Gene Therapy

The safety of all forms of gene therapy using retroviral vectors has been reviewed (Temin, 1990). One of the principal risks is that insertion of the provirus might alter the function of a vital gene or deregulate expression of other genes located in the insertion site. From the reported experiences with gene therapy in animals as well as the brief experience with humans, there have not been any side effects as a result of retroviral gene transfer (Anderson, 1992). In cultured keratinocytes, integration of a retroviral vector that lacks an oncogene has not resulted in immortalization or transformation of the transduced cells (Barrandon *et al.*, 1989). There is, however, a reported case of T-cell lymphoma in three immunodeficient monkeys after lethal irradiation and reconstitution with retroviral vector-transduced marrow stem cells (Kolberg, 1992).

In addition to insertional mutagenesis and deregulation of host genes, helper virus generated by recombination with endogenous sequences in the producer cells used to manufacture the retrovector can pose a problem of infection and systemic spread (Temin, 1990). This can be controlled by screening culture supernatants from producer cells, but generation of helper virus in the target cells may be more difficult to control. Currently all approved gene therapy protocols include assays to screen for helper virus.

Ideally, an introduced gene must be regulated and repressed if necessary. In keratinocyte gene therapy, uncontrolled expression of the introduced gene within the graft could be addressed and interrupted by removal of graft tissue. This point is one of the intrinsic safety advantages of altered epidermal tissue over other tissues considered for gene therapy.

Outlook for Keratinocyte Gene Therapy

It is likely that the initial trials of gene therapy will be limited to life-threatening disorders until the full range of safety concerns is investigated. At the present time, investigations are underway that will permit such trials to take place. There is, for example, substantial experience both in grafting of cultured autologous keratinocytes and transferring genes by retroviral transduction.

However, other aspects of gene therapy will need further investigation. Stable long-term gene expression must be achieved and promoters capable of regulated expression developed. It is also vital that epidermal stem cells be identified to gain an understanding of which transduced populations will survive. Ideally, vectors will be developed that will direct expression specifically to keratinocytes and to specific epidermal strata. The goal would be that these vectors would transfer genetic material at high efficiencies both *ex vivo* and *in vivo* to keratinocytes. Finally, the fate of genetically altered keratinocytes in epithelial grafts needs to be thoroughly investigated to appreciate the boundaries of migration and the fate of genetically modified keratinocytes.

Keratinocyte gene therapy is still in an embryonic phase, but does offer hope for ameliorating a range of local and systemic disorders. Its full impact on health has yet to be gauged. In addition to acting as a vehicle for gene therapy, genetically altered keratinocytes will provide models for diseased tissue and for the study of cutaneous biology.

References

Anderson, W. F. (1992). Human gene therapy. *Science,* **256**, 808–813.

Aneskievich, B. A., and Taichman, L. B. (1988). Epithelium-specific response of cultured keratinocytes to infection with adenovirus type 2. *J. Invest. Dermatol.* **91**, 309–314.

Bailleul, B., Surani, M. A., White, S., Barton, S. C., Brown, K., Blessing, M., Jorcano, J., and Balmain, A. (1990). Skin hyperkeratosis and papilloma formation in transgenic mice expressing a *ras* oncogene from a suprabasal keratin promoter. *Cell* **62**, 697–708.

Barrandon, Y., and Green, H. (1987). Three clonal types of keratinocytes with different capacities for multiplication. *Proc. Natl. Acad. Sci. U.S.A.* **84**, 2302–2306.

Barrandon, Y., Morgan, J. R., Mulligan, R. C., and Green, H. (1989). Restoration of growth potential in paraclones of human keratinocytes by a viral oncogene. *Proc. Natl. Acad. Sci., U.S.A.* **86**, 4102–4106.

Benvenisty, N., and Reshef, L. (1986). Direct introduction of genes into rats and expression of the genes. *Proc. Natl. Acad. Sci. U.S.A.* **83**, 9551–9555.

Bowtell, D. L., Cory, S., Johnson, G. R., and Gonda, T. J. (1988). Comparison of expression in hemopoietic cells by retroviral vectors carrying two genes. *J. Virol.* **62**, 2464–2473.

Broker, T. R., and Botchan, M. (1986). Papillomaviruses: Retrospectives and prospectives. *Cancer Cells* **4**, 17–36.

Cepko, C. (1989). Lineage analysis and immortalization of neural cells via retrovirus vectors. *In*

"Neuromethods" (A. A. Boulton, G. B. Baker, and A. T. Campagnoni, eds.). pp. 177–219. The Humana Press, Clifton, New Jersey.

Compton, C. C., Gill, J. M., Bradford, D. A., Regauer, S., Gallico, G. G., and O'Connor, N. E. (1989). Skin regenerated from cultured epithelial autografts on full-thickness burn wounds from 6 days to 5 years after grafting: A light, electron microscopic and immunohistochemical study. *Lab. Invest.* **60,** 600–612.

Culver, K. W., Ram, Z., Wallbridge, S., Ishii, H., Oldfield, E. H., and Blaese, R. M. (1992). *In vivo* gene transfer with retroviral vector-producer cells for treatment of experimental brain tumors. *Science* **256,** 1550–1552.

Davies, J., and Jimenez, A. (1980). A new selective agent for eukaryotic cloning vectors. *Am. J. Trop. Med. Hyg.* **29,** 1089–1092.

DiMaio, D., Treisman, R., and Maniatis, T. (1982). Bovine papillomavirus vector that propagates as a plasmid in both mouse and bacterial cells. *Proc. Natl. Acad. Sci. U.S.A.* **79,** 4030–4034.

Emerman, M., and Temin, H. M. (1984). Genes with promoters in retrovirus vectors can be independently suppressed by an epigenetic mechanism. *Cell* **39,** 459–467.

Emerman, M., and Temin, H. M. (1986). Quantitative analysis of gene suppression in integrated retrovirus vectors. *Mol. Cell. Biol.* **6,** 792–800.

Evans, R. M. (1988). The steroid and thyroid receptor superfamily. *Cell* **240,** 889–895.

Fenjves, E. S., Gordon, D. A., Pershing, L. K., Williams, D. L., and Taichman, L. B. (1989). Systemic distribution of apolipoprotein E secreted by grafts of epidermal keratinocytes: Implications for epidermal function and gene therapy. *Proc. Natl. Acad. Sci. U.S.A.* **86,** 8803–8807.

Fenjves, E. S., Lee, J. I., Garlick, J. A., Gordon, D. A., Williams, D. L., and Taichman, L. B. (1990). Prospects for epithelial gene therapy. *In* "DNA Damage and Repair in Human Tissues" (B. M. Sutherland, and A. D. Woodhead, eds.). pp. 215–223. Plenum Press.

Finzi, E., Fleming, T., and Pierce, J. H. (1990). Retroviral expression of transforming growth factor-alpha does not transform fibroblasts or keratinocytes. *J. Invest. Dermatol.* **95,** 382–387.

Flowers, M. E. D., Stockschlaeder, M. A. R., Schuening, F. G., Niederwieser, D., Hackman, R., Miller, A. D., and Storb, R. (1990). Long-term transplantation of canine keratinocytes made resistant to G418 through retrovirus-mediated gene transfer. *Proc. Natl. Acad. Sci. U.S.A.* **87,** 2349–2353.

Gallico, G. G. III, O'Connor, N. E., Compton, C. C., Kehinde, O., and Green, H. (1984). Permanent coverage of large burn wounds with autologous cultured human epithelium. *N. Engl. J. Med.* **311,** 448–451.

Garlick, J. A., Katz, A. B., Fenjves, E. S., and Taichman, L. B. (1991). Retrovirus-mediated transduction of cultured epidermal keratinocytes. *J. Invest. Deramtol.* **97,** 824–829.

Gilboa, E., Eglitis, M. A., Kantoff, P. W., and Anderson, W. F. (1986). Transfer and expression of cloned genes using retroviral vectors. *BioTechniques* **4,** 504–512.

Gordon, D. A., Fenjves, E. S., Williams, D. L., and Taichman, L. B. (1989). Synthesis and secretion of apolipoprotein E by cultured human keratinocytes. *J. Invest. Dermatol.* **92,** 96–99.

Jiang, C.-K., Connolly, D., and Blumenberg, M. (1991). Comparison of methods for transfection of human epidermal keratinocytes. *J. Invest. Dermatol.* **97,** 969–973.

Kolberg, R. (1992). Gene transfer virus contaminant linked to monkeys' cancer. As reported in *J.N.I.H. Res.* **4,** 43–44.

Kupper, T. S. (1989). Mechanisms of cutaneous inflammation. *Arch. Dermatol.* **125,** 1406–1412.

Luger, T. A., and Schwarz, T. (1990). Evidence for an epidermal cytokine network. *J. Invest. Dermatol.* **95,** 100s–104s.

Miller, A. D. (1990). Retroviral packaging cells. *Hum. Gene Ther.* **1,** 5–14.

Miller, A. D., and Rosman, G. J. (1989). Improved retroviral vectors for gene transfer and expression. *BioTechniques* **7**, 980–990.

Miller, A. D., Ong, E. S., Rosenfeld, M. G., Verma, I. M., Evans, R. M. (1984). Infectious and selectable retrovirus containing an inducible rat growth hormone minigene. *Science* **225**, 993–998.

Miller, A. D., Bender, M. A., Harris, E. A. S., Kaleko, M., and Gelinas, R. E. (1988). Design of retrovirus vectors for transfer and expression of the human β-globin gene. *J. Virol.* **62**, 4337–4345.

Miller, D. G., Adam, M. A., and Miller, A. D. (1990). Gene transfer by retrovirus vectors occurs only in cells that are actively replicating at the time of infection. *Mol. Cell. Biol.* **10**, 4239–4242.

Morgan, J. R., Barrandon, Y., Green, H., and Mulligan, R. C. (1987). Expression of an exogenous growth hormone gene by transplantable human epidermal cells. *Science* **237**, 1476–1479.

Morgan, J. R., and Eden, C. A. (1991). Retroviral-mediated gene transfer into transplantable human epidermal cells. *In* "Clinical and Experimental Approaches to Dermal and Epidermal Repair: Normal and Chronic Wounds" (Barbul, A., Caldwell, M. D., Eaglstein, W. H., Hunt, T. K., Pines, E., and Skover, G., eds.) pp. 417–428. Wiley-Liss, Inc.

Mungal, S., Steinberg, B. M., and Taichman, L. B. (1992). Episomal replication of plasmid-derived HPV11 DNA in cultured keratinocytes. *J. Virol.* **66**, 3220–3224.

Nicolau, C., Le Pape, A., Soriano, P., Fargette, F., and Juhel, M.-F. (1983). *In vivo* expression of rat insulin after intravenous administration of the liposome-entrapped gene for rat insulin I. *Proc. Natl. Acad. Sci. U.S.A.* **80**, 1068–1072.

Palmer, T. D., Rosman, G. J., Osborne, W. R. A., and Miller, A. D. (1991). Genetically modified skin fibroblasts persist long after transplantation but gradually inactivate introduced genes. *Proc. Natl. Acad. Sci. U.S.A.* **88**, 1330–1334.

Potten, C. S. (1975). Epidermal transit times. *Br. J. Dermatol.* **93**, 649–658.

Rheinwald, J. G., and Green, H. (1975). Serial cultivation of strains of human epidermal keratinocytes; the formation of keratinocyte colonies from single cells. *Cell* **6**, 331–344.

Rosenberg, S. A. (1990). TNF/TIL Human gene therapy clinical protocol. *Hum. Gene Ther.* **1**, 443–462.

Rosenfeld, M. A., Yoshimura, K., Trapnell, B. C., Yoneyama, K., Rosenthal, E. R., Dalemans, W., Fukayama, M., Bargon, J., Stier, L. E., Stratford-Perricaudet, L., Perricaudet, M., Guggino, W. B., Pavirani, A., Lecocq, J.-P., and Crystal, R. G. (1992). *In vivo* transfer of the human cystic fibrosis transmembrane conductance regulator gene to the airway epithelium. *Cell* **68**, 143–155.

Rosenthal, D. S., Steinert, P. M., Chung, S., Huff, C. A., Johnson, J., Yuspa, S. H., and Roop, D. R. (1991). A human epidermal differentiation-specific keratin gene is regulated by calcium but not negative modulators of differentiation in transgenic mouse keratinocytes. *Cell Growth and Differentiation* **2**, 107–113.

Sanes, J. R., Rubenstein, J. L. R., and Nicolas, J.-F. (1986). Use of a recombinant retrovirus to study post-implantation cell lineage in mouse embryos. *EMBO J.* **5**, 3133–3142.

Sugden, B., Marsh, K., and Yates, J. (1985). A vector that replicates as a plasmid and can be efficiently selected in B-lymphoblasts by Epstein-Barr virus. *Mol. Cell. Biol.* **5**, 410–413.

Temin, H. M. (1990). Safety considerations in somatic gene therapy of human disease with retrovirus vectors. *Hum. Gene Ther.* **1**, 111–123.

Teumer, J., Lindahl, A., and Green, H. (1990). Human growth hormone in the blood of athymic mice grafted with cultures of hormone-secreting human keratinocytes. *FASEB J.* **4**, 3245–3250.

Turksen, K., Kupper, T., Degenstein, L., Williams, I., and Fuchs, E. (1992). Interleukin 6: Insights to its function in skin by overexpression in transgenic mice. *Proc. Natl. Acad. Sci. U.S.A.* **89**, 5068–5072.

Varmus, H. (1988). Retroviruses. *Science* **240**, 1427–1435.

Vassar, R., and Fuchs, E. (1991). Transgenic mice provide new insights into the role of TGF-α during epidermal development and differentiation. *Genes & Dev.* **5**, 714–727.

Vassar, R., Rosenberg, M., Ross, S., Tyner, A., and Fuchs, E. (1989). Tissue-specific and differentiation-specific expression of a human K14 keratin gene in transgenic mice. *Proc. Natl. Acad. Sci. U.S.A.* **86**, 1563–1567.

Wagner, E., Zatloukal, K., Cotten, M., Kirlappos, H., Mechtler, K., Curiel, D. T., and Birnstiel, M. L. (1992). Coupling of adenovirus to transferrin–polylysine/DNA complexes greatly enhances receptor-mediated gene delivery and expression of transfected genes. *Proc. Natl. Acad. Sci. U.S.A.* **89**, 6099–6103.

Williams, R. S., Johnston, S. A., Riedy, M., DeVit, M. J., McElligott, S. G., and Sanford, J. C. (1991). Introduction of foreign genes into tissues of living mice by DNA-coated microprojectiles. *Proc. Natl. Acad. Sci. U.S.A.* **88**, 2726–2730.

Wolff, J. A., Malone, R. W., Williams, P., Chong, W., Acsadi, G., Jani, A., and Felgner, P. L. (1990). Direct gene transfer into mouse muscle *in vivo*. *Science* **247**, 1465–1468.

Wong, P. M. C., Chung, S.-W., and Nienhuis, A. W. (1987). Retroviral transfer and expression of interleukin-3 gene in hemopoietic cells. *Genes. & Dev.* **1**, 358–365.

Index